高 等 职 业 教 育 教 材

Instrumental Analysis

仪器分析

（中英对照版）

夏德强　主编

于娇娇　副主编

化学工业出版社

·北京·

内容简介

本书是顺应“走出去”企业人才培养的需要和推动现代职业教育高质量发展的要求而编写，体系完整、内容新颖、插图清新、中英文对照、数字资源丰富，既富有普通高等教育的学科特点，又突出了职业教育的类型特征，具有较强的创新性、先进性、实用性。全书重点介绍紫外可见分光光度法、原子吸收分光光度法、电化学分析法、气相色谱分析法、高效液相色谱分析法的基本原理和仪器结构；关于典型工业产品分析方面，主要介绍石油产品分析中所涉及的基本概念及相关指标的测定意义。

本书既可作为本专科职业院校资源环境、生物化工、食品药品等相关学科或专业的双语教材，也能作为普通本科学校相关专业教材，以及相关“走出去”企业员工本地化培养或“引进来”企业员工的培训教材，还可供其他专业师生及分析检验工作者参考。

图书在版编目（CIP）数据

仪器分析/夏德强主编；于娇娇副主编. —北京：化学工业出版社，2022.9

ISBN 978-7-122-41777-0

Ⅰ.①仪… Ⅱ.①夏…②于… Ⅲ.①仪器分析-教材 Ⅳ.①O657

中国版本图书馆CIP数据核字（2022）第110156号

责任编辑：刘心怡
责任校对：张茜越
装帧设计：关　飞

出版发行：化学工业出版社
（北京市东城区青年湖南街13号　邮政编码100011）
印　　装：大厂聚鑫印刷有限责任公司
787mm×1092mm　1/16　印张14¾　字数393千字
2022年11月北京第1版第1次印刷

购书咨询：010-64518888
售后服务：010-64518899
网　　址：http：//www.cip.com.cn
凡购买本书，如有缺损质量问题，本社销售中心负责调换。

定　　价：46.00元　

前言

习近平总书记于2013年提出了"一带一路"(The Belt and Road，缩写B&R)倡议。"十三五"期间，中国石油和化工企业"走出去"步伐不断加速，中国石化企业市场份额进一步增大，在全球产业链、供应链、价值链中发挥了重要作用。然而，"走出去"企业尤其是石油化工企业的员工本土化培养培训迫切需要相关的教学资源。2021年10月中共中央办公厅、国务院办公厅印发《关于推动现代职业教育高质量发展的意见》，明确提出要通过提升中外合作办学水平、拓展中外合作交流平台、推动职业教育走出去等措施打造中国特色职业教育品牌，推出一批具有国际影响力的专业标准、课程标准、教学资源。基于此，为顺应"走出去"企业人才培养的需要和国家职业教育改革发展的要求，本教材编写组以新修订实施的《中华人民共和国职业教育法》为依据，以科学性、先进性、适用性为目标，突出职业教育类型特色，紧扣实际、力求实用，解构重构了《分析化学》内容，编写了《化学分析》(中英对照版)和《仪器分析》(中英对照版)。

《仪器分析》(中英对照版)共分7章。第1章为绪论，第2章、第3章分别为紫外可见分光光度法和原子吸收分光光度法，第4章为电化学分析法，第5章、第6章分别为气相色谱分析法和高效液相色谱分析法，第7章为石油产品分析。

本书既可作为职业本科院校及高等职业专科学校资源环境、生物化工、食品药品等相关学科或专业的双语教材，也能作为普通本科学校相关专业教材，以及相关"走出去"企业员工本地化培养、"引进来"企业员工培训的教材，还可供其他专业师生及分析检验工作者参考。

本书前6章中英文内容由兰州石化职业技术大学夏德强教授编写，第7章由兰州石化职业技术大学于娇娇副教授和兰州石化职业技术大学汪永丽副教授编写。第7章的英文部分由兰州石化职业技术大学毛建梅翻译。兰州石化职业技术大学郑晓明、李晓婷、于娇娇、汪永丽、代学玉参与了本书有关章节的审核校对、实验实训项目选题及课后习题的编写工作。全书由夏德强统稿并担任主编，于娇娇任副主编，兰州石化职业技术大学甘黎明任主审。

本书的编写参考了大量的相关教材、专著、论文、规范及标准等，尤其是书中部分插图引用了相关的国外教材，在此对本书所引用成果的单位和个人表示衷心感谢。

由于编者的知识和能力水平有限，书中不足之处在所难免，恳请广大师生、读者、专家批评指正，以便今后进一步修订。

夏德强

2022年4月

PREFACE

In 2013, General Secretary Xi Jinping proposed the "*The Belt and Road*" initiative. During the period of the 13th Five-Year Plan, these companies accelerated the pace of globalization and kept enlarging their market share, so they played a crucial part in industrial chain, supply chain and value chain all over the world. However, the staff in the global-oriented enterprises, especially for those who work in petrochemical industry, urgently needs relevant teaching materials for training. Moreover, in October 2021, General Office of the CPC Central Committee and the General Office of the State Council issued the document entitled *Opinions on Promoting the High-quality Development of Modern Vocational Education,* where it was clearly stated that a brand of vocational education with Chinese characteristics should be created by promoting Sino-foreign cooperation level in running schools, expanding exchange platforms between China and other countries as well as urging vocational education to go abroad, and then a number of professional curriculum standards and teaching resources with great international influence should be made. In view of the above-mentioned, in order to meet the needs of enterprise talent training and the requirements of the national vocational education reform and development, followed by the *Vocational Education Law of the People's Republic of China*, writers and editors compiled the textbook of *Chemical Analysis* (Chinese-English Edition) & *Instrumental Analysis* (Chinese-English Edition) with the aims of highlighting vocational education's features and emphasizing pragmatic under the guide of scientificity, advancement and applicability.

Instrumental Analysis (Chinese-English Edition) is split up into 7 chapters. Certain related methods of instrumental analysis mainly include UV-VIS spectrophotometry, atomic absorption spectrophotometry, electrochemical methods, gas chromatography and liquid chromatography are talked about from chapter 2 to 6; The last chapter covers analysis of petroleum products.

This textbook can be used as a training material both for staff in local petrochemical industry with globalized intention and for employees of foreign enterprises in China. It can also be used as bilingual textbook for biology and chemical engineering, environmental protection and other related majors in Vocational Institutions for Undergraduates, Higher Vocational Colleges and Normal Undergraduate Universities.

The Chinese and English of chapters 1 to 6 were written by Professor Xia Deqiang from Lanzhou Petrochemical University of Vocational Technology. Associate Professor Yu Jiaojiao and Wang Yongli of Lanzhou Petrochemical University of Vocational Technology wrote Chinese of Chapter 7. Mao Jianmei of Lanzhou Petrochemical University of Vocational Technology translated Chapter 7. Xia Deqiang also served as the final editor, compiler and editor-in-chief of the book; Yu Jiaojiao served as deputy editors. Gan Liming, Zheng Xiaoming, Li Xiaoting, Yu Jiaojiao, Wang Yongli and Dai Xueyu participated in the proofreading, selection of experimental training projects and compilation of after-class exercises.

The compilation of this coursebook takes a large number of related textbooks, papers, specifications and standards for reference, and I would like to extend my sincere and heartfelt thanks to the copyholders for generous permission to make use of the pieces here.

Suggestions for improvement will be gratefully received.

Xia Deqiang

2022.04

CONTENTS

Chapter 5 Gas Chromatography / 125

Chapter 6 High Performance Liquid Chromatography / 172

Chapter 7 Analysis of Petroleum Products / 199

Reference / 227

Chapter 1 Introduction

第1章 概 论

Study Guide 学习指南

Instrumental analysis is based on the physical or physical and chemical properties of the substance, and seeks the internal relationship and rules of the analysis signals in the analysis process between the analysis signals and the composition of the analyzed material, and then conducts qualitative, quantitative, morphological and structural analysis. In this chapter, we shall focus on the development history, characteristics and classification of instrumental analysis, and introduces the main performance indicators of instruments and the development trend of modern instrumental analysis.

仪器分析是以物质的物理或物理化学性质为基础，探求这些性质在分析过程中所产生的分析信号与被分析物质组成的内在关系和规律，进而对其进行定性、定量、形态和结构分析的一类测定方法。本章重点介绍仪器分析的发展历程、特点和分类，介绍仪器的主要性能指标和现代仪器分析的发展趋势。

Section 1 Classification of Analytical Methods

第1节 分析方法的分类

Analytical methods are often classified as being either chemical or instrumental. Chemical methods, sometimes called classical methods, preceded instrumental methods by a century or more.

分析方法通常分为化学分析法和仪器分析法。化学分析法，又称经典分析法，比仪器分析法早一个世纪或更久。

1. Classical Methods

In the early years of chemistry, most analyses were carried out by separating the components of interest (the analytes) in a sample by precipitation, extraction, or distillation. For qualitative analyses. the separated components were then

1. 化学分析法

在化学发展的早期，大多数分析都是采用沉淀、萃取或蒸馏分离出待测物后进行测定。对于定性分析，将分离后的组分用试剂处理，然后通过颜色、沸

treated with reagents that yielded products that could be recognized by their colors, their boiling or melting points, their solubilities in a series of solvents, their odors, their optical activities, or their refractive indexes. For quantitative analyses, the amount of analyte was determined by **gravimetric** or by **volumetric** measurements.

点、熔点，以及一系列溶剂中的溶解度、气味、光学活性或折射率等来鉴别它们。对于定量分析，则是通过**重量分析法**或**容量分析法**来确定分析物的量。

In gravimetric measurements, the mass of the analyte or some compound produced from the analyte was determined. In volumetric, also called *titrinietric,* procedures, the volume or mass of a standard reagent required to react completely with the analyte was measured.

重量分析法是测定被分析物质量或由被分析物通过化学反应测定某种组分的质量。容量分析法，也称为滴定分析，是测定与被分析物完成化学反应所需标准试剂的体积或质量。

These classical methods for separating and determining analytes are still used in many laboratories. However, because of the increasing need for determining small amounts of analytes at low concentrations, the use of chemical methods has decreased with the passage of time and the advent of instrumental methods to supplant them.

这些经典的分离和分析方法仍在许多实验室中使用。然而，由于在低浓度下测定少量分析物的需求不断增加，加之仪器分析方法的不断涌现，化学分析方法的使用会随时间的推移逐渐减少。

2. Instrumental Methods

2. 仪器分析方法

Early in the twentieth century, scientists began to exploit phenomena other than those used for classical methods for solving analytical problems. Thus, measurements of such analyte physical properties as conductivity, electrode potential, light absorption or emission, mass-to-charge ratio, and fluorescence began to be used for quantitative analysis. Furthermore, highly efficient chromatographic and electrophoretic techniques began to replace distillation, extraction, and precipitation for the separation of components of complex mixtures prior to their qualitative or quantitative determination. These newer methods for separating and determining chemical species are known collectively as **instrumental methods of analysis**.

20世纪早期，科学工作者开始探索使用经典方法以外的其他现象解决分析问题，即分析物质的物理性质，如用电导、电位、光吸收或发射、质荷比和荧光等，进行物质的定量分析。此外，高效色谱和电泳技术开始取代蒸馏、萃取和沉淀，用于定量或定性分析前进行复杂混合物组分的分离。这类分离和确定化合物种类的新方法统称为**仪器分析方法**。

Many of the phenomena underlying instrumental methods have been known for a century or more. Their application by most scientists, however, was delayed by a lack of reliable and simple instrumentation. In fact, the growth of modern instrumental methods of analysis has paralleled the development of the electronics and computer industries.

仪器分析法的原理在一个世纪或更早已为人知。然而，由于缺乏可靠和简单的仪器，推迟了它们的应用。事实上，现代仪器分析方法的发展与电子和计算机工业的发展是并行的。

3. Development History of Instrumental Analysis

Generally, the development history of analytical chemistry is divided into three stages or three revolutions, two of which involve instrumental analysis. At the beginning of the 20th century, the determination of the four reaction equilibrium theories in solution laid the theoretical foundation of analytical chemistry, turned analytical chemistry from an operation technology into a science, and formed the first revolution of analytical chemistry. However, until the 1940s, chemical analysis played a leading role in analytical chemistry, with few instrumental analysis methods and low accuracy.

After the 1940s, instrumental analysis was in a period of great development due to the development of physics and electronics the needs of semiconductor material industry and atomic energy industry. A series of major scientific discoveries in this period also laid a theoretical foundation for the establishment and development of instrumental analysis. For example, the award of Rabi et al in 1944 laid the foundation for the creation of nuclear magnetic resonance spectroscopy; the award of Martin et al in 1952 greatly promoted the rapid development of chromatography The rapid development of instrumental analysis triggered the second revolution of analytical chemistry. During this period, the automation of instrument analysis was low, the instrument operation was mostly manual, and the spectrum analysis mostly depended on experience.

In the early 1980s, the third revolution in analytical chemistry, which was marked by computer application, realized the collection and processing of analytical data, information mining and three-dimensional image display under the control of computer. The analysis process turned to continuous, fast, real-time and intelligent. At the same time, new computer-based instruments were constantly emerging, such as Fourier transform infrared spectrometer and chromatography- mass spectrometry, making the computer an integral part of modern analytical instruments. At present, instrumental analysis is developing towards high sensitivity, high

3. 仪器分析发展历程

通常将分析化学的发展历程分为三个阶段或三次变革，其中两次涉及了仪器分析。20 世纪初，溶液中四大反应平衡理论的确定，奠定了分析化学的理论基础，使分析化学由一门操作技术变成一门科学，形成了分析化学的第一次变革。20 世纪 40 年代以前，化学分析在分析化学中占据着主导地位，仪器分析方法很少且精度较低。

20 世纪 40 年代以后，物理学、电子学的发展，及半导体材料工业和原子能工业生产的需要，使仪器分析处于大发展时期。这一时期的一系列重大科学发现，也为仪器分析的建立和发展奠定了理论基础。1944 年 Rabi 等的研究为核磁共振波谱分析法的创立奠定了基础；1952 年 Martin 等的研究极大地推动了色谱分析法的迅速发展。仪器分析的快速发展引发了分析化学的第二次变革。在这一时期，仪器分析的自动化程度较低，仪器操作多为手动操作，谱图解析多靠经验。

20 世纪 80 年代初，出现了以计算机应用为标志的分析化学的第三次变革，实现了计算机控制下的分析数据采集与处理、信息挖掘及三维图像显示。分析过程转向了连续、快速、实时和智能化，同时以计算机为基础的新仪器不断出现，如傅里叶变换红外光谱仪，色谱 - 质谱联用仪等，使计算机成为现代分析仪器不可分割的一部分。目前，仪器分析呈现出向高灵敏度、高选择性、自动化、智能化、信息化和微型化方向发展的趋势，建立了原位、活体、实

selectivity, automation, intelligence, informatization and miniaturization. The analysis methods of in-situ, living, real-time and online dynamic analysis and multivariate and multi parameter detection have been established. Table 1-1 and table 1-2 respectively show the future development path of instrumental analysis and the future demand for instrumental analysis, respectively.

时、在线的动态分析及多元多参数检测的分析方法。表 1-1 和表 1-2 分别为仪器分析的未来发展途径及未来对仪器分析的需求。

Table 1-1 Future development of instrumental analysis

表 1-1 仪器分析的未来发展途径

Automation and Robotics	自动化和机器人
Real intelligent instrument	真正智能仪器
On-line sensors and miniaturization system	在线传感器和微型化系统
Instrument network	仪器网络
Advanced remote sensing	高级遥感
More complex data compression	更复杂的数据压缩

Table 1-2 Future requirements for instrumental analysis

表 1-2 未来对仪器分析的需求

Higher sensitivity or selectivity	更高的灵敏度或选择性
Combination with more innovative analysis methods	更具创新性的分析方法联用
Advanced 3 D trace, nano, and subsurface analysis	高级三维微量、纳米和亚表面分析
A deeper understanding of measurement science	对测量科学更深入的理解
The ability to perform the analysis under more demanding in situ conditions	在更苛刻的原位条件下进行分析的能力
Energy analysis of the direct detection of molecules, transition states, and reaction kinetics	直接探测分子、过渡态和反应动力学的能量分析
Interpreting raw analytical data by an expert system	用专家系统解释原始分析数据

The development history and three revolutions of analytical chemistry show that instrumental analysis plays a role in connecting the past and the future. It is the most widely-used method and technology in modern analytical chemistry, and also the cutting-edge topic of analytical chemistry research today.

分析化学发展历程和三次变革说明，仪器分析起到承前启后作用，是现代分析化学应用最广泛的方法、技术，也是当今分析化学研究的前沿。

4. Characteristics of Instrumental Analysis

4. 仪器分析的特点

Instrumental analysis promotes the rapid development of analytical chemistry. Compared with chemical analysis, instrumental analysis has a series of characteristics, mainly including:

仪器分析推动分析化学迅速发展，与化学分析比较，仪器分析具有一系列特点，主要有：

(1) The sample dosage is small, suitable for trace, semi-trace and even ultra-trace analysis.

(1) 试样用量少，适用于微量、半微量乃至超微量分析。由化学分析的 mL、mg 级降到 μL、μg 级，甚至更低的 ng 级。

(2) Higher sensitivity, and the minimum detection amount and detection concentration are greatly reduced. From the 10^{-6} g of chemical analysis to 10^{-12} g, the minimum has reached 10^{-18} g, which is suitable for trace, super-trace composition determination.

(2) 检测灵敏度高，最低检出量和检出浓度大大降低。由化学分析的 10^{-6}g 级降至 10^{-12}g，最低已达 10^{-18}g 级，适用于痕量、超痕量成分测定。

(3) Better reproducibility, faster analysis speed, easy operation, easy to realize automation, information and online detection.

(3) 重现性好，分析速度快，操作简便，易于实现自动化、信息化和在线检测。

(4) Chemical analysis is carried out in solution, and the sample needs to be dissolved or decomposed; instrumental analysis can be analyzed in the original state of material, realizing non-destructive analysis and surface, microzone and morphology.

(4) 化学分析在溶液中进行，试样需要溶解或分解；仪器分析可在物质原始状态下分析，可实现试样非破坏性分析及表面、微区、形态等分析。

(5) Component separation, identification or structure determination of complex mixtures are realized by instrumental analysis. But the general chemical analysis methods can not.

(5) 可实现复杂混合物成分分离、鉴定或结构测定，一般化学分析方法难以实现。

(6) Generally, the relative error of chemical analysis is less than 0.3%, which is suitable for constant and high-content composition analysis. The relative error of instrumental analysis is high, which is 3%-5%. It is not suitable for constant and high content component analysis.

(6) 化学分析一般相对误差小于 0.3% 左右，适用于常量和高含量成分分析。仪器分析一般相对误差较高，为 3% ～ 5%，较不适宜常量和高含量成分分析。

(7) It requires expensive instruments and equipment with more complex structure, and the analysis cost is generally higher than that of chemical analysis.

(7) 仪器分析需要结构较复杂的昂贵仪器设备，分析成本一般比化学分析高。

5. Innovative Achievements of Analytical Chemistry

5. 分析化学创新成就

The Nobel Prize for science, which embodies the highest scientific spirit and highest scientific achievements such as innovation, realism and dedication, reflects the landmark scientific inventions and technological progress in analytical chemistry in the past 100 years, especially instrumental analysis. Table 1-3 lists some winners of Nobel Prize and their contributions to the establishment of some modern instrumental analysis methods.

作为体现创新、求实、献身等最高意义科学精神和最高科学成就的诺贝尔科学奖反映了近 100 年来分析化学，尤其是仪器分析发展中里程碑式的科学发明和技术进步。表 1-3 列出了与建立现代仪器分析方法有关的、获得诺贝尔奖的科学家及其贡献。

Table 1-3 Nobel Prize winners related to the development of analytical instruments

表1-3 与分析仪器发展相关的诺贝尔奖获得者

Number 序号	Year 年份	Nobel Prize winners 诺贝尔奖获得者	Outstanding contributions 突出贡献		Award category 奖项类别
1	1901	Röntgen W.C.(Germany)	Discovery of X-rays	发现 X 射线	Physics award
2	1907	Michelson A.A.(U.S.A.)	Manufacturing optical precision instruments and Spectral Research on celestial bodies	制造光学精密仪器及对天体所作的光谱研究	Physics award
3	1915	Bragg W.H.(England) & Bragg W.L.(England)	Analysis of crystal structure by X-ray technique	采用 X 射线技术对晶体结构分析	Physics award
4	1922	Aston F.W.(England)	Discovery of mass spectrometry and determination of isotopes	发明质谱技术并用来测定同位素	Chemical award
5	1923	Pregl F. (Austria)	Invention of microanalysis of organic matter	发明有机物质微量分析法	Chemical award
6	1924	Sieghahn M. (Sweden)	Discovery and research in X-ray instruments	在 X 射线仪器方面的发现及研究	Physics award
7	1931	Raman C.V. (India)	Discovery of Raman effect	发现 Raman（拉曼）效应	Physics award
8	1944	Rabi I.I. (U.S.A.)	Recording the resonance of atomic nucleus by resonance method.	用共振方法记录了原子核的共振	Physics award
9	1948	Tiselius A.W.K. (Sweden)	Separation of protein components in human serum by electrophoresis and adsorption	采用电泳及吸附分离人血清中蛋白质组分	Chemical award
10	1952	Bloch F.(U.S.A.) & Purcell E.N.(U.S.A.)	Development of fine measurement methods for nuclear magnetic resonance	发展核磁共振的精细测量方法	Physics award
11	1952	Martin A.J.P.(England) & Synge R.L.M.(England)	Invention of distribution chromatography	发明分配色谱法	Chemical award
12	1959	Heyrovsky J. (Czech)	Development of polarographic analyzer and analysis method	首先发展极谱分析仪及分析方法	Chemical award
13	1977	Yalow R. (U.S.A.)	Invention of Radioimmunoassay	开创放射免疫分析法	Physiological medicine award
14	1981	Sieghahn K.M. (Sweden)	High-resolution electron spectroscopy and instruments for chemical analysis	发展高分辨电子能谱学、仪器并用于化学分析	Physics award

续表

Number 序号	Year 年份	Nobel Prize winners 诺贝尔奖获得者	Outstanding contributions 突出贡献		Award category 奖项类别
15	1986	Binnig G. (Germany) & Roher H. (Switzerland)	Invention of tunnel scanning microscope	发明隧道扫描显微镜	Physics award
16	1991	Ernst R.R. (Switzerland)	Development of high resolution nuclear magnetic resonance analysis	对高分辨核磁共振分析的发展	Chemical award
17	2002	Wüthrich K. (Switzerland), Fenn J. B. (U.S.A.) & Tanaka K. (Japan)	A major breakthrough in the field of biological macromolecular analysis by nuclear magnetic resonance and mass spectrometry	核磁共振、质谱生物大分子分析研究领域的重大突破	Chemical award

Section 2 Types of Instrumental Methods

第2节 仪器分析的主要类型

1. Chemical and Physical Properties Used in Instrumental Methods

Let us first consider some of the chemical and physical characteristics that are useful for qualitative or quantitative analysis. Table 1-4 lists many of the characteristic properties that are currently used for instrumental analysis. Most of the characteristics listed in the table require a source of energy to stimulate a measurable response from the analyte. For example, in atomic emission an increase in the temperature of the analyte is required first to produce gaseous analyte atoms and then to excite the atoms to higher energy states. The excited-state atoms then emit characteristic electromagnetic radiation, which is the quantity measured by the instrument. Sources of energy may take the form of a rapid thermal change as in the previous example; electromagnetic radiation from a selected region of the spectrum; application of an electrical quantity, such as voltage, current, or charge; or perhaps subtler forms intrinsic to the analyte itself.

1. 仪器分析方法中使用的化学和物理性质

首先应了解定性或定量分析常用的化学或物理性质。表 1-4 列出了目前用于仪器分析的特征性质，其中大多数特征都需要能量来激发待测物质产生可测信号。例如，在原子发射光谱中，首先需要提高待测物质的温度产生气态待测物质的基态原子，然后再将基态原子激发到更高能态（激发态）。激发态原子跃迁为基态时会发射特征电磁辐射，这是仪器测量的基础。能量可来源于快速热变化、特定频段的电磁辐射、施加电量（如电压、电流或电荷），亦或是分析物本身固有的更微妙的形式。

Table 1-4 Chemical and Physical Properties Used in Instrumental Methods

Characteristic Properties	Instrumental Methods
Emission of radiation	Emission spectroscopy (X-ray, UV, visible, electron, Auger), fluorescence, phosphorescence, and luminescence (X-ray, UV, and visible)
Absorption of radiation	Spectrophotometry and photometry (X-ray, UV, visible, IR), photoacoustic spectroscopy, nuclear magnetic resonance and electron spin resonance spectroscopy
Scattering of radiation	Turbidimetry, nephelometry, Raman spectroscopy
Refraction of radiation	Refractometry, interferometry
Diffraction of radiation	X-ray and electron diffraction methods
Rotation of radiation	Polarimetry, optical rotary dispersion, circular dichroism
Electrical potential	Potentiometry, chronopotentiometry
Electrical charge	Coulometry
Electrical current	Amperometry, polarography
Electrical resistance	Conductometry
Mass	Gravimetry (quartz crystal microbalance)
Mass-to-charge ratio	Mass spectrometry
Rate of reaction	Kinetic methods
Thermal characteristics	Thermal gravimetry and titrimetry, differential scanning calorimetry, differential thermal analyses, thermal conductometric methods
Radioactivity	Activation and isotope dilution methods

表 1-4 仪器分析方法中使用的化学和物理性质

特征性质	仪器分析方法
辐射的发射	发射光谱（X 射线、紫外、可见、电子能谱、俄歇电子能谱），荧光，磷光和化学发光（X 射线、紫外、可见）
辐射的吸收	分光光度法和光度法（X 射线、紫外、可见、红外），光声光谱，核磁共振，电子自旋共振谱
辐射的散射	比浊法，浊度测定法，拉曼光谱
辐射的折射	折射法，干涉衍射法
辐射的衍射	X 射线，电子衍射法
辐射的旋转	偏振测定法，旋光散射法，圆二色谱
电位	电位法，计时电位分析法
电荷	库仑法
电流	安培法，极谱法
电阻	电导法
质量	重量法（石英晶体微天平）
质荷比	质谱法
反应速率	动力学方法
热性质	热重量和热滴定法；差示扫描量热法；差热分析法；热导法
放射性	放射化学分析法

Note that the first six entries in table 1-4 involve interactions of the analyte with electromagnetic radiation. In the first property, radiant energy is produced by the analyte; the next five properties involve changes in electromagnetic radiation brought about by its interaction with the sample. Four electrical properties then follow. Finally, five miscellaneous properties are grouped together: mass, mass-to-charge ratio, reaction rate, thermal characteristics, and radioactivity.

表 1-4 中的前六项涉及被分析物与电磁辐射的相互作用。在第一个性质中，辐射能由被分析物产生；接下的五个性质涉及其与样品相互作用所引起的电磁辐射的变化；再接下来是四种电学性质。最后将五种其他性质分组，即质量、质荷比、反应速率、热性质和放射性。

The second column in table 1-4 lists the instrumental methods that are based on the various physical and chemical properties. Be aware that it is not always easy to select an optimal method from among available instrumental techniques and their classical counterparts. Many instrumental techniques are more sensitive than classical techniques, but others are not. With certain combinations of elements or compounds, an instrumental method may be more selective, but with others, a gravimetric or volumetric approach may suffer less interference. Generalizations on the basis of accuracy, convenience, or expenditure of time are equally difficult to draw. Nor is it necessarily true that instrumental procedures employ more sophisticated or more costly apparatus.

表 1-4 中第二列为基于各种物理和化学性质建立的仪器分析方法。要注意，从仪器分析方法和经典分析方法中选择一个最优方案并不容易。大多数仪器分析方法比经典分析方法更灵敏，但少数也未必。对具有特定组成的元素或化合物进行分析，仪器方法可能选择性更好，但其他方法，如重量法或容量法可能受到的干扰更少。基于准确、方便或耗时考虑，对分析方法的综合判断同样很难。况且，仪器分析也不一定都用更复杂或更昂贵的仪器。

As noted earlier, in addition to the numerous methods listed in the second column of table 1-4, there is a group of instrumental procedures that are used for separation and resolution of closely related compounds. Most of these procedures are based on chromatography, solvent extraction, or electrophoresis. One of the characteristics listed in table 1-4 is usually used to complete the analysis following chromatographic separations. Thus, for example, thermal conductivity, ultraviolet and infrared absorption, refractive index, and electrical conductance are often used for this purpose.

如前所述，除了表 1-4 第二列中列出的分析方法外，还有一类仪器分析方法，常用于分离分析相关化合物。这类方法主要以色谱技术、溶剂萃取技术及电泳技术等为代表。表 1-4 中列出的特征性质常用于完成色谱分离后的检测，如导热性、紫外和红外吸收、折射率和电导等。

2. Optical Methods of Analysis

2. 光学分析法

Optical analysis method is a kind of analysis method based on the change of electromagnetic radiation signal after the energy acts on the matter or after the interaction between the electromagnetic radiation and the matter.

光学分析法是基于能量作用于物质后产生电磁辐射信号或电磁辐射与物质相互作用后产生辐射信号的变化而建立起来的一类分析方法。光学分析法分为

Optical analysis methods are divided into spectral analysis methods and non-spectral analysis methods.

光谱分析法和非光谱分析法。

Spectral analysis is an optical analysis method based on the absorption, emission and Raman scattering of light, which is established by detecting the changes of wavelength and intensity of spectrum after interaction. Qualitative analysis can be performed according to the wavelength of the characteristic spectrum of the matter, and quantitative analysis can be performed according to the change of spectral intensity.

光谱分析法是基于物质对光的吸收、发射和拉曼散射等作用，通过检测相互作用后光谱的波长和强度变化而建立的光分析方法。根据物质特征光谱的波长可进行定性分析，根据光谱强度变化可进行定量分析。

Spectroscopy can be divided into atomic spectroscopy and molecular spectroscopy, mainly including atomic emission spectroscopy, atomic absorption spectroscopy, X-ray spectroscopy, molecular fluorescence and phosphorescence, chemiluminescence, UV-vis spectroscopy, infrared spectroscopy, Raman spectroscopy, nuclear magnetic resonance spectroscopy, etc. Among them, infrared spectroscopy, Raman spectroscopy and nuclear magnetic resonance spectroscopy are commonly used for structural analysis of compounds, while others are mostly used for quantitative analysis.

光谱法又可分为原子光谱法和分子光谱法两大类，主要包括原子发射光谱法、原子吸收光谱法、X 射线光谱法、分子荧光和磷光法、化学发光法、紫外 - 可见光谱法、红外光谱法、拉曼光谱法、核磁共振波谱法等。其中，红外光谱法、拉曼光谱法、核磁共振波谱法常用于化合物的结构分析，其他多用于定量分析。

Non-spectral method is an optical analysis method established by measuring the changes of light reflection, refraction, interference, diffraction and polarization, including refraction method, interference method, rotation method, X-ray diffraction method, etc. The development of new high intensity, short pulse, tunable light source and complex spectral analysis are cutting-edge topics of optical analysis.

非光谱法是指通过测量光的反射、折射、干涉、衍射和偏振等变化所建立的分析方法，包括：折射法、干涉法、旋光法、X 射线衍射法等。新型高强度、短脉冲、可调谐光源的研制及复杂光谱解析等都是光学分析法的前沿领域。

3. Electroanalytical methods

3. 电化学分析法

Electroanalytical chemistry encompasses a group of qualitative and quantitative analytical methods that are based upon the electrical properties of solution of the analyte when it is made part of an electrochemical cell. Electroanalytical techniques are capable of producing low detection limits and a wealth of characterization information describing electrochemically accessible

电分析化学包括定性和定量分析方法，其是基于被分析物溶液成为电化学电池的组成部分时的电学性质。电分析技术具有较低检测限，能够提供丰富的描述电化学可访问系统的表征信息。这些信息包括界面电荷转移的化学计量学和速率、传质速率、吸附或化学吸附程度以及化学反应的速率和平衡常数。

systems. Such information includes the stoichiometry and rate of interfacial charge transfer, the rate of mass transfer, the extent of adsorption or chemisorption, and the rates and equilibrium constants for chemical reactions.

Electroanalytical methods have certain advantages. First, electrochemical measurements are often specific for a particular oxidation state of an element. For example, in electrochemical methods it is possible to determine the concentration of each of the species in a mixture of cerium (Ⅱ) and cerium (Ⅳ), whereas most other analytical methods can reveal only the total cerium concentration. A second important advantage of electrochemical methods is that the instrumentation is relatively inexpensive. Most electrochemical instruments cost less than $30,000. An important feature of certain electrochemical methods, which may be an advantage or a disadvantage, is that they provide information about activities rather than concentrations of chemical species. Generally, in physiological studies, for example, activities of ions such as calcium and potassium are more important than concentrations.

电化学分析法有一定的优势。首先，电化学测量通常针对元素的特定氧化状态。例如，电化学分析法可以确定铈（Ⅱ）和铈（Ⅳ）混合物中每种物质的浓度，而大多数其他分析方法只能显示总铈浓度。电化学方法的第二个重要优点是仪器相对便宜。大多数电化学分析仪器的价格不到20万元。某些电化学分析法还有一个重要特点，它们提供的信息是活度，而不是化学物质的浓度，这可能是优点也可能是缺点。例如，一般来说，在生理学研究中，钙和钾等离子的活度比浓度更重要。

4. Separation analysis method

Separation analysis method refers to the instrument method integrating separation and determination, mainly represented by gas chromatography, high performance liquid chromatography, capillary electrophoresis and its separation and analysis technology combined with the above instruments. Chromatographic analysis includes separation and detection. Chromatographic separation is based on the difference of physical and chemical properties such as adsorption, vapor pressure, solubility, hydrophobicity, ion exchange and molecular volume of substances on adsorbents, separation media or separation materials, and is not included in the characteristic properties in table 1-4. After separation, the detection of each component can be based on the physical and chemical properties of the substance, including some characteristic properties in table 1-4. Although the

4. 分离分析法

分离分析法是指融分离与测定于一体的仪器分析法，主要是以气相色谱、高效液相色谱、毛细管电泳等为代表的分离分析方法及其与上述仪器联用的分离分析技术。色谱分析包括分离和检测两部分。色谱分离基于物质在吸附剂、分离介质或分离材料上的吸附、蒸气压、溶解度、疏水性、离子交换、分子体积等多种物理化学性质差异，未包含在表1-4特征性质中。色谱分离后，各组分的检测是基于物质的物理化学性质，包括表1-4某些特征性质。尽管色谱检测器与一般分析仪器原理相似，但设计、结构相差很大。分离分析法可用于混合物，特别是各种复杂混合物的分离测定。

principle of chromatographic detector is similar to that of general analytical instruments, its design and structure are quite different. The separation analysis method is used for the separation and determination of mixtures, especially various complex mixtures.

5. Other instrumental analysis

Other instrumental analysis methods are mainly based on the last four feature properties in table 1-4. Including mass spectrometry, that is, the substance is ionized in the ion source to form charged ions, which are measured in the mass analyzer according to the ion mass charge ratio (m/z); thermal analysis method is based on the relationship between mass, volume, thermal conductivity or reaction heat and temperature, radiochemical analysis by radioactive isotopes, etc.

5. 其他仪器分析方法

其他仪器分析方法主要基于表 1-4 中的最后四个特征性质。包括质谱法，即物质在离子源中被电离形成带电离子，在质量分析器中根据离子质荷比（m/z）进行测定；热分析法，基于物质的质量、体积、热导率或反应热等与温度之间关系的测定方法；以及利用放射性同位素进行分析的放射化学分析法等。

Section 3 Instruments for Analysis

第 3 节 分析仪器

1. Basic Structural Unit of Instrument

An instrument for chemical analysis converts information about the physical or chemical characteristics of the analyte to information that can be manipulated and interpreted by a human. Thus, an analytical instrument can be viewed as a communication device between the system under study and the investigator. To retrieve the desired information from the analyte, it is necessary to provide a stimulus, which is usually in the form of electromagnetic, electrical, mechanical, or nuclear energy, as illustrated in figure 1-1 and figure 1-2. The stimulus elicits a response from the system under study whose nature and magnitude are governed by the fundamental laws of chemistry and physics. The resulting information is contained in the phenomena that result from the interaction of the stimulus with the

1. 分析仪器的基本结构单元

分析仪器的原理是分析物质的有关物理或化学性质转换为能为人们操控和解释的信息。因此，分析仪器可看成被研究体系与研究者之间的通信设备。如图 1-1 和图 1-2 所示，为了从分析物中检索所需信息，必须提供通常以电磁辐射、电能、机械能或核能形式存在的探测能源。该探测能源引发研究系统的反应，其性质和大小遵守化学或物理基本定律。一个常见的例子是：将一个狭窄波长的可见光通过样品，并测量其被待测物质吸收的程度，分别测定光与样品相互作用前后的强度变化，即可确定分析物的浓度。

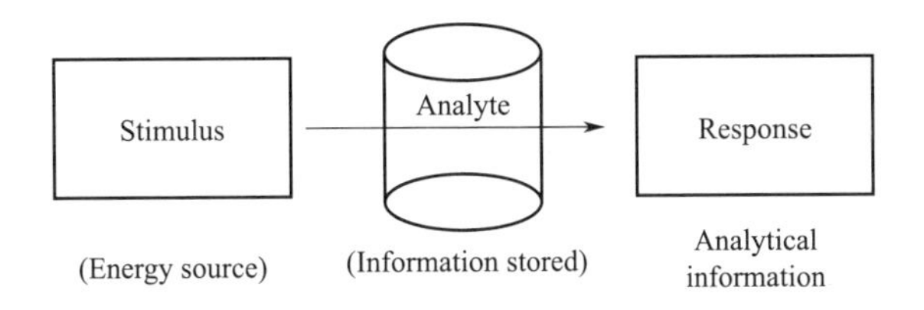

Fig.1-1 Block diagram showing the overall process of an instrumental measurement.

图 1-1 分析仪器基本组成框图

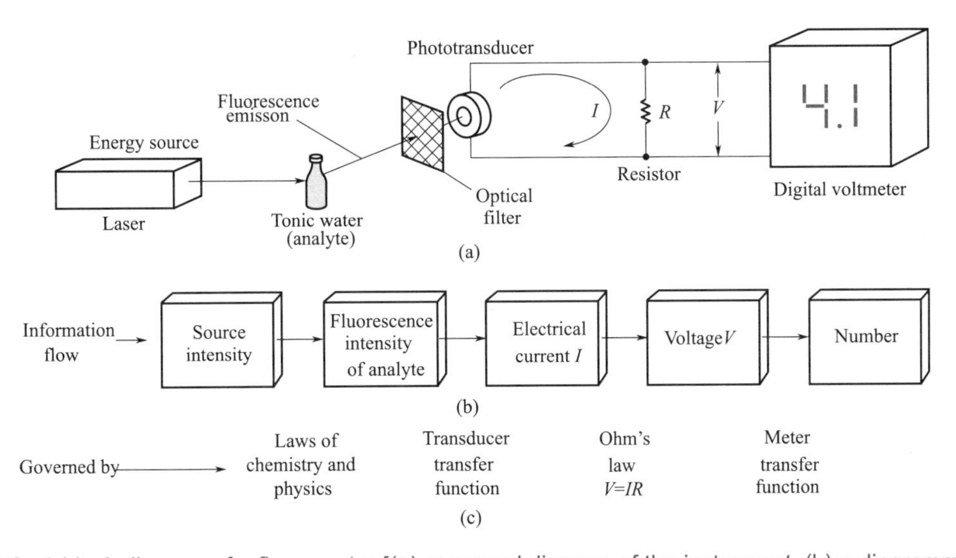

Fig.1-2 A block diagram of a fluorometer [(a) a general diagram of the instrument, (b) a diagrammatic representation of the flow of information through various data domains in the instrument, and (c) the rules governing the data-domain transformations during the measurement process]

图 1-2 荧光仪的框图 [(a) 仪器的流程图，(b) 通过仪器中不同数据域的信息流的图解表示，(c) 测量过程中数据转换]

analyte. A familiar example is passing a narrow band of wavelengths of visible light through a sample to measure the extent of its absorption by the analyte. The intensity of the light is determined before and after its interaction with the sample, and the ratio of these intensities provides a measure of the analyte concentration.

Generally, instruments for chemical analysis comprise just a few basic components, some of which are listed in table 1-5. To understand the relationships among these instrument components and the flow of information from the characteristics of the analyte through the components to the numerical or graphical output produced by the instrument, it is instructive to explore how the information of interest can be represented and transformed.

通常，化学分析仪器包括几个基本组件，如表 1-5 所示。仪器组件之间的关系以及从待测物质特征到仪器产生的数值或输出的图形信息，对如何表示和转换为有用信息具有指导意义。

Table 1-5 Some Examples of Instrument Components

表 1-5 一些分析仪器组件案例

Instrument 仪器	Energy Source (stimulus) 能源	Analytical Information 分析信息	Information Sorter 信息分拣	Input Transducer 输入转换器	Data Domain of Transduced Information 信息转换的数据域	Signal Processor/ Readout 信息处理 / 输出
Photometer 光度计	Tungsten lamp 钨灯	Attenuated light beam 衰减光束	Filter 滤光片	Photodiode 光电管	Electrical current 电流	Amplifier, digitizer, LED display 放大器，数字化仪，LED 显示屏
Atomic emission spectrometer 原子发射光谱仪	Inductively coupled plasma 电感耦合等离子体	UV or visible radiation 紫外 - 可见光	Monochromator 单色器	Photomulti-plier tube 光电倍增管	Electrical current 电流	Amplifier, digitizer, digital display 放大器，数字化仪，数字显示器
Coulometer 库伦仪	Direct-current source 直流电源	Charge required to reduce or oxidize analyte 还原或氧化被分析物所需的电荷量	Cell potential 电池电位	Electrodes 电极	Time 时间	Amplifier, digital timer 放大器，数字定时器
pH meter pH 计	Sample/glass electrode 样品 / 玻璃电极	Hydrogen ion activity 氢离子活度	Glass electrode 玻璃电极	Glass-Calomel electrodes 玻璃 - 甘汞电极	Electrical voltage 电压	Amplifier, digitizer, digital display 放大器，数字化仪，数字显示器
Mass spectrometer 质谱仪	Ion source 离子源	Mass-to-charge ratio 质荷比	Mass analyzer 质量分析仪	Electron multiplier 电子倍增管	Electrical current 电流	Amplifier, digitizer, computer system 放大器，数字化仪，计算机系统
Gas chromatograph with flame ionization 火焰离子化气相色谱仪	Flame 火焰	Ion concentration vs. time 离子浓度与时间的关系	Chromatographic column 色谱柱	Biased electrodes 加偏压的电极	Electrical current 电流	Electrometer, digitizer, computer system 电能表、数字化仪、计算机系统

2. Performance Characteristics of Instruments

Table 1-6 lists quantitative instrument performance parameters used to determine the suitability of a given instrument method to solve the corresponding analytical problem. These characteristics are expressed in numerical terms called figures of merit, which permit us to narrow the choice of instruments for a given analytical problem to a relatively few. Then, the instrument analysis method is further selected according to qualitative performance criteria such as analysis speed, operation convenience, technical requirements for operators, equipment cost and analysis capability, and per-sample cost, etc.

2. 分析仪器的性能指标

表 1-6 列出了仪器定量性能参数，用于确定给定的仪器分析方法是否适合解决相应的分析问题。这些性能用数字术语表示，称为品质因数。品质因数可以将某一特定分析问题的仪器选择范围缩小到相对较少的几个。然后，再根据分析速度、操作方便性、对操作者的技术要求、设备成本和分析能力、样品前处理成本等定性参数进一步选择仪器分析方法。

Table 1-6 Numerical Criteria for Selecting Analytical Methods

表 1-6 分析方法选择的参数标准

Criterion 标准	Figure of Merit 品质因数
1. Precision 1. 精密度	Absolute standard deviation, relative standard deviation, coefficient of variation, variance 绝对标准偏差，相对标准偏差 , 变异系数，方差
2. Error 2. 误差	Absolute systematic error, relative systematic error 绝对系统误差，相对系统误差
3. Sensitivity 3. 灵敏度	Calibration sensitivity, analytical sensitivity 校准灵敏度，分析灵敏度
4. Detection limit 4. 检出限	Blank plus three times standard deviation of the blank 以基质空白产生的背景信号平均值加上 3 倍的均数标准差
5. Dynamic range 5. 动态范围	Concentration limit of quantitation (LOQ) to concentration limit of linearity (LOL) 定量测定限至线性响应限的区间
6. Selectivity 6. 选择性	Coefficient of selectivity 选择性系数

(1) Precision The precision of analytical data is the degree of mutual agreement among data that have been obtained in the same way. Precision provides a measure of the random, or indeterminate, error of an analysis. Figures of merit for precision include **absolute standard deviation, relative standard deviation, standard error of the mean, coefficient of variation, and variance.** These terms are defined in table 1-7.

（1）精密度 分析数据的精密度是指以相同方式获得数据之间的一致程度。精密度是表征分析的随机或不确定误差的指标。精密度的指标包括绝对标准差、相对标准差、均数标准差、变异系数和方差，如表 1-7。

Table 1-7 Figures of Merit for Precision of Analytical Methods

表 1-7　分析方法的精密度指标

No.	Terms	Definition①	Equation
1	Absolute standard deviations, s 绝对标准差	$s=\sqrt{\dfrac{\sum_{i=1}^{N}(x_1-\bar{x})^2}{N-1}}$	(1.1)
2	Relative standard deviation (RSD) 相对标准差	$RSD=\dfrac{s}{\bar{x}}$	(1.2)
3	Standard error of the mean, s_m 均数标准差	$s_m=\dfrac{s}{\sqrt{N}}$	(1.3)
4	Coefficient of variation (CV) 变异系数	$CV=\dfrac{s}{\bar{x}}\times 100\%$	(1.4)
5	Variance 方差	s^2	(1.5)

① x_i=numerical value of the ith measurement; $\bar{x}$=mean of N measurements=$\dfrac{\sum_{i=1}^{N}x_i}{N}$.

(2) Error There are two ways to express error: absolute error (E) and relative error (E_r). The absolute error refers to the difference between the measured value (x_i) and the true value (μ), shown in formula (1.6); while the relative error is the percentage of the absolute error equivalent to the true value, which is shown in formula (1.7).

（2）误差　误差有两种表示方法：绝对误差（E）和相对误差（E_r）。绝对误差是测量值（x_i）与真实值（μ）之间的差值，即式（1.6）；相对误差是绝对误差对于真实值的百分比，即式（1.7）。

$$E=x_i-\mu \tag{1.6}$$

$$E_r=\frac{E}{\mu}\times 100\%=\frac{x_i-\mu}{\mu}\times 100\% \tag{1.7}$$

① Systematic errors This kind of error is caused by some special reason. It has unidirectionality, that is, positive and negative, the size has a certain regularity. Systematic errors reappear when the measurement is repeated. If the cause can be found and corrected, the systematic error can be eliminated, so it is also called measurable error.

① 系统误差　这类误差由某种固定的原因造成的，它具有单向性，即正负、大小都有一定的规律性。当重复进行测定时，系统误差会重复出现。若能找出原因，并设法加以校正，系统误差就可以消除，因此也称为可测误差。

The main causes for systematic errors are as follows.

系统误差产生的主要原因如下：

a. Method error refers to the error caused by the analysis methods itself. For example, in the titration analysis, the titration end point determined by the indicator does not completely coincide with the chemometric point and

a. 方法误差指分析方法本身所造成的误差。例如滴定分析中，由指示剂确定的滴定终点与化学计量点不完全符合以及副反应的发生等，都将系统地使测

side reactions occur, which make measurement result higher or lower completely.

定结果偏高或偏低。

b. Instrument error is mainly caused by the instrument itself, inaccurate or not calibrated. For example, when the balance, scale and vessel have errors, the measurement results will be inaccurate in the process of working.

b. 仪器误差主要是仪器本身不够准确或未经校准所引起的。如天平、砝码和容量器皿不准等，在使用过程中就会使测定结果产生误差。

c. Reagent error is error caused by impure reagent or trace impurities in distilled water.

c. 试剂误差是指由于试剂不纯或蒸馏水中含有微量杂质引起的误差。

d. Operation error is caused by unreasonable operation for the operator. For example, some people are sensible for judgment of the color change of the end point, but others are dull;the dropper reading is high or low.

d. 操作误差是由于操作人员的主观原因造成。例如，对终点颜色变化的判断，有人敏锐，有人迟钝；滴管读数偏高或偏低等。

② Random error Random error refers to the error caused by the random variation of various factors for measured result, such as temperature, humidity, pressure fluctuations, small changes for the instrumental performance, slight deviations in operation, which will cause the analysis results to fluctuate in a certain range, errors happen. Since the random error depends on a series of random factors in the measurement process, whose size and direction are not fixed, it cannot be measured and corrected. Therefore, the random error is also called unmeasurable error.

② 随机误差　随机误差是指测定值受各种因素的随机变动而引起的误差，如温度、湿度、气压的波动，仪器性能的微小变化，操作稍有出入等，都将使分析结果在一定范围内波动，从而造成误差。由于随机误差取决于测定过程中一系列随机因素，其大小和方向都不固定，因此无法测量，也不可能校正，所以机误差又称为不可测误差。

Random error inevitably happens, which objectively exists, and it is difficult to detect and control, there seems to be no regularity from its appearance, but after eliminating the systematic error and measuring it for many times under the same conditions, it can be found that the distribution of random errors is also regular. Generally speaking, it obeys the statistical discipline of normal distribution.

随机误差不可避免，客观存在，难以觉察，难以控制，从表面上似乎没有规律，但是消除系统误差后，在同样条件下多次测定，则可发现随机误差的分布也是有规律的，一般服从正态分布统计规律。

(3) Sensitivity There is general agreement that the sensitivity of an instrument or a method is a measure of its ability to discriminate between small differences in analyte concentration. Two factors limit sensitivity: the slope of the calibration curve and the reproducibility or precision of the measuring device. Of two methods that have equal precision, the one that has the steeper calibration curve will be the more sensitive. A corollary to this statement is that if two methods have calibration

（3）灵敏度　仪器或分析方法的灵敏度是指区别具有微小浓度差异分析物能力的指标。灵敏度决定于两个因素：即校准曲线的斜率和仪器设备的重现性或精密度。在相同精密度的两种方法中，校准曲线斜率越大，方法越灵敏。同样，在校准曲线斜率相等的两种方法中，精密度好的则灵敏度高。

curves with equal slopes, the one that exhibits the better precision will be the more sensitive.

The quantitative definition of sensitivity that is accepted by the International Union of Pure and Applied Chemistry (IUPAC) is calibration sensitivity, which is the slope of the calibration curve at the concentration of interest. Most calibration curves that are used in analytical chemistry are linear and may be represented by the equation (1.8).

根据IUPAC规定，灵敏度用校准灵敏度表示，即测定浓度范围内校准曲线的斜率。分析化学中，大多数校准曲线是线性的，可以用式（1.8）表示。

$$S=mc+S_{bl} \tag{1.8}$$

where S is the measured signal, c is the concentration of the analyte, S_{bl} is the instrumental signal for a blank, and m is the slope of the straight line. The quantity S_{bl} is the y-intercept of the straight line. With such curves, the calibration sensitivity is independent of the concentration c and is equal to m. The calibration sensitivity as a figure of merit suffers from its failure to take into account the precision of individual measurements.

式中，S为响应信号，c为待测物浓度，S_{bl}为仪器的本底空白信号，m为直线的斜率。S_{bl}是校准曲线在纵坐标上的截距。用这种校准曲线，校准灵敏度与浓度c无关，且等于m。在不考虑单个测定精密度时，将校准灵敏度作为性能指标具有一定优势。

Mandel and Stiehler recognized the need to include precision in a meaningful mathematical statement of sensitivity and proposed the following definition for analytical sensitivity γ:

Mandel和Stiehler认为，灵敏度在具有重要价值的数学处理中，需要包括精密度。因而提出分析灵敏度γ，如式（1.9）所示。

$$\gamma=m/s_s \tag{1.9}$$

Here, m is again the slope of the calibration curve, and s_s is the standard deviation of the measurement.

式中m仍为校准曲线斜率，s_s为测量的标准偏差。

The analytical sensitivity offers the advantage of being relatively insensitive to amplification factors. For example, increasing the gain of an instrument by a factor of five will produce a fivefold increase in m. Ordinarily, however, this increase will be accompanied by a corresponding increase in s_s, thus leaving the analytical sensitivity more or less constant. A second advantage of analytical sensitivity is that it is independent of the measurement units for s. A disadvantage of analytical sensitivity is that it is often concentration dependent because s_s may vary with concentration.

分析灵敏度具有的优点是对仪器放大系数相对不敏感。例如，用5倍的放大系数提高仪器增益，则可增加5倍的m。然而，通常这种增加会伴随s_s的增加，从而保持分析灵敏度相对恒定。分析灵敏度的第二个优点是与测定s的单位无关。分析灵敏度的缺点是与浓度的相关性，因s_s可能随浓度变化。

(4) Detection limit The most generally accepted qualitative definition of **detection limit** is that it is the minimum

（4）检出限　最普遍接受的**检出限**的定义是在已知置信水平下检出的待测

concentration or mass of analyte that can be detected at a known confidence level. This limit depends on the ratio of the magnitude of the analytical signal to the size of the statistical fluctuations in the blank signal. That is, unless the analytical signal is larger than the blank by some multiple k of the variation in the blank due to random errors, it is impossible to delect the analytical signal with absolute certainty. Thus, as the limit of detection is approached, the analytical signal and its standard deviation approach the blank signal S_{bl} and its standard deviation s_{bl}. The minimum distinguishable analytical signal S_m is then taken as the sum of the mean blank signal $\overline{S}_{bl}$ plus a multiple k of the standard deviation of the blank s_{bl}. That is,

物的最小量或最低浓度，它取决于待测物产生的信号与本底空白信号波动或噪声统计平均值之比。也就是说，只有当待测物信号大于由随机误差导致的空白信号一定倍数 k 时，待测物才可能被检出。因此，检出限的分析信号及其标准偏差接近于空白信号 S_{bl} 及其标准偏差 s_{bl}。最小检出信号 S_m 应为空白信号的平均值$\overline{S}_{bl}$加上 k 倍的空白信号标准偏差 s_{bl}，见式（1.10）。

$$S_m=\overline{S}_{bl}+ks_{bl} \tag{1.10}$$

Experimentally, S_m can be determined by performing twenty to thirty blank measurements, preferably over an extended period of time. The resulting data are then treated statistically to obtain $\overline{S}_{bl}$ and s_{bl}. Finally, the slope from equation (1.8) is used to convert S_m to c_m, which is defined as the detection limit. The detection limit is then given by

实验中，计算 S_m 时可以进行一定时间内 20 ～ 30 次的空白测定，然后对结果数据进行统计处理，得出$\overline{S}_{bl}$和 s_{bl}, 最后用式（1.8）中的斜率将 S_m 转换为 c_m，即为最低检测浓度，见式（1.11）。

$$c_m=\frac{S_m-\overline{S}_{bl}}{m}=\frac{ks_{bl}}{m} \tag{1.11}$$

Kaiser argues that a reasonable value for the constant is k=3. He points out that it is wrong to assume a strictly normal distribution of results from blank measurements and that when k=3, the confidence level of detection will be 95% in most cases. He further argues that little is to be gained by using a larger value of k and thus a greater confidence level.

Kaiser 认为常数 k 的合理值是 3。他指出，假设空白测量结果严格正态分布是错误的，且当 k=3 时，大多数情况下检测置信水平为 95%；k 值进一步增加，却难以获得更高的检测置信水平。

Example 1-1

A least-squares analysis of calibration data for the determination of lead based on its flame emission spectrum yielded the equation

$$S=1.12\ c_{Pb}+0.312$$

where c_{Pb} is the lead concentration in μL/L and S is a measure of the relative intensity of the lead emission line. The following replicate data were then obtained：

Concentration/(μL/L)	No. of Replications	Mean Value of S	s
10.0	10	11.62	0.15
1.00	10	1.12	0.025
0.000	24	0.0296	0.0082

Calculate (a) the calibration sensitivity, (b) the analytical sensitivity at 1 μL/L and 10 μL/L of Pb, and (c) the detection limit.

Solution

(a) By definition, the calibration sensitivity is the slope m=1.12.

(b) At 10 μL/L, $\gamma=m/s$=1.12/0.15=7.5; At 1 μL/L, γ=1.12/0.025=45.

Note that the analytical sensitivity is quite concentration dependent. Because of this, it is not reported as often as the calibration sensitivity.

(c) Applying equation (1.10), $S_m=0.0296+3\times0.0082=0.054$

Substituting into equation (1.11) gives $c_m=\dfrac{0.054-0.0296}{1.12}$=0.022(μL/L)

(5) Dynamic range Figure 1-3 illustrates the definition of the dynamic range of an analytical method, which extends from the lowest concentration at which quantitative measurements can be made (limit of quantitation, or LOQ) to the concentration at which the calibration curve departs from linearity by a specified amount (limit of linearity, or LOL). Usually, a deviation of 5% from linearity is considered the upper

（5）动态范围　图 1-3 是分析方法动态范围的定义，即定量测定最低浓度（即定量测定限，LOQ）扩展到校准曲线偏离线性响应（即线性响应限，LOL）的浓度范围。通常偏离线性的 5% 被认为是上限。由于非理想的检测器响应或化学效应，在高浓度下，偏差线性是常见的。定量测定下限一般取

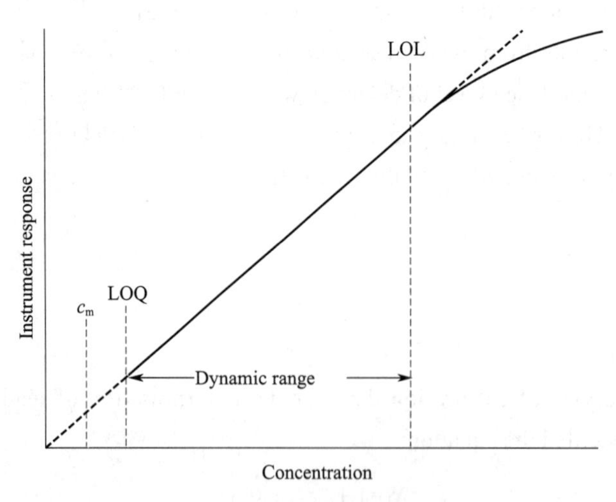

Fig.1-3 Useful range of an analytical method(LOQ =limit of quantitative measurement; LOL= limit of linear response.)

图 1–3　仪器分析方法适用线性范围（LOQ 为定量测定限；LOL 为线性响应限。）

limit. Deviations from linearity are common at high concentrations because of nonideal detector responses or chemical effects. The lower limit of quantitative measurements is generally taken to be equal to ten times the standard deviation of repetitive measurements on a blank, or $10s_{bl}$. At this point, the relative standard deviation is about 30% and decreases rapidly as concentrations become larger.

10倍的空白重复测定标准差，或$10s_{bl}$。此时，相对标准差约为30%，并随浓度增加而迅速降低。

To be very useful, an analytical method should have a dynamic range of at least a few orders of magnitude. Some analytical techniques, such as absorption spectrophotometry, are linear over only one to two orders of magnitude. Other methods, such as mass spectrometry and molecular fluorescence, may exhibit linearity over four to five orders of magnitude.

分析方法应具有至少几个数量级的动态范围。有些分析技术，如吸收分光光度法，仅在一到两个数量级上是线性的；有些分析方法，如质谱法和分子荧光法，可能存在超过4～5个数量级的线性。

(6) Selectivity　Selectivity of an analytical method refers to the degree to which the method is free from interference by other species contained in the sample matrix. Unfortunately, no analytical method is totally free from interference from other species, and frequently steps must be taken to minimize the effects of these interferences.

（6）选择性　仪器分析方法的选择性是指试样中含有的其他组分对测定组分干扰的程度。没有一种分析方法能完全避免其他组分干扰，因而尽量降低干扰是分析测试中常需要的步骤。

Exercises

1-1　What is a transducer in an analytical instrument?

1-2　What is the information processor in an instrument for measuring the color of a solution visually?

1-3　What is the detector in a spectrograph in which spectral lines are recorded photographically?

1-4　What is a data domain?

1-5　List four input transducers and describe how they are used.

1-6　What is a figure of merit?

1-7　A 25.0-mL sample containing Cu^{2+} gave an instrument signal of 25.2 units (corrected for a blank). When exactly 0.500 mL of 0.0275 mol · L^{-1} $Cu(NO_3)_2$ was added to the solution, the signal increased to 45.1 units. Calculate the molar concentration of Cu^{2+} assuming that the signal was directly proportional to the analyte concentration.

(6.66×10^{-4}mol · L^{-1})

1-8 The data in the following table were obtained during a colorimetric determination of glucose in blood serum.

Glucose concentration/mmol · L^{-1}	Absorbance, A
0.0	0.002
2.0	0.150
4.0	0.294
6.0	0.434
8.0	0.570
10.0	0.704

(a) Assuming a linear relationship, find the least-squares estimates of the slope and intercept.

(m=0.0701; b=0.0083)

(b) Use the linest function in Excel to find the standard deviations of the slope and intercept. What is the standard error of the estimate?

(s_m=0.0007; s_b=0.0040)

(c) Determine the 95% confidence intervals for the slope and intercept.

(m=0.0701±2.78×0.0007=0.070±0.002)

(b= 0.0083±2.78×0.004=0.08±0.01)

(d) A serum sample gave an absorbance of 0.350. Find the glucose concentration and its standard deviation.

[(4.87±0.09) mmol · L^{-1}]

1-9 Exactly 5.00-mL aliquots of a solution containing phenobarbital were measured into 50.00-mL volumetric flasks and made basic with KOH. The following volumes of a standard solution of phenobarbital containing 2.000 μg · mL^{-1} of phenobarbital were then introduced into each flask and the mixture was diluted to volume: 0.000 mL, 0.500 mL, 1.00 mL, 1.50 mL, and 2.00 mL. The fluorescence of each of these solutions was measured with a fluorometer, which gave values of 3.26, 4.80, 6.41, 8.02, and 9.56, respectively.

(a) Plot the data.

(b) Using the plot from (a), calculate the concentration of phenobarbital in the unknown.

(0.410μg · mL^{-1})

(c) Derive a least-squares equation for the data.

(S=3.16V_s+3.25)

(d) Find the concentration of phenobarbital from the equation in (c).

(0.410μg · mL^{-1})

(e) Calculate a standard deviation for the concentration obtained in (d).

Chapter 2 Ultraviolet and Visible Spectrophotometry

第 2 章 紫外可见分光光度法

Study Guide 学习指南

Absorptiometry is one of the typical instrumental analysis methods based on the selective light absorption properties of molecules (or atoms) of the substance in interest, including ultraviolet and visible (UV-vis.) spectrophotometry and infrared spectroscopy. This chapter will focus on the introduction of the principle, characteristics, type of commonly used instruments and industrial applications of UV-vis. spectrophotometry. The target of this chapter is to familiarize with the principle, determination procedure and result calculation of UV-vis spectrophotometry.

吸光光度法是经典的仪器分析方法之一，它是基于被测物质的分子（或原子）对光具有选择吸收的特性而建立的分析方法，包括紫外可见分光光度法和红外光谱法等。本章重点介绍紫外可见分光光度法的原理、特点、常用仪器的类型及工业应用等内容。通过本章的学习，掌握紫外可见分光光度法的原理、测定过程及结果计算。

Section 1 Introduction

第 1 节 概述

When a sample solution with a specific concentration is continuously irradiated by the light of multiple wavelengths, it shows different absorption intensity corresponding to the different wavelength. If a graph is arranged to show wavelength (λ) on the horizontal axis and absorption intensity (A) on the vertical, an absorption spectrum of the substance can be plotted. The qualitative and quantitative analysis by using the absorption spectrum

在吸光光度法中，将不同波长的光连续地照射到一定浓度的样品溶液时，便可得到与不同波长相对应的吸收强度。如以波长（λ）为横坐标，以吸光度（A）为纵坐标，就可绘出该物质的吸收光谱曲线。利用该曲线进行物质定性、定量的分析方法，称为吸光光度法，也称为吸收光谱法。用紫外光源

is known as absorptiometry or absorption spectroscopy. UV spectrophotometry means a UV light source is used to determine the colourless substances, while visible spectrophotometry means a visible light source is used to determine the coloured substances. Similar to colorimetry, the two methods are based on Beer-Lambert's Law. UV and visible regions stated hereinabove are commonly used. The applicable regions in absorptiometry include UV region (200-400 nm), visible region (400-780 nm) and infrared region (0.78-2.5 μm).

测定无色物质的方法，称为紫外分光光度法；用可见光光源测定有色物质的方法，称为可见光光度法。它们与比色法一样，都以朗伯·比尔定律为基础。上述的紫外光区与可见光区是常用的，但吸光光度法的应用光区包括紫外光区（波长 200 ～ 400nm）、可见光区（400 ～ 780nm）与红外光区（0.78 ～ 2.5μm）。

1. Properties of Electromagnetic Radiation

Electromagnetic radiation is a form of energy that is transmitted through space at enormous velocities. Electromagnetic radiation can be described as a wave with properties of wavelength, frequency, velocity, and amplitude (fig.2-1). In contrast to sound waves, light requires no supporting medium for its transmission; thus, it readily passes through a vacuum. Electromagnetic radiation can be treated as discrete packets of energy or particles called photons or quanta. These dual views of radiation as particles and waves are not mutually exclusive but complementary.

Wavelength (fig.2-2 shows the wavelengths of various types of electromagnetic radiation), **λ**, is the crest-to-

1. 电磁辐射的特性

电磁辐射是一种以很高速度通过空间的能量传播形式。电磁辐射可以描述为具有波长、频率、速度和振幅特性的波（图 2-1）。与声波相比，光不需要传播介质，因此，它能很容易通过真空。电磁辐射可以被视为离散的能量包或称为光量子的粒子。这些将辐射作为粒子和波的双重观点并不是相互排斥的，而是互补的。

波长 λ，即波峰的距离（各类型电磁辐射的波长见图 2-2）。**频率 υ** 是波

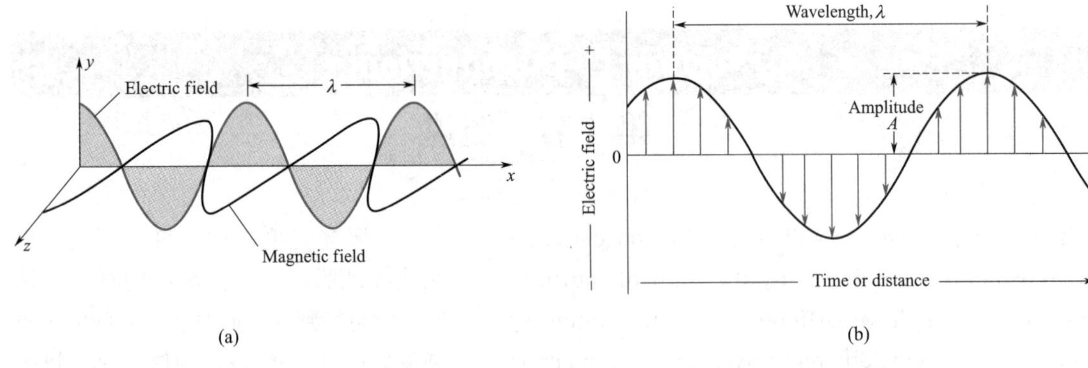

Fig.2-1 Plane-polarized electromagnetic radiation of wavelength λ, propagating along the x-axis. The electric field of plane-polarized light is confi ned to a single plane. Ordinary, unpolarized light has electric field components in all planes parallel to the direction of travel.

图 2-1 波长为 λ 的沿 x 轴传播的平面偏振电磁辐射。平面偏振光的电场集中在一个平面内。普通的非偏振光在所有平行于传播方向的平面上都有电场分量。

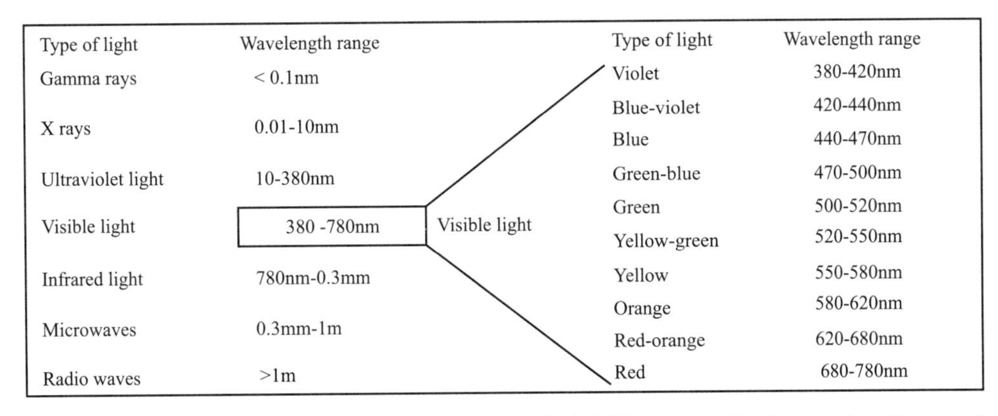

Fig.2-2 Various types of electromagnetic radiation, or "light" (The approximate wavelengths are shown for each type of electromagnetic radiation, with the scale on the right showing an enlarged view of the wavelengths that make up visible light.)

图 2-2 各种类型的电磁辐射或“光”。（明确了每种电磁辐射的大致波长，右边为可见光的波长范围放大图。）

crest distance between waves. **Frequency, υ,** is the number of complete oscillations that the wave makes each second. The unit of frequency is 1/second. One oscillation per second is called one hertz (Hz). A frequency of 10^6 s^{-1} is therefore said to be 10^6 Hz, or 1 megahertz (MHz). The relation between frequency and wavelength is

每秒产生的完全振荡的次数，频率单位为 1/ 秒，即每秒一次振荡被称为 1Hz（赫兹）。因此，$10^6 s^{-1}$ 的频率被称为 10^6Hz，或 1MHz（兆赫）。频率和波长之间的关系见式（2.1）。

$$c=\lambda\upsilon \tag{2.1}$$

where c is the speed of light (3.0×10^8 m/s in vacuum). With regard to energy, it is more convenient to think of light as particles called photons.
Each photon carries energy E given by

其中 c 为光速（在真空中为 3.0×10^8m/s）。关于能量，我们可以把光看作是光子，每个光子所携带的能量 E 见式（2.2）。

$$E=h\upsilon \tag{2.2}$$

where h is Planck's constant (6.626×10^{-34} J · s).

Equation (2.2) states that energy is proportional to frequency. Combining equation (2.1) and equation (2.2), we can write

其中 h 为普朗克常数（6.626×10^{-34} J · s）。

式（2.2）表示能量与频率成正比。结合式（2.1）和式（2.2），可得式（2.3）。

$$E=h\cdot\frac{c}{\lambda}=hc\sigma \tag{2.3}$$

where $\sigma=\frac{1}{\lambda}$ is called **wavenumber**. Energy is inversely proportional to wavelength and directly proportional to wavenumber. Red light, with a longer wavelength than blue light, is less energetic than blue light. The most common unit of wavenumber is cm^{-1}, read "reciprocal centimeters" or "wavenumbers".

其中 $\sigma=\frac{1}{\lambda}$，**为波数**。能量与波长成反比，与波数成正比。红光比蓝光波长长，能量不如蓝光。最常见的波数单位是 cm^{-1}，即“倒数厘米”或“波数”。

2. Principle of UV-Vis Spectrophotometry

2. 紫外可见分光光度法原理

(1) Transmittance and absorbance As shown in fig.2-3, when a beam of parallel light passes through a homogenous solution, part of the light is absorbed and part of the light is reflected by container. If the incident light intensity is I_0, absorbed light intensity is I_a, transmitted light intensity is I_t, and reflected light intensity is I_r, then

（1）透射比和吸光度 如图 2-3 所示，当一束平行光通过均匀的溶液介质时，光的一部分被吸收，一部分被器皿反射。设入射光强度为 I_0，吸收光强度为 I_a，透射光强度为 I_t，反射光强度为 I_r，则

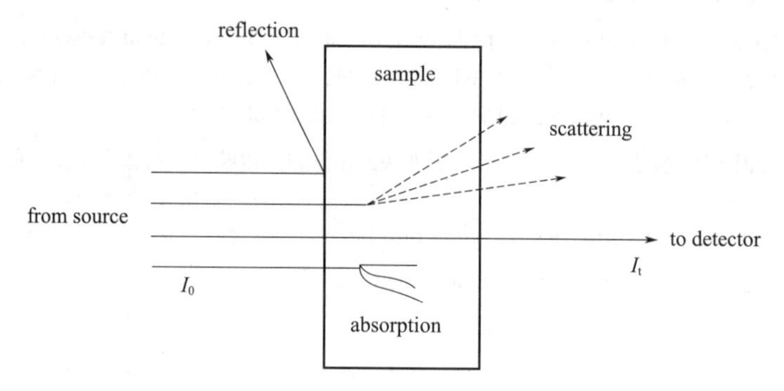

Fig.2-3 Intensity loss of a light beam of intensity I_0 by reflection, scattering and absorption

图 2-3 强度为 I_0 的光束经过反射、散射和吸收后的强度损失

$$I_0 = I_a + I_t + I_r \tag{2.4}$$

In absorption spectrophotometric analysis, the sample solution and the reference solution are contained in two absorption cells of the same material and thickness. Emit a beam of monochromatic light with intensity I_0 through the two absorption cells respectively, adjust the zero absorption point of the instrument with a reference cell, then re-measure the intensity of the transmitted light. Hence the influence of the reflected light is eliminated by the reference solution. Thus the above equation can be simplified into:

在进行吸收光谱分析中，被测溶液和参比溶液是分别放在同样材料及厚度的两个吸收池中，让强度同为 I_0 的单色光分别通过两个吸收池，用参比池调节仪器的零吸收点，再测量被测量溶液的透射光强度，所以反射光的影响可以从参比溶液中消除，见式（2.5）。

$$I_0 = I_a + I_t \tag{2.5}$$

The ratio of the intensity of transmitted light and the intensity of incident light is called the transmittance, expressed as T, then

透射光强度（I_t）与入射光强度（I_0）之比称为透射比（亦称透射率），用 T 表示。

$$T = \frac{I_t}{I_0} \tag{2.6}$$

The greater is the T of the solution, the weaker is the absorption of light, and vice versa. To explicitly illustrate the co-relationship of the absorption intensity of solution and its physical quantity, absorbance (A) is most often used to express the degree that a substance absorbs light:

溶液的 T 越大，表明它对光的吸收越弱；反之，T 越小，表明它对光的吸收越强。为了更明确地表明溶液的吸光强弱与表达物理量的相应关系，常用吸光度（A）表示物质对光的吸收程度。

From the equation, it is known that the larger is A, the stronger the substance absorbs light. Both T and A are measures showing the degree that a substance absorbs light. Transmittance is usually expressed as a percentage and called percentage transmittance $T\%$, while absorbance A is a dimensionless quantity. They can be inter-converted as equation (2.7).

A 值越大，表明物质对光吸收越强。T 及 A 都是表示物质对光吸收程度的一种量度，透射比常以百分率表示，称为百分透射比，$T\%$，吸光度 A 为一个无量纲的量，两者可通过式（2.7）互相换算。

$$A = \lg\frac{1}{T} = \lg\frac{I_0}{I_t} \tag{2.7}$$

(2) Lamber-Beer's Law　The significance of Lamber-Beer's Law in physics is that when a beam of parallel monochromatic light passes through a solution perpendicularly, the absorbance A is directly proportional to the concentration (c) and the path length (b) of solution. When the path length (b) and concentration (c) have units of centimeter and mol · L^{-1} respectively, the extinction coefficient K is shown as ε and known as molar absorptivity. Then Lamber-Beer's Law is expressed as follows, where the unit of A is L · mol^{-1} · cm^{-1}.

（2）朗伯 - 比尔定律　朗伯 - 比尔定律的物理意义是，当一束平行单色光垂直通过某溶液时，溶液的吸光度 A 与吸光物质的浓度 c 及液层厚度 b 成正比。当液层厚度 b 以 cm、吸光物质浓度 c 以“mol · L^{-1}”为单位时，系数 K 就以 ε 表示，称为摩尔吸光系数。此时朗伯 - 比尔定律表示如下。式中摩尔吸光系数单位为 L · mol^{-1} · cm^{-1}。

$$A = \varepsilon bc \tag{2.8}$$

Example 2-1

Find the absorbance and transmittance of a 0.00240 mol · L^{-1} solution of a substance with a molar absorptivity of 313 L · mol^{-1} · cm^{-1} in a cell with a 2.00 cm pathlength.

Solution

Equation (2.5) gives us the absorbance.

$A=\varepsilon bc=313\text{L} \cdot \text{mol}^{-1} \cdot \text{cm}^{-1}\times 0.00240\text{mol} \cdot \text{L}^{-1}\times 2.00\text{cm}=1.50$

$A=\lg\frac{1}{T}=-\lg T$

$\Rightarrow T=10^{\lg T}=10^{-A}=10^{-1.5}=0.0316$

Just 3.16% of the incident light emerges from this solution.

(3) Absorption curve It can be seen from figure 2-4 that

（3）光吸收曲线 由图2-4可知，

① The same substance has different absorbance of light for different wavelengths. The wavelength corresponding to the maximum absorbance is called the maximum absorption wavelength λ_{max}.

① 同一种物质对不同波长光的吸光度不同。吸光度最大处对应的波长称为最大吸收波长 λ_{max}。

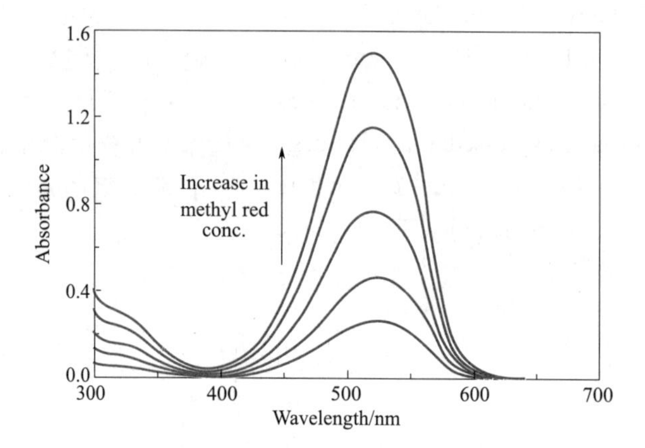

Fig.2-4 Absorption curve of the same substance at different concentrations.

图2-4 不同浓度下同一物质的吸收曲线

② For the same substance with different concentrations, the shape of absorption curve is similar and λ_{max} remains unchanged. But for different substances, their shape of absorption curve and λ_{max} are different

② 不同浓度的同一种物质，其吸收曲线形状相似 λ_{max} 不变。而对于不同物质，它们的吸收曲线形状和 λ_{max} 则不同。

③ Absorption curves can provide information on the structure of substances and serve as one of the basis for qualitative analysis of substances.

③ 吸收曲线可以提供物质的结构信息，并作为物质定性分析的依据之一。

④ For the same substance with different concentrations, absorbance A is different at a certain wavelength, and the difference in absorbance A is greatest at λ_{max}, so the characteristic can be used as a basis for quantitative analysis of substances.

④ 不同浓度的同一种物质，在某一定波长下吸光度 A 有差异，在 λ_{max} 处吸光度 A 的差异最大。此特性可作为物质定量分析的依据。

⑤ At λ_{max} the absorbance varies most with the concentration, so measurement is the most sensitive. Absorption curve is an important basis for selecting the wavelength of incident light in quantitative analysis.

⑤ 在 λ_{max} 处吸光度随浓度变化的幅度最大，所以测定最灵敏。吸收曲线是定量分析中选择入射光波长的重要依据。

3. Characteristics of UV-Vis Spectrophotometry

3. 紫外可见分光光度法特点

Spectrophotometry is one of the most useful analytical methods for analyzers. Nearly all the laboratories

分光光度法对于分析人员来说，可以说是最有用的工具之一。几乎每一个

are equipped with UV-Vis spectrophotometers. The characteristics of UV-Vis spectrophotometry are:

分析实验室都离不开紫外可见分光光度计。分光光度法的主要特点为：

(1) Wide application　Various inorganic compounds and organic compounds absorb light in the UV or visible regions, so these compounds can be determined by UV-Vis Spectrophotometric method. Of all the analysis-related papers published internationally, those subjected on spectrophotometry occupy 28%. And this figure reaches 33% in China. So far, nearly all the elements on the periodic table (except a few radioactive elements and inert gases) can be determined by this method. In APHA (American Public Health Association), the application of spectrophotometry is about 37.4% in Standard Methods for the Examination of Water and Wasterwater, and this figure is 35% in China's monitoring and analysis of water and wastewater, such as the determination of zinc ion, all silica and turbidity of recycle cooling water; the determination of trace iron, phosphate radical and trace copper in the boiler feed water.

（1）应用广泛　由于各种各样的无机物和有机物在紫外可见区都有吸收，因此均可借此法加以测定。在国际上发表的有关分析的论文总数中，光度法约占28%，我国发表的论文约占所发表论文总数的33%。到目前为止，几乎周期表上的所有元素（除少数放射性元素和惰性气体外）均可采用本法测定。美国水和废水标准检验方法中光度法占37.4%，我国水和废水监测分析方法中光度法占35.0%。如：循环冷却水中锌离子、全硅、浊度等的测定；锅炉给水中微量铁、磷酸根、微量铜等的测定。

(2) High sensitivity　Thanks to the development of relevant scientific fields, the synthesis of and the research on the new organic chromogenic agents have made significant progress, and the sensitivity in the determination of elements is greatly improved thereby. In particular, the application of multicomponent complex and various surfactants has promoted the molar absorptivity of many elements from tens of thousands to hundreds of thousands. Comparing with other trace analysis methods, the precision and accuracy of spectrophotometry is highly acknowledged. Furthermore, spectrophotometry is not only widely used in the actual applications but also valued in the researches of standard reference substances. A series of spectrophotometric methods have been drawn up into standards.

（2）灵敏度高　由于相应学科的发展，使新的有机显色剂的合成和研究取得可喜的进展，从而对元素测定的灵敏度大大提高了一步。特别是由于多元络合物和各种表面活性剂的应用研究，使许多元素的摩尔吸光系数由原来的几万提高到几十万。相对于其他痕量分析方法而言，光度法的精密度和准确度公认是比较高的。不但在实际工作中光度法被广泛采用，在标准参考物质的研制中，它更受重视，很多光度分析法已制定成为标准方法。

(3) Good selectivity　Till now, some elements such as Co, U, Ni, Cu, Ag and Fe can be determined via the spectrophotometric method as long as there are appropriate and controllable colour conditions.

（3）选择性好　目前，已有些元素只要利用控制适当的显色条件就可直接进行光度法测定，如钴、铀、镍、铜、银、铁等元素的测定，已有比较满意的方法了。

(4) High accuracy　For general spectrophotometric

（4）准确度高　对于一般的分光

methods, the relative error of measured concentration falls in the range of 1%-3%. If differential spectrophotometry is employed, the relative error often drops to three significant figures.

光度法来说，其浓度测量的相对误差在1%～3%范围内，如采用示差分光度法测量，则误差往往可减少到千分之几。

(5) Broad concentration range   The concentration of solutions can vary from normal quantity (1%-50%) (differential spectrophotometry is most applicable) to trace quantity (10^{-6} %-10^{-8} %) (after being preenriched).

（5）适用浓度范围广　可从检测常量（1%～50%）（尤其是使用示差法）到痕量（10^{-6} %～10^{-8} %）（经预富集后）的物质含量。

(6) UV-Vis spectrophotometry is low in cost, easy with operation and rapid in process.

（6）分析成本低、操作简便、快速。

Section 2 UV-Vis Spectrophotometer

第2节　紫外可见分光光度计

1. Basic Structure

(1) Sources of light   The light sources could emit continuous light in UV and visible regions, with sufficient radiation intensity, good stability and long life time.

Visible region:

A tungsten lamp (fig.2-5) is an excellent source of continuous visible and near-infrared radiation. A typical tungsten filament operates at a temperature near 3000K and produces useful radiation in the range **320 nm to 2500 nm** (fig.2-6).

Ultraviolet region:

Ultraviolet spectroscopy normally employs a deuterium arc lamp in which a controlled electric discharge causes D_2 to dissociate and emit ultraviolet radiation from **180 nm to 400 nm**.

(2) Monochromator   A monochromator disperses light into its component wavelengths and selects a narrow band of wavelengths to pass on to the sample or detector. The monochromator in figure 2-7 consists of entrance and exit slits, mirrors, and a grating to disperse

1. 基本结构

Operation of UV7504
UV7504 型紫外可见分光光度计的基本操作

（1）光源　在整个紫外光区或可见光谱区可以发射连续光谱，具有足够的辐射强度、较好的稳定性、较长的使用寿命。

可见光区及近红外光区常用钨灯（图2-5）作为光源，其辐射波长范围在320～2500nm。

紫外区常用氢灯或氘灯作光源，发射180～400nm的连续光谱（图2-6）。

（2）单色器　单色器是将光源发射的复合光分解成单色光，并从中选出一个较窄波长带的单色光传递给样品或检测器。图2-7中的单色器由入口和出口狭缝、凹面镜和一个分散光的光栅组成。在较旧的仪器中，人们使用了棱镜来代替光栅，见图2-8。

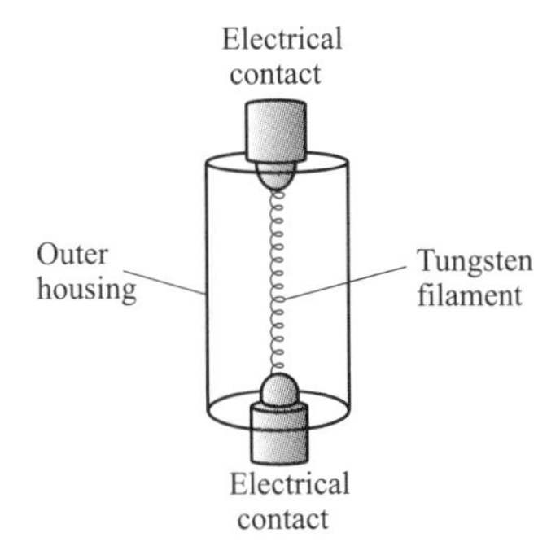

Fig.2-5 The general design of a tungsten lamp(A tungsten/halogen lamp has a similar design, but includes some I_2 in the chamber that surrounds the tungsten filament)

图 2-5 钨灯结构示意图［卤钨灯结构类似，只是在填充气体中含有卤族元素或卤化物（如 I_2）］

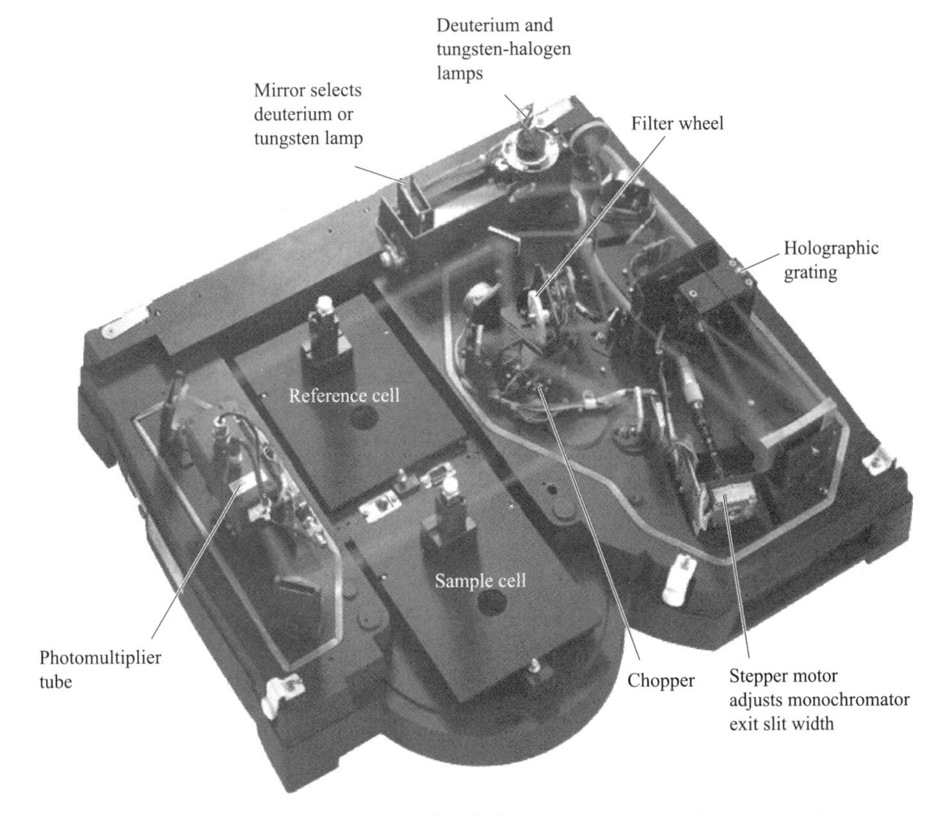

Fig.2-6 Optical train of Evolution 600, showing layout of components

图 2-6 Evolution 600 型紫外可见分光光度计光学系统及其布局

the light. Prisms (fig.2-8) were used instead of gratings in older instruments.

① Entrance slit: light from source enters into monochromator through this slit.

① 入射狭缝：光源的光由此进入单色器；

② Collimating device: a lens or a reflector that changes incident lights into a parallel light beam.

② 准光装置：透镜或反射镜使入射光成为平行光束。

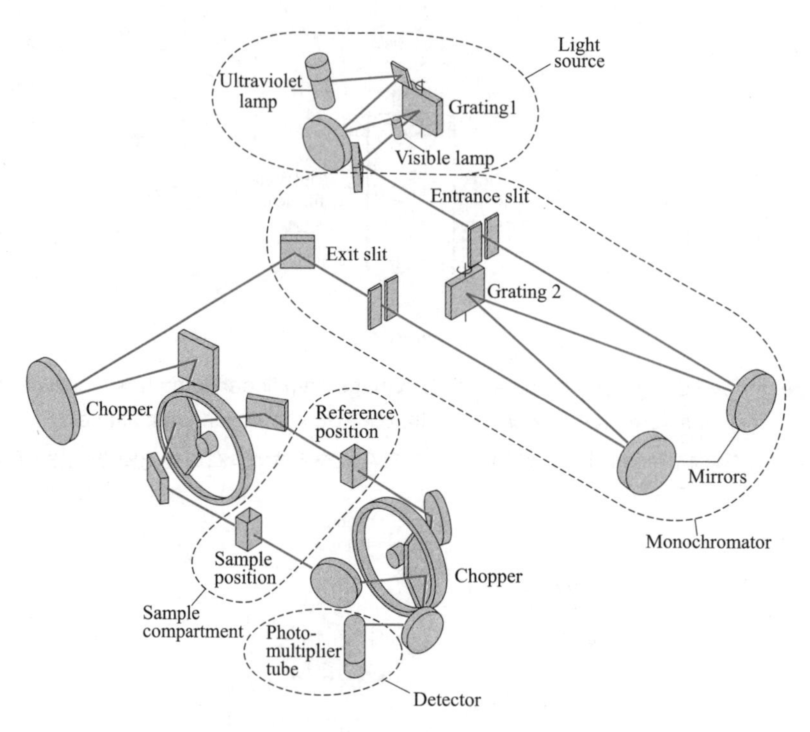

Fig.2-7 Optical train of Varian Cary 3E ultraviolet-visible double-beam spectrophotometer

图 2-7 Varian Cary 3E 型双光束紫外可见分光光度计光学系统

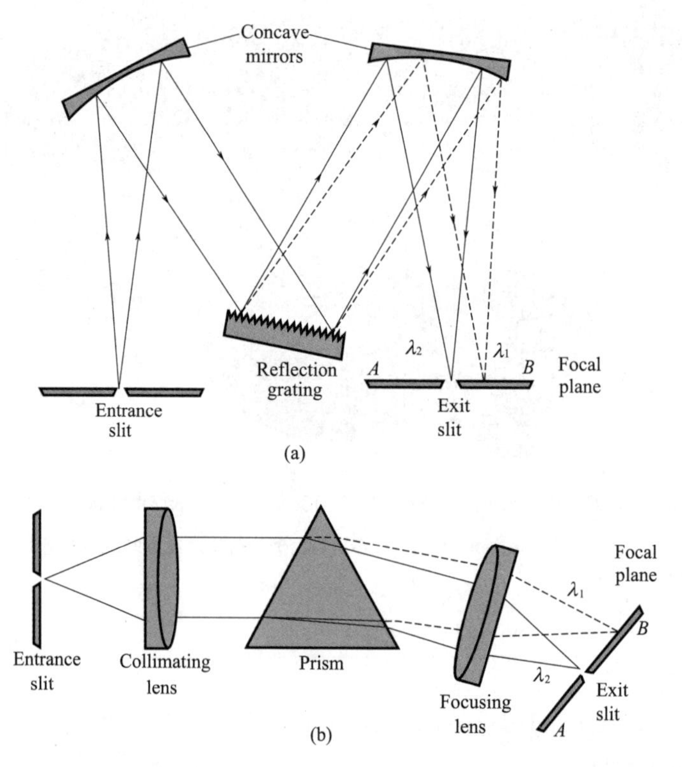

Fig.2-8 Two types of monochromators (a) Czerney-Turner grating monochromator, (b) Bunsen prism monochromator (In both instances, $\lambda_1 > \lambda_2$.)

图 2-8 两种典型的单色器（a）切尔尼 - 特纳光栅单色器；（b）本生棱镜单色器（在这两个实例中，$\lambda_1 > \lambda_2$。）

③ Dispersive element: a prism or a grating which used to separate polychromatic lights into monochromatic lights.

④ Focusing device: a lens or a concave reflector, focusing monochromatic light onto the exit slit.

⑤ Exit slit.

(3) Sample cell (absorption cell)　The sample chamber is placed with various types of absorption cells (cuvettes) and their cell rack accessories. There are two main types of absorption cell: quartz cell and glass cell. A quartz cell (glass can absorb ultraviolet light) must be used in the ultraviolet region, and a glass cell is often used in the visible region. Figure 2-9 shows common cuvets for visible and ultraviolet spectroscopy.

③ 色散元件：将复合光分解成单色光；棱镜或光栅。

④ 聚焦装置：透镜或凹面反射镜，将分光后所得单色光聚焦至出射狭缝。

⑤ 出射狭缝。

Compatibility test of sample cell
吸收池配套性检验

（3）样品室（吸收池）　样品室放置各种类型的吸收池（比色皿）和相应的池架附件。吸收池主要有石英池和玻璃池两种。在紫外区须采用石英池（因为玻璃能够吸收紫外光），可见区一般用玻璃池。图 2-9 展示了常见的紫外可见分光光度计吸收池。

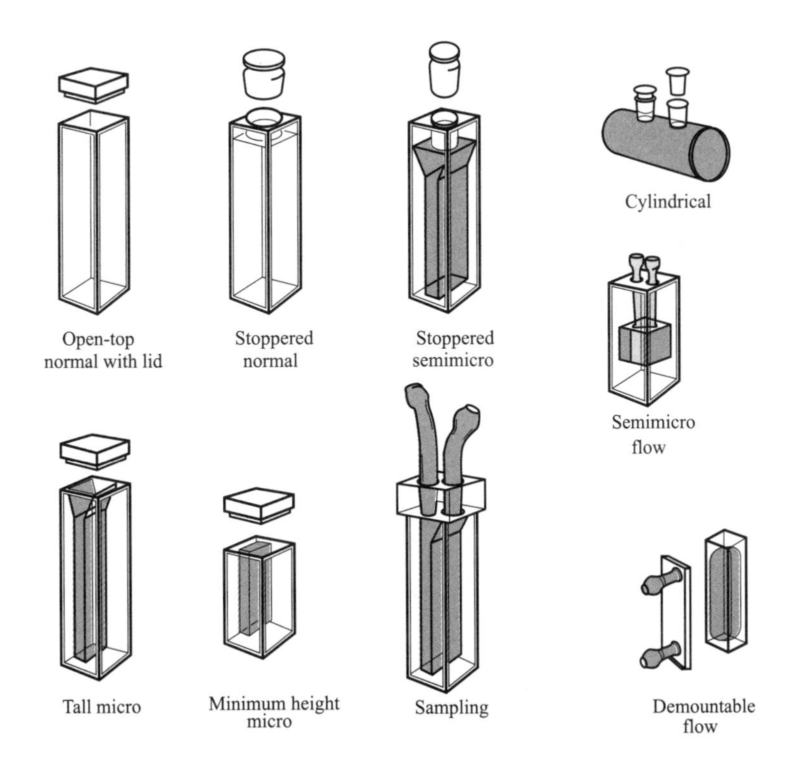

Fig.2-9　Common cuvets for visible and ultraviolet spectroscopy(Flow cells permit continuous flow of solution through the cell. In the thermal cell, liquid from a constant temperature bath flows through the cell jacket to maintain a desired temperature.)

图 2-9　常见的紫外可见分光光度计吸收池（流动池允许溶液连续地流过该池。在恒温样品池中，来自恒温槽的液体流过样品池夹套，保持所需的温度。）

(4) Detector Detectors change the optical signals passed through absorption cell into electric signals by photo effect. The most often used detectors are photocell, phototube or photomultiplier tube. A detector produces an electric signal when it is struck by photons. For example, a phototube emits electrons from a photosensitive, negatively charged surface (the cathode) when struck by visible light or ultraviolet radiation. Electrons flow through a vacuum to a positively charged collector whose current is proportional to the radiation intensity (figure 2-10-figure 2-12).

（4）检测器 利用光电效应将透过吸收池的光信号变成可测的电信号，常用的有光电池、光电管或光电倍增管。当检测器被光子轰击时，会产生电信号。例如，当被可见光或紫外光照射时，光电管从光敏的、带负电荷的表面（阴极）发射电子。电子通过真空流入一个带正电荷的集电器，其电流与辐射强度成正比，见图 2-10 ～图 2-12。

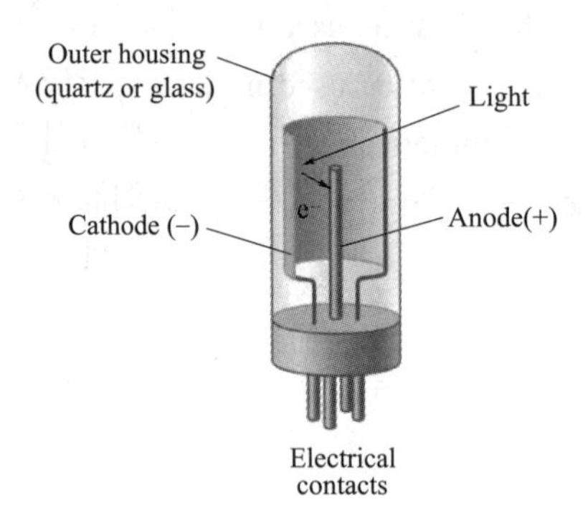

Fig.2-10 The basic components of a phototube(As light enters this device and strikes the cathode, electrons are given off by a photoemissive material on the cathode's surface. These electrons then travel to the more positive anode. This produces a current that is related to the intensity of light entering the phototube and striking the cathode.)

图 2-10 光电管的基本组件

（当光照到阴极的光敏材料时，阴极发射出电子，被阳极收集而产生光电流，该电流与光强度有关。）

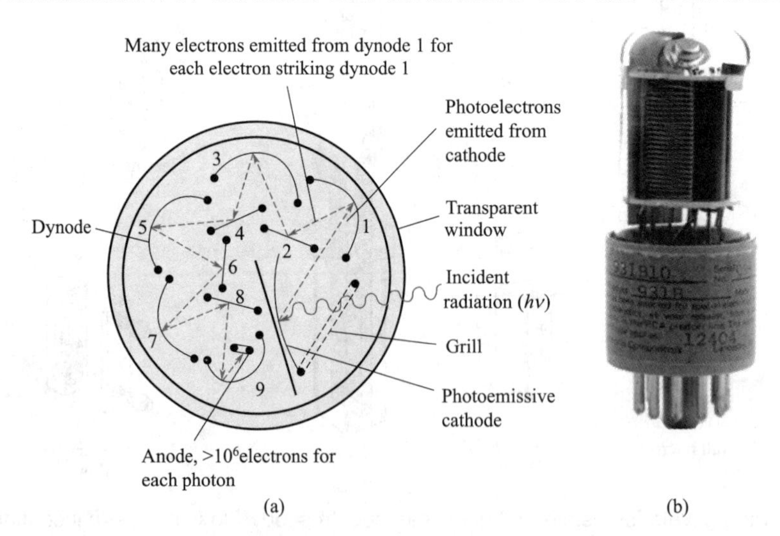

Fig.2-11 Photomultiplier tube with nine dynodes(Amplification of the signal occurs at each dynode, which is approximately 90V more positive than the previous dynode.)

图 2-11 具有 9 个电子发射器的光电倍增管 [每个次级电子发射器（电子倍增级）产生的信号比前一发射器放大 90V。]

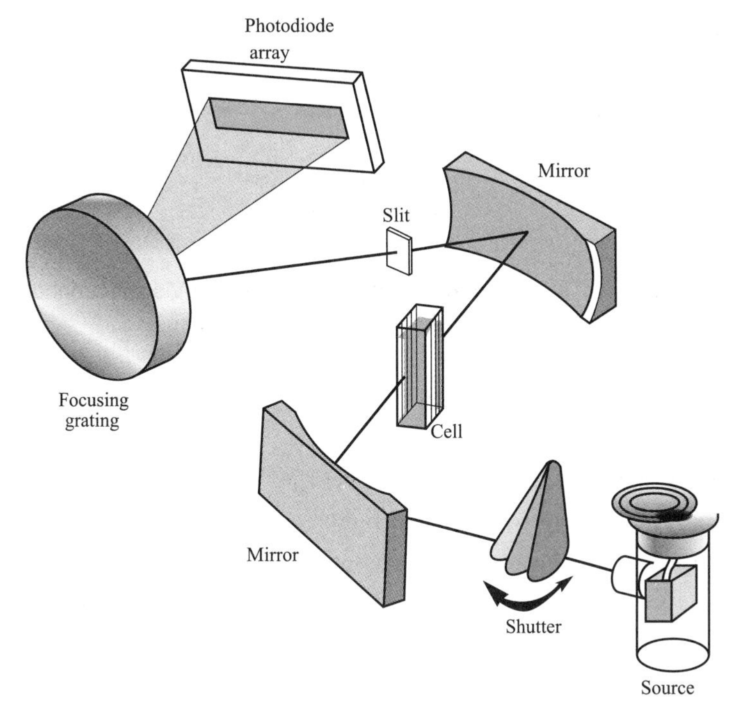

Fig.2-12 Design of photodiode array spectrophotometer

图 2-12 光电二极管阵列分光光度计

(5) Readout recording system Galvanometer, digital indicator, automat ic control and data processing by computers.

（5）结果显示记录系统 检流计（721 型分光光度计）、数字显示（7230，UV-1801 等型分光光度计），并以计算机进行仪器自动控制和结果处理。

2. Type of Spectrophotometer

2. 仪器类型

(1) Single wavelength single beam spectrophotometer Diagram of a single wavelength single beam UV-Vis spectrophotometer is shown in figure 2-13. After the lights are splitted by a monochromator, a beam of parallel light passes through the reference solution and the sample solution in order to determine the absorbance. This type of spectrophotometer is simple in structure, easy in operation and maintenance, applicable to routine analysis. The determination of transmittance involves three successive steps that are separated in time: (1) the T (0%) setting with a shutter in place, (2) the T (100%) adjustment with the solvent in the light path, and (3) the measurement of T with the sample in place.

（1）单波长单光束分光光度计 其光路示意图如图 2-13 所示，经单色器分光后的一束平行光，轮流通过参比溶液和样品溶液，以进行吸光度的测定。这种简易型分光光度计结构简单，操作方便，维修容易，适用于常规分析。透射率的测定包括三个连续步骤：（1）透射率 T(0%) 调节；（2）透射率 T(100%) 调节；（3）样品的透射率测定。

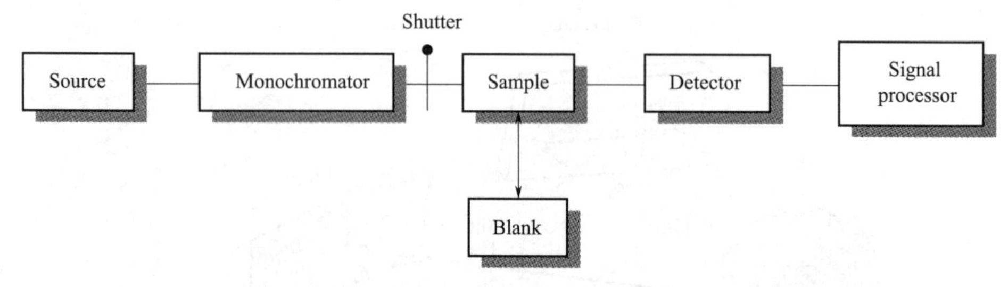

Fig.2-13 Diagram of a single wavelength single beam UV-Vis spectrophotometer

图 2-13 单波长单光束紫外可见分光光度计示意图

(2) Single wavelength dual beam spectrophotometer Diagram of a single wavelength dual beam UV-Vis spectrophotometer is shown in figure 2-14. The light is splitted by a reflector into two beam of light of same intensity, in which one beam of light passes through reference cell and another passes through sample cell. The photometer automatically compares the intensities of the two beams, the ratio of which is the transmittance of the sample. The logarithm of this ratio is the transmittance, which is recorded as a function of wavelength. Generally, dual beam spectrophotometers could record the absorption curce automatically. Since two beams pass through reference cell and sample cell at the same time, the measuring error caused by the change of light intensity can be corrected.

（2）单波长双光束分光光度计 其光路示意图见图 2-14。经单色器分光后经反射镜分解为强度相等的两束光，一束通过参比池，另一束通过样品池，光度计能自动比较两束光的强度，此比值即为试样的透射比，经对数变换将它转换成吸光度并作为波长的函数记录下来。双光束分光光度计一般都能自动记录吸收光谱曲线。由于两束光同时分别通过参比池和样品池，还能自动消除光源强度变化所引起的误差。

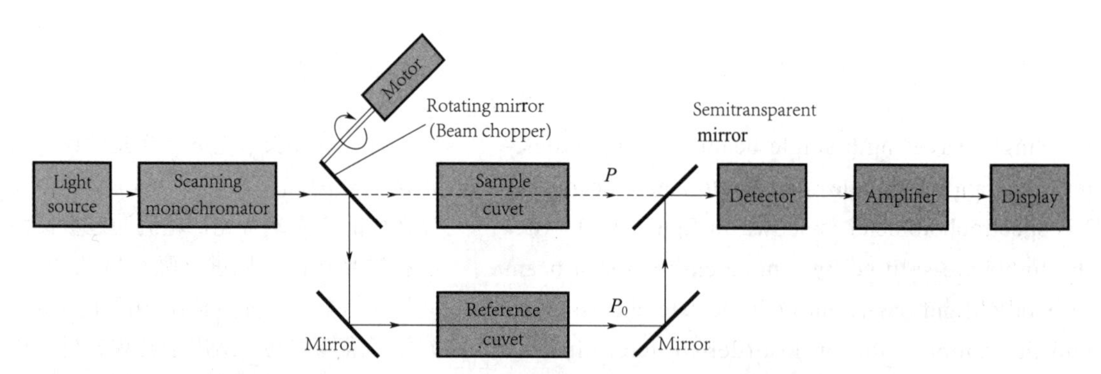

Fig.2-14 Schematic diagram of a single wavelength double beam scanning spectrophotometer(The rotating beam chopper passes the incident beam alternately through the sample and reference cuvets.)

图 2-14 单波长双光束扫描分光光度计原理图（旋转斩光器使入射光束交替地通过样品池和参比池。）

(3) Dual wavelength Dual beam spectrophotometer Diagram of a dual wavelength dual beam UV-Vis spectrophotometer is shown in figure 2-15. The light emitted from a single source is splitted

（3）双波长双光束分光光度计 其基本光路如图 2-15 所示。由同一光源发出的光被分成两束，分别经过两个单色器，得到两束不同波长（λ_1 和 λ_2）的

into two beams. The two beams pass through two monochromators to become two monochromatic lights with different wavelength (λ_1 and λ_2). The two beams of monochromatic lights pass through a beam chopper, which blocks one beam at a time. Thus, two beams of light pass through the same absorption cell, photomultiplier tube and and electronic control system alternatively. Finally the indicator will show the absorbance difference of the two wavelengths. The use of dual wavelength spectrophotometer often can improve the sensitivity and selectivity for multicomponent samples and muddy samples (e.g. tissue fluid), or if background interference or absorption interference of co-components exist. The derivative spectrum can be obtained by using dual wavelength spectrophotometer. By the optical conversion, dual wavelength UV-Vis spectrophotometer can easily work in the way of the single wavelength instrument. If a curve of the absorbance change with time is recorded and plotted at λ_1 and λ_2, the kinetics of chemical reactions can also be studied.

单色光；利用切光器使两束光以一定的频率交替照射同一吸收池，然后经过光电倍增管和电子控制系统，最后由显示器显示出两个波长处的吸光度差值。对于多组分混合物、混浊试样（如生物组织液）分析，以及存在背景干扰或共存组分吸收干扰的情况下，利用双波长分光光度法，往往能提高方法的灵敏度和选择性。利用双波长分光光度计，能获得导数光谱。通过光学系统转换，使双波长分光光度计能很方便地转化为单波长工作方式。如果能在 λ_1 和 λ_2 处分别记录吸光度随时间变化的曲线，还能进行化学反应动力学研究。

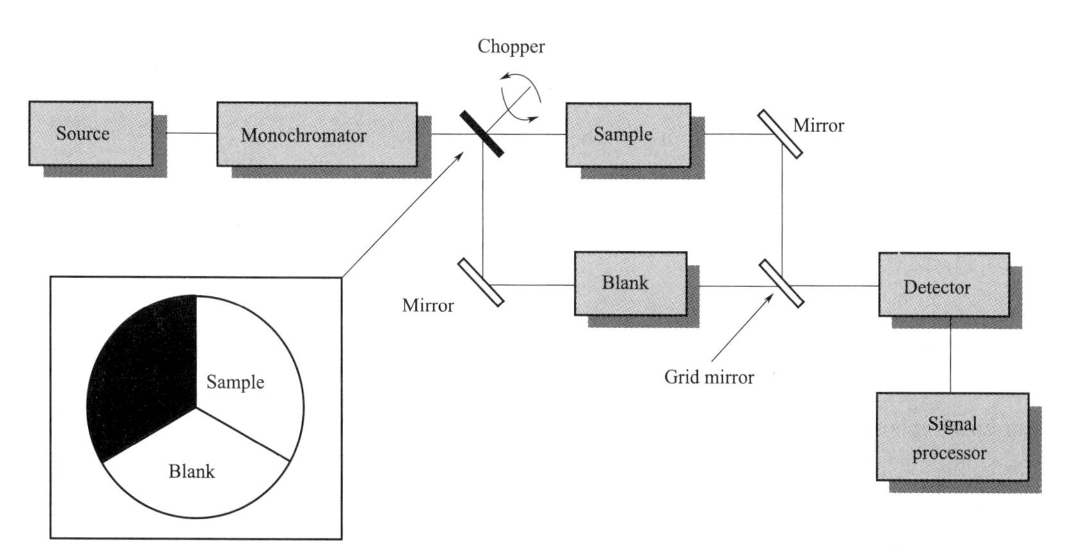

Fig.2-15 Diagram of a dual wavelength dual beam UV-Vis spectrophotometer

图 2-15 双波长双光束紫外可见分光光度计示意图

Section 3 Selection of Instrument Measuring Conditions

第 3 节 仪器测量条件的选择

The instrument measuring conditions include the measuring wavelength, proper absorbance range, the slit width and reference solution.

仪器测量条件的选择包括测量波长、适宜吸光度范围、仪器狭缝宽度及参比溶液的选择。

1. Selection of Measuring Wavelength

In general, the maximum absorption wavelength of the most intensified absorption band is selected as the measuring wavelength to obtain the highest analytical sensitivity, which is called the principle of maximum absorption. Around the maximum absorption wavelength, the absorbance has only slight changes along with the change of wavelength, so that a high measuring precision is secured. However, it's better to choose the absorption peak wavelength of low sensitivity when determining highly concentrated substances in order to make sure the calibration curve has enough linear range. If the absorption peak is too spiked, the wavelength of low sensitivity can be chosen on the premise that the sensitivity meets the measuring requirements to lower down the deviation in Beer-Lambert's Law.

1. 测量波长的选择

通常都是选择最强吸收带的最大吸收波长作为测量波长，称为最大吸收原则，以获得最高的分析灵敏度。而且在最大吸收波长附近，吸光度随波长的变化一般较小，可得到较好的测定精密度。但在测量高浓度组分时，宁可选用灵敏度低一些的吸收峰波长（ε 较小）作为测量波长，以保证校正曲线有足够的线性范围。如果所处吸收峰太尖锐，则在满足分析灵敏度前提下，可选用灵敏度低一些的波长进行测量，以减少比尔定律的偏差。

Wavelength calibration
分光光度计波长校正

2. Selection of Proper Absorbance Range

All the spectrophotometers have measuring errors due to the instability of light sources, inaccuracy of readouts or the accidental changes in experiment conditions during measurements. In Beer-Lambert's Law, the transmittance T is negatively logarithmic to concentration c. It is known from the relation of negative logarithm that the same transmittance readout errors cause different relative errors of concentration in different concentration ranges. When the concentration is too high or too low, the relative error

2. 适宜吸光度范围的选择

任何光度计都有一定的测量误差，这是由于测量过程中光源的不稳定、读数的不准确或实验条件的偶然变动等因素造成的。由于吸收定律中透射率 T 与浓度 c 是负对数的关系，从负对数的关系曲线可以看出，相同的透射比读数误差在不同的浓度范围中，所引起的浓度相对误差不同，当浓度较大或浓度较小时，相对误差都比较大。**因此，要选**

is high. **Therefore, proper absorbance ranges have to be selected to lower down the relative errors of results.** In absorptiometric analysis, the measuring range of A is generally from 0.2 to 0.8 (T is from 65% to 15%). If the readout error (ΔT) of absorbance is 1%, the relative error of measuring result ($\Delta c/c$) is about 3%. In fig.2-16, the relative error of measuring was minimal when T=36.8%, namely A=0.434. In practical work, we can adjust the concentration of sample solution or choose appropriate thickness of absorption cell to help the absorbance fall in the required range.

择适宜的吸光度范围进行测量，以降低测定结果的相对误差。在吸光分析中，一般选择 A 的测量范围为 0.2 ～ 0.8（T 为 65% ～ 15%），此时如果仪器透射率读数误差（ΔT）为 1% 时，由此引起的测定结果相对误差（$\Delta c/c$）约为 3%。图 2-16 中，当 T=36.8%，即 A=0.434 时，测量的相对误差最小。在实际工作中，可通过调节待测溶液的浓度或选用适当厚度的吸收池的方法，使测得的吸光度落在所要求的范围内。

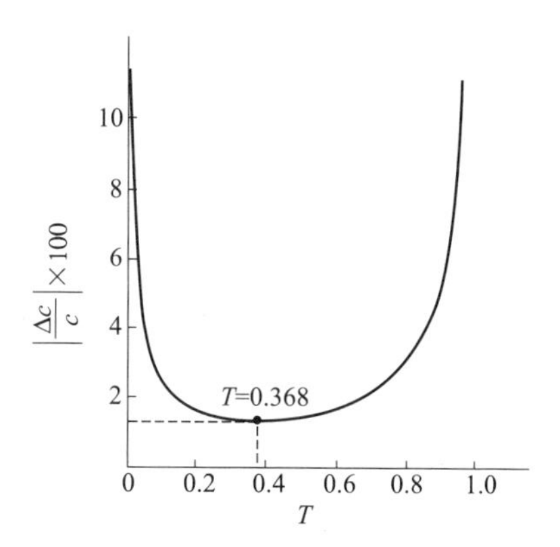

Fig.2-16 The relationship between measuring errors and transmittance

图 2-16 测量误差与透射率的关系

3. Selection of Slit Width

The width of slit directly affects the sensitivity of measurement and the linear range of calibration curve. If the slit is too wide, the monochromatic light of incident light is reduced, the calibration curve will deviate from Beer-Lambert's Law and the sensitivity will be low. If the slit is too narrow, light will become weak, the instrument gain has to be increased which consequently results in loud noise, which is not good for the measurement. To select a proper slit width, the change of absorbance with the change of slit width should be measured. When the slit width is within a certain range, the absorbance does not change. However, if the slit width exceeds a limit, the absorbance will decrease. Therefore, the largest slit width within that limit is the most proper slit width for the measurement.

3. 仪器狭缝宽度的选择

狭缝的宽度会直接影响到测定的灵敏度和校准曲线的线性范围。狭缝宽度过大时，入射光的单色光降低，校准曲线偏离朗伯 - 比尔定律，灵敏度降低；狭缝宽度过窄时，光强变弱，势必要提高仪器的增益，随之而来的是仪器噪声增大，不利于测量。选择狭缝宽度的方法是：测量吸光度随狭缝宽度的变化。狭缝的宽度在一个范围内，吸光度是不变的，当狭缝宽度大到某一程度时，吸光度开始减小。因此，在不减小吸光度时的最大狭缝宽度，即是所要选取的合适的狭缝宽度。

4. Selection of Reference Solution

When measuring the absorbance of a sample solution, the transmittance of the sample solution should be adjusted to 100% with a reference solution first in order to eliminate the errors caused by other components in the solution, absorption cell as well as the reflection and absorption of light by solvent. Hence it is important to select a reference solution with proper components according to the properties of the sample solution.

(1) Solvent as reference solution　If the composition of the sample solution is simple with few co-existing components which seldom absorbs measuring light and the chromogenic agent does not absorb light, solvent (e.g. distilled water) can be utilized as the reference solution to get rid of the influence by solvent and absorption cell.

(2) Reagent as reference solution　If chromogenic agent or other reagents absorb light in the measuring range of wavelength, add the reagent and solvent as reference solution under the same conditions as the colour reaction to eliminate the influence of reagents absorption in sample solution.

(3) Sample solution as reference solution　If the sample solution matrix absorbs light in the measuring range of wavelength but does not react with chromogenic agent, the sample can be treated in the same way as in the colour reaction, only that the chromogenic agent is not added. This kind of reference solution is applicable to the situations in which sample solutions have many co-existing components, little amount of chromogenic agent is added and the chromogenic agent does not absorb light in that wavelength range.

(4) Sample solution of parallel test as reference solution　Handle a sample solution that does not contain analyte under the same conditions as for the sample solution to obtain the parallel test sample solution as the reference solution.

4. 参比溶液的选择

测量试样溶液的吸光度时，先要用参比溶液调节透射率为100%，以消除溶液中其他成分的吸收以及吸收池和溶剂对光的反射和吸收所带来的误差。根据试样溶液的性质，选择合适组分的参比溶液是很重要的。

（1）溶剂参比　当试样溶液的组成较为简单，共存的其他组分很少且对测定波长的光几乎没有吸收，而且显色剂也没有吸收时，可采用溶剂（如蒸馏水等）作为参比溶液，这样可消除溶剂、吸收池等因素的影响。

（2）试剂参比　如果显色剂或其他试剂在测定波长有吸收，按显色反应相同的条件，只是不加入试样，同样加入试剂和溶剂作为参比溶液。这种参比溶液可消除样品溶液中试剂吸收产生的影响。

（3）试样参比　如果试样基体在测定波长有吸收，而与显色剂不起显色反应时，可按与显色反应相同的条件处理试样，只是不加显色剂。这种参比溶液适用于试样中有较多的共存组分，加入的显色剂量不大，且显色剂在测定波长无吸收的情况。

（4）平行操作溶液参比　用不含被测组分的试样，在相同条件下与被测试样同样进行处理，由此得到平行操作参比溶液。

Section 4 Quantitative Analytical Methods of UV-Vis Spectrophotometry

第4节　紫外可见分光光度法的定量分析方法

1. Quantitative Analysis of Single Component

If only one component of a sample is to be determined, and the other components in the sample does not interfere with the analyte in the range of measuring wavelength, the quantitative analysis of this single component is simple. There are usually two methods, one is standard comparison and another is standard curve (fig.2-17).

1. 单组分的定量分析

如果在一个样品中只要测定一种组分，且在选定的测量波长下，试样中其他组分对该组分不干扰，这种单组分的定量分析较简单。一般有标准对照法和标准曲线法两种（图 2-17）。

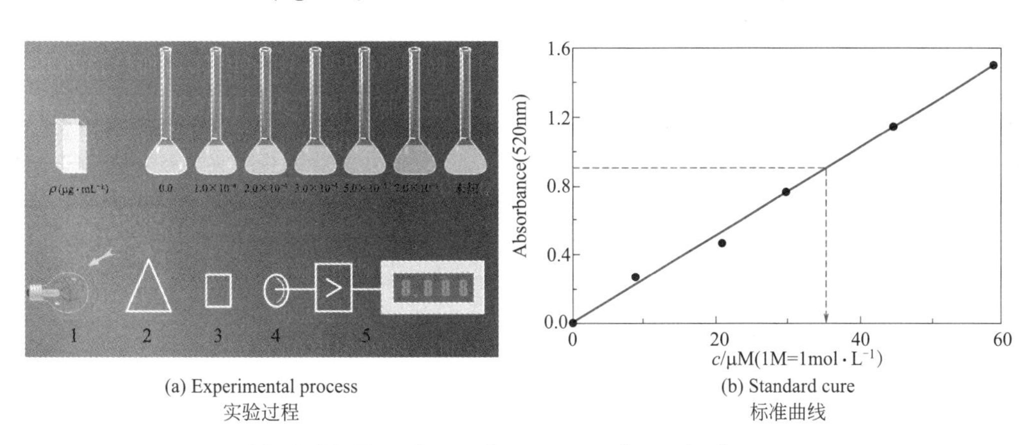

(a) Experimental process
实验过程

(b) Standard cure
标准曲线

Fig.2-17　Experimental process and standard curve

图 2-17　标准曲线法及其实验过程

1—光源；2—单色器；3—吸收池；4—接收器；5—测量系统

2. Quantitative Analysis of Multiple Components

If the absorption spectra of all the components do not overlap, the quantitative analytic method of single component can be employed; If the absorption spectra of two components overlap (fig.2-18), measure the absorbances in multiple wavelengths according to the additivity of absorbance and get the result by simultaneous equation.

2. 多组分的定量分析

当各组分的吸收光谱不重叠时，可以用单组分的定量分析方法分别测定；当两种组分的吸收光谱有重叠时（图 2-18），可以根据吸光度的加和性，在多个波长下测定吸光度，并利用解联立方程的方法求解。

$$\begin{cases} A_1=\varepsilon_{a1}bc_a+\varepsilon_{b1}bc_b \\ A_2=\varepsilon_{a2}bc_a+\varepsilon_{b2}bc_b \end{cases}$$

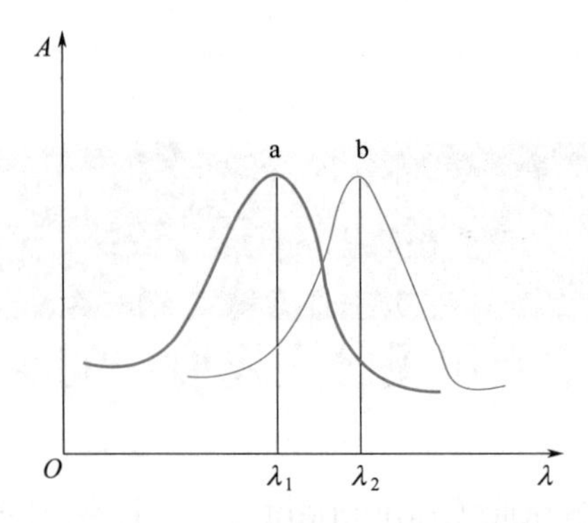

Fig.2-18 Two-component absorption curve (overlapping)

图 2-18 双组分吸收曲线（重叠）

3. Dual Wavelength Spectrophotometry

If the absorption spectra of two components in the sample overlap greatly, there will be large errors to determine the contents of the two components by simultaneous equation. At this time, dual wavelength spectrophotometric method (fig.2-19) can be used, which can determine a analyte under interference or determine two analytes at the same time. There are two methods available in determining two components by dual wavelength spectrophotometry, equal absorption point of dual wavelength and ratio.

3. 双波长分光光度法

当试样中两组分的吸收光谱重叠较为严重时，用解联立方程的方法测定两组分的含量可能误差较大，这时可以用双波长分光光度法（图 2-19）测定。它可以在有其他组分干扰时，测定某组分的含量，也可以同时测定两组分的含量。双波长分光光度法定量测定两混合物组分的主要方法有等吸收波长法和系数倍率法两种。

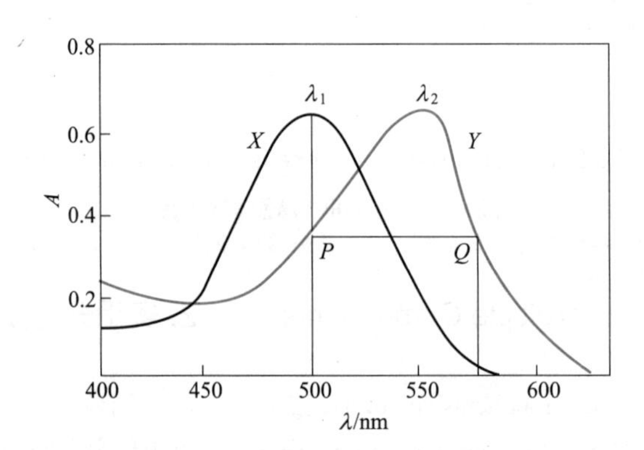

Fig.2-19 Dual Wavelength Spectrophotometry

图 2-19 双波长分光光度法

4. Differential Spectrophotometry

The determination of absorbance of extremely diluted or concentrated solution by common spectrophotometry generates large errors. Select a standard solution of

4. 示差分光光度法

用普通分光光度法测定很稀或很浓溶液的吸光度时，测量误差都很大。若用一已知合适浓度的标准溶液作为参比

known concentration as reference solution, adjust the transmittance of instrument to 100% (that is absorbance to 0), then measure the transmission ratio of sample solution to standard solution. It is found that the accuracy of absorbance is improved (fig.2-20). This method is called differential spectrophotometry. The principle is that the standard solution concentration used as a reference shall be c_s, and the solution concentration to be tested shall be c_x, and $c_x>c_s$. According to Lamber-Beer's Law:

溶液，调节仪器的 100% 透射率点（即 0 吸光度点），测量样品溶液对该已知标准溶液的透射率，则可以改善测量吸光度的精确度（图 2-20），这种方法称为示差分光光度法。其原理为：设用作参比的标准溶液浓度为 c_s，待测试液浓度为 c_x，且 $c_x > c_s$。根据朗伯比尔定律，则有：

$$A_s=\varepsilon bc_s \qquad A_x=\varepsilon bc_x$$
$$\Delta A=A_x-A_s=\varepsilon bc_x-\varepsilon bc_s=\varepsilon b\Delta c \tag{2.9}$$

The formula shows that the difference in the absorbance is proportional to the difference in the concentration of these two solutions. Draw the ΔA-Δc standard curve, find the corresponding Δc according to the measured ΔA, and then find the concentration of the liquid to be calculated from $c_x=c_s+\Delta c$.

该式表明，吸光度之差与这两种溶液的浓度差成正比。作 ΔA-Δc 标准曲线，根据测得的 ΔA 即可求出对应的 Δc，再从 $c_x=c_s+\Delta c$ 可求出待测试液的浓度。

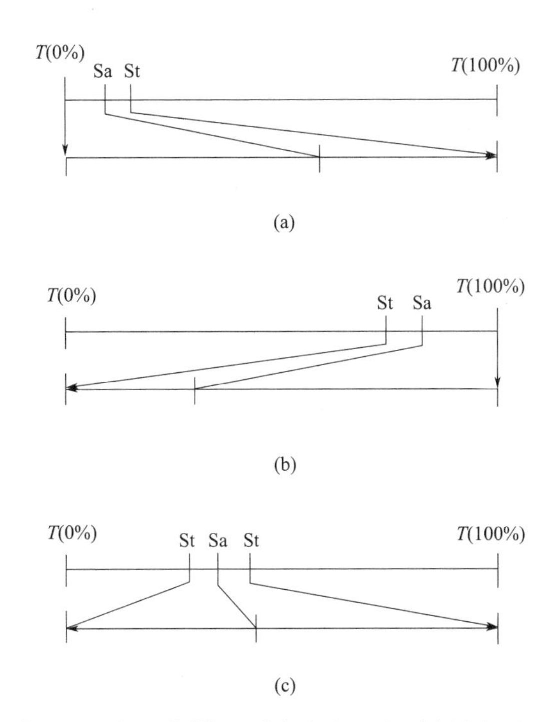

Fig.2-20 Principle of scale extension of differential photometry (a) high-absorbance method, (b) low-absorbance method, and (c) maximum precision method (Abbreviations: Sa=sample; St=standard.)

图 2-20 示差分光度法的尺度扩展原理（a）高吸光度法；（b）低吸光度法；（c）最大精度法（缩写：Sa= 样本；St= 标准。）

Experiment 1
Determination of trace iron by *o*-diphenanthrene spectrophotometry

1. Purposes of the experiment

(1) Familiar with the use of visible spectrophotometer.

(2) Practice mapping absorption curves.

(3) Master the basic method and calculation of spectrophotometric determination of trace components by using standard curve.

2. The experimental principle

In the solution of pH 2-9, *o*-diphenanthrene (abbreviated as phen) and Fe^{2+}occur the following chromogenic reactions:

$$Fe^{2+}+3phen \longrightarrow [Fe(phen)_3]^{2+}$$

The resulting orange-red complex is very stable, $\lg K_{stable}=21.3$ (20℃), in which has the maximum absorption peak at 510 nm with molar absorption coefficient $\varepsilon_{510}=1.1\times10^4\ L\cdot mol^{-1}\cdot cm^{-1}$. Trace iron can be determined by the above reaction.

The suitable pH range of color reaction is wide (2-9). The reaction rate is slow when the acidity is too high (pH<2). Fe^{2+} will hydrolyze, when the acidity is too low. It is usually measured in HOAc-NaOAc buffer media with a pH of about 5.

The selectivity of the reaction between o-diphenanthrene and Fe^{2+} is very high. Co^{2+}, Cu^{2+} which are equal to 5 times of iron content, Cr^{3+}, Mn^{2+}, PO_4^{3-}, V (Ⅴ), which are equal to 20 times of iron content and even Al^{3+}, Ca^{2+}, Mg^{2+}, which are equal to 40 times of iron content;Besides, SiO_3^{2-}, Sn^{2+} and Zn^{2+}do not interfere with the determination.

In this experiment, hydroxylamine hydrochloride or ascorbic acid can be used as reducing agent to reduce Fe^{3+} to Fe^{2+}. And quantitative determination will be carried out by standard curve method.

Due to the high selectivity of the reaction between o-diphenanthrene and Fe^{2+}, the good stability of colored complex and good reproducibility of testing results, the spectrophotometric method of o-phenanthrene coloration is used to determine the content of iron in iron and steel, tin, lead solder, lead ingot and other metallurgical products and industrial sulfuric acid, industrial sodium carbonate, alumina and other chemical products according to China's national standards.

3. Instruments and reagents

(1) The instrument　Visible spectrophotometer or ultraviolet visible spectrophotometer，glass cuvette (3 cm)，volumetric flasks (50 mL, 250 mL), several pipettes (5 mL, 10 mL), pipette(25 mL).

(2) Reagents and samples

① Standard solution of iron salt: 0.01000 $mg\cdot mL^{-1}$.

Accurately weigh several grams (self-calculation) of superior pure $NH_4Fe(SO_4)_2 \cdot 12H_2O$ in a small beaker, add water to dissolve it, add 5 mL of 6 mol · L^{-1} HCl solution, transfer the acidified solution to 250 mL volumetric flask, dilute it to scale with distilled water, and stir well. The resulting solution contains 0.100 mg iron permL. Then absorb the above solution into a 25.00 mL volumetric flask and add 5 mL of 6 mol · L^{-1}HCl solution, and dilute to the scale with distilled water and shake well, the obtained solution contains 0.0100 mg iron · mL^{-1}.

Preparation of solution

② An aqueous solution of *o*-diphenanthrene 1 g · L^{-1}. Weigh 0.5 g of *o*-diphenanthrene in a small beaker, add 2-3 mL of 95% ethanol solution, and dilute to 500 mL with water.

③ Aqueous solution of hydroxylamine hydrochloride 10 g · L^{-1}.

④ HAc-NaAc buffer solution (pH=4.6). Weigh 136 g superior pure sodium acetate, add 120 mL glacial acetic acid, dissolve with water, and dilute to 500 mL.

⑤ 3 mol · L^{-1} HCl solution.

⑥ Sample: limestone.

4. Experimental steps

(1) To measure Fe^{2+}-Phen absorption curve Use pipettes to draw 0.0100 mg · mL^{-1}standard iron solutions 0 mL, 2.0 mL and 4.0 mL into three 50 mL volumetric flasks, add 2.5 mL hydroxylamine hydrochloride solution to each, and shake well, then add 5 mL HAc-NaAc buffer solution and 5 mL *o*-diphenanthrene solution, respectively. Dilute with distilled water to the scale, shake well and place for 10 minutes. The absorbance of the solution is measured with a spectrophotometer in the wavelength range of 420-600 nm by using a 3 cm cuvette and the blank solution (e.g. the above-mentioned the standard solution without iron) as the reference solution. Generally, a point is measured at an interval of 20 nm, and the measurement points near 510 nm must be made more densely.

Draw a absorption curve

(2) To draw a standard curve Use pipettes to draw 0.0100 mg · mL^{-1} standard iron solutions 0 mL, 1.0 mL, 2.0 mL, 3.0 mL, 4.0 mL, 5.0 mL, 6.0 mL and 7.0 mL into eight 50 mL volumetric flasks, add 2.5 mL hydroxylamine hydrochloride solution and shake well, then add 5 mL HAc-NaAc buffer solution and 5 mL *o*-diphenanthrene solution, respectively. Dilute with distilled water to the scale, shake well and set aside for 10 minutes. The absorbance of each solution will be measured at the maximum absorption wavelength obtained in step 1 with a 3 cm cuvette and blank reagent solution as the reference solution.

Draw a standard curve

(3) To determine trace iron in limestone samples Accurately weigh the sample 0.4-0.5 g (if the iron content is high, reduce the weight appropriately) in a small beaker, add a small amount of distilled water to wet and cover the surface dish. Carefully add 3 mol · L^{-1} HCl solution to dissolve the sample, transfer the sample to a 50 mL volumetric flask, rinse the beaker with a small amount

of distilled water several times, and then transfer as a whole to the volumetric flask. Then follow the same method in experiment step 2 to develop color and measure absorbance.

5. Data and processing

(1) Fill in the following table with the measurement results

① Absorption curve

Wavelength λ/nm		420	440	460	480	…	580	600
Absorbance A	2.0 mL Fe^{2+}							
	4.0 mL Fe^{2+}							

② The standard curve

$V_{Fe^{2+}}$/mL	1.0	2.0	3.0	4.0	5.0	6.0	7.0
$m_{Fe^{2+}}$/mL							
A							

③ Sample No.: ________，Sample quality: ________(g)，Measured absorbance: ________

(2) Drawing and calculation

① Taking wavelength as abscissa and absorbance as ordinate, draw Fe^{2+}-Phen absorption curve and find the wavelength λ_{max} of the maximum absorption peak. λ_{max} is generally used as the measurement wavelength of spectrophotometry.

② Taking the mass of iron in the solution after color development as abscissa and the absorbance as ordinate, draw the standard curve of iron determination, and work out the regression equation and linear regression coefficient.

③ According to the absorbance of the sample, calculate the mass fraction of iron in the sample.

6. Questions

① What is the significance of absorption curve and standard curve?

② When we map Fe^{2+}-Phen absorption curves, why should the measurement points be more closely spaced around 510 nm?

③ Why blank reagent instead of distilled water is used as the reference solution in the experiment?

④ How many grams of $NH_4Fe(SO_4)_2 \cdot 12H_2O$ should be weighed to prepare 1 L of 100 μg · mL^{-1} standard solution?

实验1 邻二氮菲分光光度法测定微量铁

1. 目的要求

（1）熟悉可见分光光度计的使用方法。

（2）练习测绘吸收曲线的方法。

（3）掌握利用标准曲线进行微量成分分光光度测定的基本方法和有关计算。

2. 基本原理

在pH为2～9的溶液中，邻二氮菲（简写作phen）与Fe^{2+}发生下列显色反应：

$$Fe^{2+}+3phen \longrightarrow [Fe(phen)_3]^{2+}$$

生成的橙红色配合物非常稳定，$\lg K_{稳}=21.3$（20℃），其溶液在510nm有最大吸收峰，摩尔吸光系数$\varepsilon_{510}=1.1\times10^4 L\cdot mol^{-1}\cdot cm^{-1}$，利用上述反应可以测定微量铁。显色反应的适宜pH范围很宽（2～9），酸度过高（pH<2）反应速率较慢；若酸度过低，Fe^{2+}将水解。通常在pH约为5的HOAc-NaOAc缓冲介质中测定。

邻二氮菲与Fe^{2+}反应的选择性很高，相当于含铁量5倍的Co^{2+}、Cu^{2+}，20倍的Cr^{3+}、Mn^{2+}、PO_4^{3-}、V（Ⅴ），甚至40倍量的Al^{3+}、Ca^{2+}、Mg^{2+}，而且SiO_3^{2-}、Sn^{2+}和Zn^{2+}都不干扰测定。本实验以盐酸羟胺为还原剂，也可使用抗坏血酸将Fe^{3+}还原为Fe^{2+}，采用标准曲线法进行定量测定。由于邻二氮菲与Fe^{2+}的反应选择性高，显色反应所生成的有色配合物稳定，测定结果的重现性好，因此在我国的国家标准中，测定钢铁、锡、铅焊料、铅锭等冶金产品和工业硫酸、工业碳酸钠、氧化铝等化工产品的铁含量，都采用邻二氮菲显色的分光光度法。

3. 仪器与试剂

（1）仪器　可见分光光度计或紫外可见分光光度计；玻璃比色皿（3cm）；容量瓶（50mL、250mL）若干；移液管（5mL、10mL）若干；25mL吸量管。

（2）试剂和样品

① 铁盐标准溶液：0.0100 $mg\cdot mL^{-1}$。准确称取若干克（自行计算）优级纯的铁铵矾$NH_4Fe(SO_4)_2\cdot12H_2O$于小烧杯中，加水溶解，加入6 $mol\cdot L^{-1}$HCl溶液5mL，酸化后的溶液转移到250mL容量瓶中，用蒸馏水稀释至刻度，摇匀，所得溶液每毫升含铁0.100mg。然后吸取上述溶液25.00mL容量瓶中，加入6$mol\cdot L^{-1}$HCl溶液5mL，用蒸馏水稀释至刻度，摇匀，所得溶液含铁0.0100$mg\cdot mL^{-1}$。

溶液配制

② 1$g\cdot L^{-1}$邻二氮菲水溶液。称取0.5g邻二氮菲于小烧杯中，加入2～3mL 95%乙醇溶液，再用水稀释到500mL。

③ 10$g\cdot L^{-1}$盐酸羟胺水溶液。

④ HAc-NaAc缓冲溶液（pH=4.6）。称取136g优级纯醋酸钠，加入120mL冰醋酸，加水溶解后，稀释至500mL。

⑤ 3$mol\cdot L^{-1}$HCl溶液。

⑥ 样品：石灰石。

4. 实验步骤

吸收曲线绘制

工作曲线绘制

（1）测量 Fe^{2+}-phen 吸收曲线　用吸量管吸取 0.0100mg · mL^{-1} 的铁标准溶液 0mL、2.0mL 和 4.0mL 分别加入三只 50mL 容量瓶中，各加入 2.5mL 盐酸羟胺溶液，摇匀，再各加入 5mL HAc-NaAc 缓冲溶液和 5mL 邻二氮菲溶液，用蒸馏水稀释至刻度，摇匀，放置 10min。以 3cm 比色皿，试剂空白溶液（即上述不加铁标准溶液的）为参比溶液，用分光光度计在 420 ～ 600nm 波长区间测定溶液的吸光度随波长的变化。一般间隔 20nm 测一个点，在 510nm 附近测量点须取得密一些。

（2）绘制标准曲线　用吸量管分别吸取 0.0100mg · mL^{-1} 铁标准溶液 0mL、1.0mL、2.0mL、3.0mL、4.0mL、5.0mL、6.0mL 和 7.0mL 于 8 只 50mL 容量瓶中，依次各加入 2.5mL 盐酸羟胺溶液、5mL HAc-NaAc 缓冲溶液、5mL 邻二氮菲溶液，用蒸馏水稀释至刻度，摇匀，放置 10min。用 3cm 比色皿，以试剂空白溶液为参比溶液，在实验步骤 1 所得到的最大吸收波长下，分别测量各溶液的吸光度。

（3）石灰石样品中微量铁的测定　准确称量样品 0.4 ～ 0.5g（如铁含量较高，则适量减少称量）于小烧杯中，加少量蒸馏水润湿，盖上表面皿，小心滴加 3 mol · L^{-1}HCl 溶液至样品溶解，转移样品至 50mL 容量瓶中，用少量蒸馏水淋洗烧杯数次，一并转入容量瓶中。然后按照实验步骤 2 中同样的方法显色和测量吸光度。

5. 数据及处理

（1）将测量结果填入下表

① 吸收曲线

波长 λ/nm		420	440	460	480	……	580	600
吸光度 A	2.0mL Fe^{2+}							
	4.0mL Fe^{2+}							

② 标准曲线

$V_{Fe^{2+}}$/mL	1.0	2.0	3.0	4.0	5.0	6.0	7.0
$m_{Fe^{2+}}$/mL							
A							

③ 样品编号：__________；样品质量：__________(g)；测得的吸光度：__________

（2）绘图及计算

① 以波长为横坐标，吸光度为纵坐标，绘制 Fe^{2+}-phen 吸收曲线，并求出最大吸收峰的波长 λ_{max}，一般选用 λ_{max} 作为分光光度法的测量波长。

② 以显色后的 50mL 溶液中铁的质量为横坐标，吸光度为纵坐标，绘制测定铁的标准曲线，并求出回归方程和线性回归系数。

③ 根据样品的吸光度，计算样品中铁的质量分数。

6. 思考题

① 吸收曲线与标准曲线各有何使用意义？

② 测绘 Fe^{2+}-phen 吸收曲线时，在 510nm 附近，测量点间隔为什么要密一些？

③ 实验所用的参比溶液为什么选用试剂空白，而不用蒸馏水？

④ 配制 1L 100μg·mL^{-1} 的铁标准溶液需称取多少克的 $NH_4Fe(SO_4)_2 \cdot 12H_2O$？

Exercises

2-1 When transmittance T%=0, absorbance A is ( ).

A. 0 B. 100 C. 1 D. ∞

2-2 The energy of a quantum of light is proportional to its ( ).

A. λ B. v C. c D. I

2-3 When the absorbance reading is in the range of ( ), the measurement is more accurate.

A. 0～1 B. 0～0.7 C. 0.2～0.8 D. 0.2～1.0

2-4 Which of the following factorsis the molar absorption coefficient related to ？( )

A. wavelength of the incident light B. color of solution

C. temperature of solution D. concentration of solution

2-5 Which of the following ways can frequency be expressed in? ( )

A. σ/c B. σc C. $1/\lambda$ D. c/σ

2-6 What are the components of a monochromator? ( )

A. Entrance slit, sources of Light, dispersive element, focusing device, exit slit.

B. Entrance slit, collimating device, dispersive element, focusing device, exit slit.

C. Entrance slit, collimating device, sample cell, focusing device, exit slit.

D. Entrance slit, collimating device, dispersive element, focusing device, detector.

2-7 What interference can be automatically eliminated by dual beam spectrophotometry? ( )

A. The measuring error caused by background interference.

B. The measuring error caused by other components in the solution.

C. The measuring error caused by the change of light intensity.

D. The measuring error caused by absorption cell as well as the reflection and absorption of light by solvent.

2-8 The relative error of measuring is minimal when T=36.8%, namely, A= ( ).

A. 0.443 B. 0.343 C. 0.334 D. 0.434

2-9 The most important component of a monochromator is ( ).

A. entrance slit B. collimating device

C. dispersive element D. focusing device

2-10 A light with a wavelength of 2.5 μm has a wave number of ( ).

A. 400 cm^{-1} B. 4000 cm^{-1} C. 40 cm^{-1} D. 0.4 cm^{-1}

2-11 What can be concluded from the light absorption curve?

2-12 What are the components and types of ultraviolet visible spectrophotometer?

2-13 How to choose the reference solution ?

2-14 The molar absorption coefficient of a ferrous complex is 12000 $L \cdot mol^{-1} \cdot cm^{-1}$. If a 1.00 cm absorption cell is used, what is the concentration range to limit the transmittance reading to between 0.200 and 0.650?

(2.0×10^{-5}-6.0×10^{-5} $mol \cdot L^{-1}$)

2-15 If the concentration of complex NiDx2 is 1.70×10^{-5} $mol \cdot L^{-1}$, the transmittance is 30.0% at 470 nm wavelength with 2.0 cm absorption cell, calculate the molar absorption coefficient of the complex at this wavelength.

($1.54 \times 10^{4} L \cdot mol^{-1} \cdot cm^{-1}$)

2-16 The alkaline solution of K_2CrO_4 has maximum absorption at 372 nm. If the concentration of alkaline K_2CrO_4 is 3.00×10^{-5} $mol \cdot L^{-1}$, the transmittance is 71.6% at this wavelength with 1.0 cm absorption cell, please calculate: (1) The absorbance of the solution; (2) The molar absorbance coefficient.

(0.145, $4.83 \times 10^{3} L \cdot mol^{-1} \cdot cm^{-1}$)

2-17 Dissolve and dilute the 500 mg sample containing color component X to 500 mL, and the absorbance of the solution is 0.900 at 400 nm in a 1.00 cm absorption cell. When 10.0 mg of pure X is dissolved in 1 L of the same solvent, the absorbance is 0.300 according to the same operation, please calculate the mass fraction of X in the original sample.

(3.0×10^{-3})

2-18 The absorbance of the solution composed of drugs A and B is measured at 295 nm and 370 nm in a 1 cm absorption cell, and the absorbance is 0.320 and 0.430, respectively. The absorbance of 0.01 $mol \cdot L^{-1}$ solution A is 0.08 and 0.90 at 295 nm and 370 nm in A 1 cm absorption cell, respectively. Under the same conditions, the absorbance of 0.01 $mol \cdot L^{-1}$ solution B is 0.67 and 0.12, respectively, please calculate the concentrations of A and B in the compound solution. (Assuming that other reagents in the compound solution do not interfere with the determination)

($c_A = 3.9 \times 10^{-3} mol \cdot L^{-1}$, $c_B = 4.3 \times 10^{-3} mol \cdot L^{-1}$)

Chapter 3 Atomic Absorption Spectrophotometry

第3章 原子吸收分光光度法

Study Guide 学习指南

Atomic absorption spectrophotometry is a spectroanalytical procedure for the quantitative determination of the concentration of a particular element (analyte) in a sample according to the attenuation of characteristic spectral lines of analyte. It employs the absorption of optical radiation (light) by ground state atoms of analyte in the gaseous state. This chapter will focus on the introduction of the principles, basic concepts, characteristics of instruments and industrial applications of atomic absorption spectrophotometry. The target of this chapter is to familiarize with the principles and concepts, main components of spectrometer and analytical methods of atomic absorption spectrophotometry.

原子吸收光谱分析是基于从光源辐射出待测元素的特征谱线通过试样蒸气时，被待测元素的基态原子吸收，由特征谱线被减弱的程度来测定试样中待测元素含量的方法。本章重点介绍原子吸收分光光度法的原理、基本概念、仪器特点及工业应用等内容。通过本章的学习，应掌握原子吸收分光光度法的基本原理及概念、了解原子吸收分光光度计的主要部件、掌握原子吸收分光光度法的分析方法。

Section 1 Principle of Atomic Absorption Spectrophotometry

第1节 原子吸收分光光度法基本原理

1. Introduction on Atomic Absorption Spectroscopy

(1) Atomic absorption spectroscopy Three forms of atomic spectroscopy are based on absorption, emission, and fluorescence (figure 3-1). The technique makes use of the attenuation of the characteristic spectral lines of

1. 原子吸收光谱分析概述

（1）原子吸收光谱分析 原子光谱法有三种形式——原子吸收、原子发射和原子荧光（图3-1）。根据测量试样所产生的原子蒸气中待测元素的基态原

radiation by the ground state atoms of analyte in the atom vapour generated by the sample to quantitatively determine the elements. It is also called atomic absorption spectrophotometry, AAS. In atomic absorption in figure 3-2, liquid sample is aspirated (sucked) into a flame whose temperature is 2000-3000 K. Liquid evaporates and the remaining solid is atomized (broken into atoms) in the flame, which replaces the cuvet of conventional spectrophotometry. The pathlength of the flame is typically 10 cm. The hollow-cathode lamp at the left in figure 3-2 has an iron cathode. When the cathode is bombarded with energetic Ne^{+}or Ar^{+} ions, excited Fe atoms vaporize and emit light with the same frequencies absorbed by analyte Fe in the flame. At the right side of figure 3-2, a detector measures the amount of light that passes through the flame.

子对光源辐射的特征谱线的吸收程度，进而定量测定化学元素的方法，称为原子吸收分光光度法，简写为AAS。在图3-2原子吸收装置图中，液体样品被吸入到温度为2000～3000K的火焰中蒸发、雾化、原子化。原子化器取代了传统分光光度法的样品池，火焰宽度通常为10cm。图3-2左侧的空心阴极灯有一个铁阴极，当阴极被高能的Ne^{+}或Ar^{+}轰击时，激发的Fe原子蒸发并发出与火焰中分析物Fe吸收的相同频率的光。在图3-2的右侧，用检测器检测通过火焰的光通量。

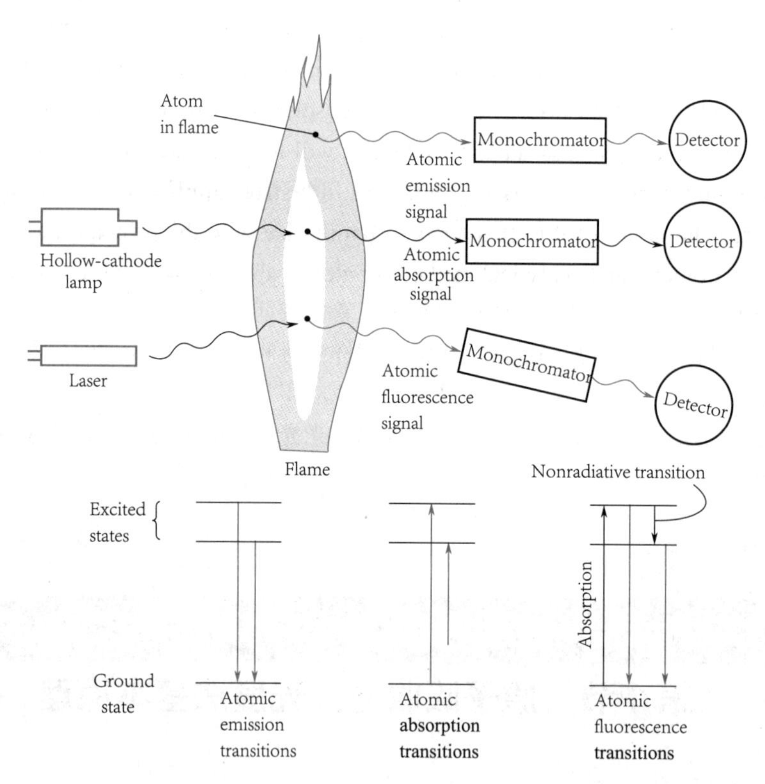

Fig.3-1 Emission, absorption, and fluorescence by atoms in a flame (In atomic absorption, atoms absorb part of the light from the source and the remainder of the light reaches the detector. Atomic emission comes from atoms that are in an excited state because of the high thermal energy of the flame. To observe atomic fluorescence, atoms are excited by an external lamp or laser. An excited atom can fall to a lower state and emit radiation.)

图3-1 原子在火焰中的发射、吸收和产生荧光（在原子吸收中，原子吸收部分来自光源的光，其余的光到达检测器。原子发射主要是因火焰的高热能使原子由基态跃迁为激发态。原子被灯或激光激发后处于激发态，瞬间又跃迁回较低的能态时，即产生原子荧光。）

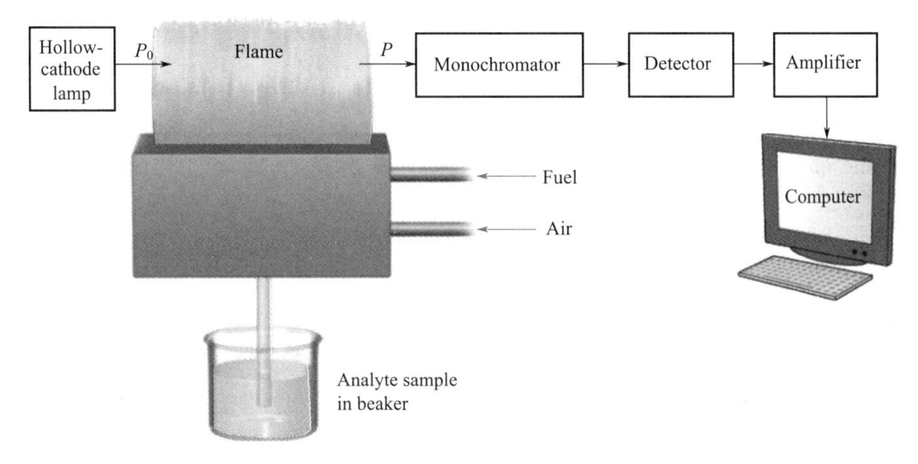

Fig.3-2 Atomic absorption experiment

图 3-2 原子吸收实验

(2) Similarities and differences between atomic absorption spectrophotometry (AAS) and UV-Vis absorption spectrophotometry (UV-Vis AS)

① Similarities

a. Both are based on the absorption of incident light by samples.

b. Both follows Beer-Lambert's Law.

c. The measuring devices mainly consist of four parts: light source, monochromator, atomizer and detector.

② Differences

a. The states of absorbing substance are different.

UV-Vis: the spectrum of molecule, ion and wide-band molecule in solution can use continuous radiation source.

AAS: the spectrum of ground state atom and narrow-band atom must use sharp line radiation source.

b. The positions of monochromator and absorption cell are different.

UV-Vis AS: radiation source→monochromator → cuvette.

AAS: radiation source →atomizer→monochromator.

(3) Characteristics of AAS

① High selectivity Utilize different element lamp for the determination of different element. There is

（2）原子吸收分光光度法（AAS）与紫外可见分光光度法（UV-Vis AS）的异同点

① 相同点

a. 都是依据样品对入射光的吸收进行测量的。

b. 两种方法都遵循朗伯 - 比尔定律。

c. 就设备而言，均由四大部分组成，即光源、单色器、原子化器、检测器。

② 不同点

a. 吸收物质的状态不同。

紫外可见分光光度法：溶液中分子、离子和宽带分子的光谱，可以使用连续光源。

原子吸收分光光度法：基态原子和窄带原子的光谱，必须使用锐线光源。

b. 单色器与吸收池的位置不同。

紫外可见分光光度法：光源→单色器→比色皿。

原子吸收分光光度法：光源→原子化器→单色器。

（3）原子吸收分光光度法的特点

① 选择性高 分析不同元素时选

little interference from the co-existing elements. Therefore, the co-existing elements do not need to be separated.

用不同元素灯，共存元素对待测元素的干扰少，一般无须分离共存元素。

② High sensitivity Applicable to micro and trace analysis. Flame atomizer: 10^{-9}g · mL^{-1}, graphite tube: 10^{-13}g · mL^{-1}.

② 灵敏度高　适合于微量及痕量分析。火焰原子化法：10^{-9}g · mL^{-1}；石墨炉：10^{-13}g · mL^{-1}。

③ Wide measuring range More than 70 elements can be determined. The treatment of sample and operation are simple. The analysis is rapid.

③ 测定的范围广　可以测定 70 多种元素。试样处理简单，操作简便，分析速度快。

(4) Applications of AAS AAS is widely used in the micro and trace analysis in the fields of environmental protection, medical science, metallurgy, geology, food, petrochemical industry and agriculture. For example,

（4）原子吸收分光光度法的应用　原子吸收分光光度法广泛应用于环保、医药卫生、冶金、地质、食品、石油化工和工农业等部门的微量和痕量分析。例如，

① *Technical Specifications for Environmental Monitoring of Groundwater*, HJ/T 164—2020.

① 地下水环境监测技术规范，HJ/T 164—2020。

② *Technical Guideline for the Development of Environmental Monitoring Analytical Method standards*, HJ 168—2020.

② 环境监测分析方法标准制订技术导则，HJ 168—2020。

③ *Standard Methods for the Examination of Water and Waste Water* (4th edition), The ministry of ecology and environment, PRC.

③ 水和废水检测分析方法（第四版），中华人民共和国生态环境部。

2. Key Concepts in Atomic Absorption Spectroscopy

2. 原子吸收分光光度法的重要概念

(1) Resonance line

（1）共振线

① Atomic energy level The atoms of all the elements consist of nuclei and extranulear electrons. The extronuclear electrons are configured layer by layer. Each layer has a definite energy, which is called atomic energy level.

① 原子能级　任何元素的原子都由原子核和核外电子组成。核外电子分层排布，每层都具有确定的能量，称为原子能级。

② Ground state All the electrons are configured regularly on the energy level. The energy of each electron depends on the level it is configured, in which the state on the lowest energy level is called the ground state (E_0=0).

② 基态　所有电子都按一定规律排布在各个能级上，每个电子的能量由它所处的能级决定。其中处于最低能量的状态，称为“基态”（E_0=0）。

③ Ground state atom A atom that is in ground state are called “ground state atom”.

③ 基态原子　原子处于基态，称为“基态原子”。

④ Excited state　When an atom is excited by absorbing the energy outside, the electrons in the outmost orbit may move into the orbits with higher energy. This state by this movement of electrons is called excited state. All the other states with higher energy are called excited state. The electrons in the excited state are extremely unstable. They will move back to ground state or the lower-energy excited state in a very short time (10^{-8}-10^{-7}s). At that moment, atoms will release energy by the form of electromagnetic wave.

④ 激发态　当原子吸收外界能量被激发时，其最外层电子可能跃迁到较高的不同能级上，原子的这种运动状态称为激发态。（即其余能量较高的状态都叫做“激发态”。）处于激发态的电子很不稳定，一般在极短的时间（10^{-8} ～ 10^{-7}s）便跃回基态（或较低的激发态），此时，原子以电磁波的形式放出能量。

$$\Delta E=E_n-E_0=h\nu=h\frac{c}{\lambda} \quad (3.1)$$

⑤ Transition　When electrons absorb outside energy, they move from the ground state to the excited state and then return to the ground state because of the unstability. The movement of electron from low energy state to higher energy state and vice verse is called transition.

⑤ 跃迁　当外界给予能量时，电子由基态跃迁到能量较高的激发态。激发态的电子是很不稳定的，随时都有返回基态的趋势，由最低能级跳到较高能级或较高能级返回较低能级的运动称“跃迁”。

⑥ Resonance absorption line　When electrons transit from the ground state to the lowest excited state (the first excited state), they absorb an amount of energy (fig.3-3). The corresponding spectral line with the absorbed energy is called resonance absorption line.

⑥ 共振吸收线　电子从基态跃迁到最低激发态（称第一激发态）时要吸收一定的能量（图 3-3），与所吸收能量对应的光谱线（即所产生的谱线）叫共振吸收线（即使电子从基态跃迁到第一激发态所产生的吸收谱线）。

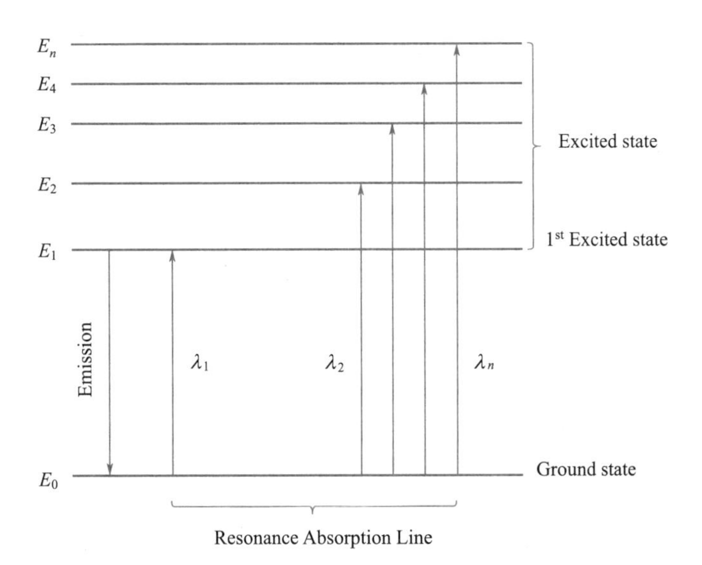

Fig.3-3　Partial electronic energy level diagram

图 3-3　部分电子能级图

⑦ Resonance emission line　When electrons transit from the the first excited state back to the ground state, they emit an amount of energy, whose corresponding spectral line is called resonance emission line.

⑧ Resonance line　Resonance absorption line and resonance emission line are called resonance line in general.

⑨ Characteristic spectral line　Every element has different atom structure and electron configuration. The atoms of different element absorb or emit different amount of energy when transitting from the ground state to the first excited state or vice verse. Therefore, each element has its unique resonance line, which is called the characteristic spectral line of the element.

⑩ Sensitive line　The transition from the ground state to the first excited state needs very little excitation energy, so the transition happens easily. For most elements, resonance line is the most sensitive line in the spectra. Thus, it is called the sensitive line of the element.

⑪ Absorptive analytical line　Atomic absorption analysis makes use of the absorption of resonance line generated by the ground state atom in the atom vapour of analyte. Therefore, the resonance line of an element is also called the absorptive analytical line.

The hollow cathode lamp containing the element of interest is used in the determination for the reason that the lamp can emit the resonance line absorbed by the atom vapor.

(2) Profile and broadening of atomic absorption spectrum

The absorption line is a curve showing the relation of intensity and frequency. When a beam of parallel lights (intensity I_0) of different frequencies pass through the atom vapor of analyte, it is observed that the light intensity attenuates at the characteristic frequency ν_0 of the element, which shows the monochromic light with frequency ν_0 is absorbed by the ground state atoms. The light intensity after absorption is I_ν, the absorption coefficient is K_ν. Figure 3-4、figure 3-5 show the I_ν-ν curve. and the K_ν-ν curve. Theoretically, the absorption line of atom should be an infinitely narrow line, but it does have a width in practice.

⑦ 共振发射线　电子从基态跃迁到第一激发态时要吸收一定频率的光，它再跃迁回基态时，则同样发射出一定频率的光（谱线），这种谱线称为共振发射线。

⑧ 共振线　共振发射线和共振吸收线都简称为共振线。

⑨ 特征谱线　各种元素的原子结构和外层电子排布不同，不同元素的原子从基态激发至第一激发态（或由第一激发态跃迁返回基态）时，吸收（或发射）的能量不同，因而各种元素都有它特有的共振线。这种特有的共振线称为元素的特征谱线。

⑩ 灵敏线　从基态到第一激发态间的跃迁所需的激发能较低，其跃迁最易发生，因此对大多数元素来说，共振线是所有谱线中最灵敏的谱线，故称共振线是元素的灵敏线。

⑪ 吸收分析线　在原子吸收分析中，就是利用待测元素原子蒸气中基态原子对光源发出的共振线的吸收来进行分析的。因此元素的共振线又称为吸收分析线。

在测定时利用被测元素制成的空心阴极灯作光源，此空心阴极灯可以发射出被该元素原子蒸气吸收的共振线。

（2）原子吸收光谱的轮廓和变宽　原子吸收线指强度随频率变化的曲线，若将一束不同频率、强度为 I_0 的平行光通过被测元素的原子蒸气，可观察到在元素的特征频率 ν_0 处的光强度减弱，表明频率为 ν_0 的单色光被基态原子吸收，吸收后光强为 I_ν。I_ν-ν、K_ν-ν 曲线见图 3-4、图 3-5。从理论上讲，原子吸收线应是一条无限窄的线，但实际上它有一定宽度。

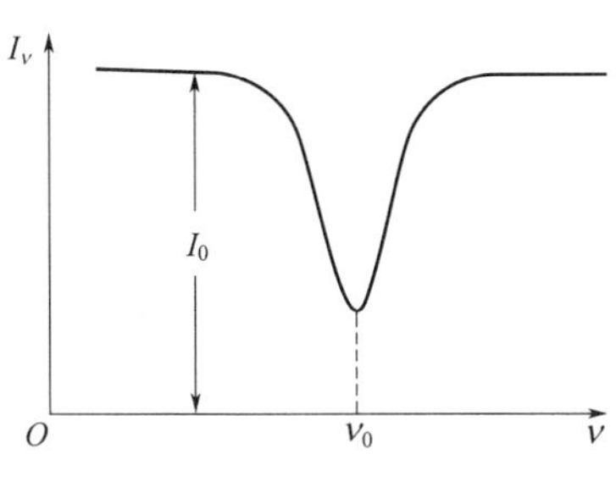

Fig.3-4 I_ν-ν curve

图 3-4 I_ν-ν 曲线

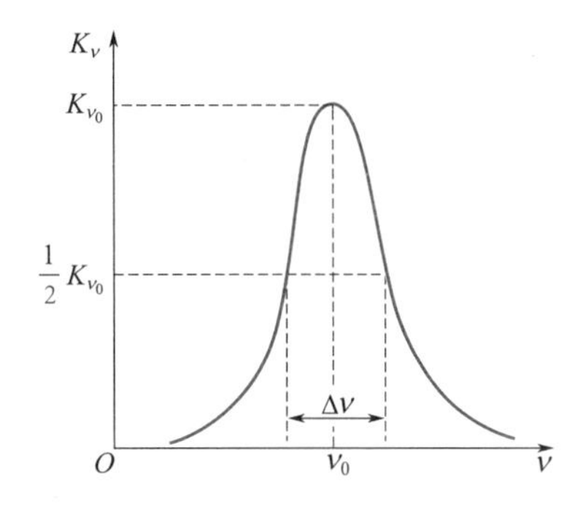

Fig.3-5 K_ν-ν curve

图 3-5 K_ν-ν 曲线

① Half width: the differential frequency of two points on the spectrum profile at half of the maximum absorptivity of the central wavelength.

② Factors affecting half width

a. Natural broadening: the natural width of spectrum line.

b. Doppler broadening: caused by thermal motion.

c. Pressure broadening: caused by the collision of atoms (Lorentz broadening, Holtsmark broadening).

③ Relationship of absorption value and concentration

a. Integrated absorption The integral of absorption coefficient in the profile of absorption curve.

b. Peak absorption The peak absorption coefficient at the central frequency of absorption curve.

At specific conditions:

① 半宽度 $\Delta\nu$：中心频率 ν_0 的吸光系数一半处谱线轮廓上两点之间的频率差。

② 影响半宽度的因素

a. 自然变宽：谱线固有宽度。

b. 多普勒变宽：由热运动引起。

c. 压力变宽：由原子间碰撞引起（劳伦兹变宽；赫鲁兹马克变宽）。

③ 原子吸收值与原子浓度的关系

a. 积分吸收 在吸收曲线的轮廓内，对吸光系数的积分。

b. 峰值吸收 吸收线中心频率处的峰值吸光系数。

在特定条件下，有：

$$A=KN_0L \quad (3.2)$$

In practice, the thickness L of atomizer is definite, the concentration c of a component in the sample is directly proportional to the total number of ground state atoms in vapor, then

在实际工作中，原子化器厚度 L 一定，试样中某组分浓度 c 与蒸气中基态原子数 N_0 成正比。

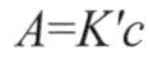

$$A=K'c \quad (3.3)$$

Where K' is a constant revelant to the experimental conditions. Then the absorbance of peak absorption is directly proportional to the concentration of analyte. The formula (3.3) is the basis of quatitative analysis in AAS.

式中 K' 是与实验条件有关的常数，即峰值吸收测量的吸光度与被测组分的浓度成正比，式（3.3）为原子吸收分光光度法进行定量分析的基本公式。

④ The essential conditions of peak absorption substituting for integrated absorption (fig.3-6):

④ 峰值吸收代替积分吸收的必要条件（图 3-6）：

a. The central frequencies of emission line and absorption are identical.

a. 发射线的中心频率与吸收线的中心频率一致。

b. The half width of emission line is much smaller than the half width of absorption line (1/10-1/5).

b. 发射线的半宽度远小于吸收线的半宽度（1/10 ～ 1/5）。

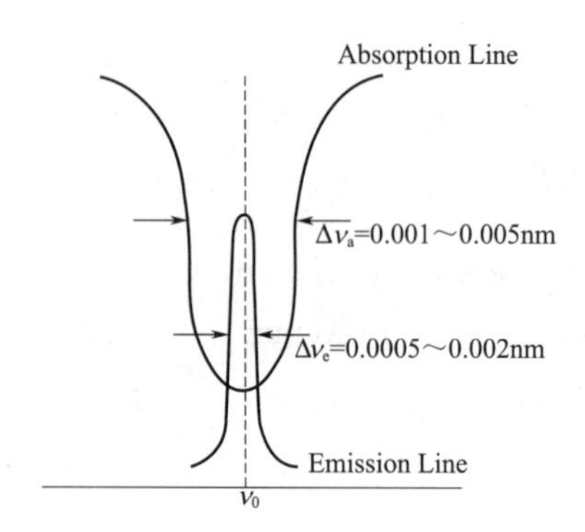

Fig.3-6 Schematic representation of the peak absorption

图 3-6 峰值吸收示意图

Section 2 Atomic Absorption Spectrometer

第 2 节 原子吸收分光光度计

There are many kinds of atomic absorption spectrometers. But the main components of these spectrometers includes light source, atomization system, optical splitting system and detection system (fig.3-7).

原子吸收分光光度计的种类很多，但都由四个主要部件组成，即光源、原子化系统、分光系统和检测系统（图 3-7）。

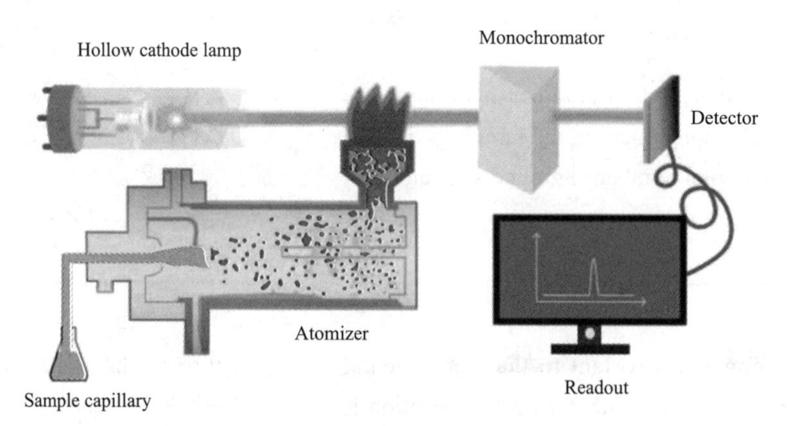

Fig.3-7 Diagram of flame AAS spectrometer

图 3-7 火焰原子吸收光谱仪示意图

1. Main Components of Spectrometer

(1) Sharp line source

① Function: emitting the characteristic spectral line of analyte–resonance line.

Sharp line source must meet the following requirements: a. Emit the resonance line of analyte; b. Emit sharp line; c. Intensified and stable radiation。

② Types of sharp line source Hollow cathode lamp, electrodeless discharge lamp, vapor discharge lamp.

③ Principle of operation The hollow-cathode lamp in figure 3-8 is filled with Ne or Ar at a pressure of about 130-700 Pa. The cathode is made of the element whose emission lines we want. When approximately 500 V is applied between the anode and the cathode, gas is ionized and positive ions are accelerated toward the cathode. After ionization occurs, the lamp is maintained at a constant current of 2-30 mA by a lower voltage. Cations strike the cathode with enough energy to "sputter" metal atoms from the cathode into the gas phase. Gaseous atoms excited by collisions with high-energy electrons emit photons. This atomic radiation has the same frequency absorbed by analyte in the flame or furnace. Atoms in the lamp are cooler than atoms in a flame, so lamp emission is sufficiently narrower

1. 仪器的主要部件

（1）锐线光源

① 锐线光源的作用是发射待测元素的特征谱线——共振线。

光源应满足要求：a. 能发射待测元素的共振线；b. 能发射锐线；c. 辐射光强度大，稳定性好。

② 锐线光源的种类　空心阴极灯、无极放电灯、蒸气放电灯。

③ 原理　图 3-8 中的空心阴极灯在约 130 ～ 700Pa 的压力下充满氖气或氩气。阴极由我们想要的发射线的元素制成。当在阳极和阴极之间施加约 500V 电压时，惰性气体电离，带正电荷的惰性气体离子加速向阴极移动并猛烈轰击阴极内壁，使阴极表面的金属原子"溅射"出来。电离发生后，空心阴极灯通过较低的电压保持 2 ～ 30mA 的恒定电流。由高能电子碰撞激发的气体原子发射光子。这种原子辐射与火焰或石墨炉中被分析元素的吸收频率相同。由于灯中的原子比火焰中的原子要冷，因此灯的发射比火焰中原子吸收线的宽度要窄得多，

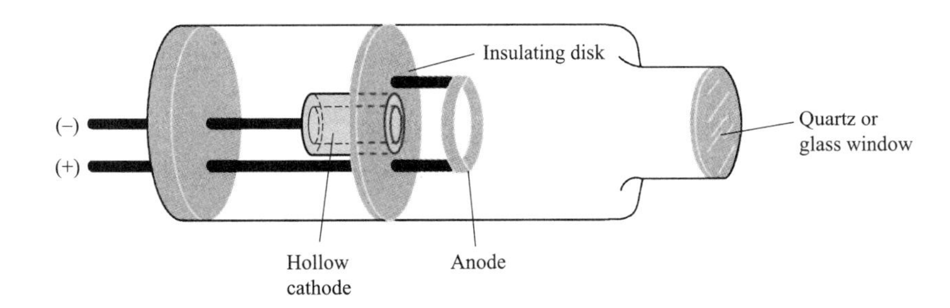

Fig.3-8　A hollow cathode lamp

图 3-8　空心阴极灯

The installation of hollow cathode lamp

空气阴极灯的安装

than the width of the absorption line of atoms in the flame to be nearly "monochromatic". The purpose of a monochromator in atomic spectroscopy is to select one line from the hollow-cathode lamp and to reject as much emission from the flame or furnace as possible. A different lamp is usually required for each element, although some lamps are made with more than one element in the cathode.

几乎是"单色的"。原子光谱法中单色器的目的是从空心阴极灯中选择一条发射线，并尽可能多地抑制火焰或石墨炉的辐射。用不同的待测元素作阴极材料可制作相应待测元素的空心阴极灯。有些灯的阴极中可有一个以上的元素。

(2) Atomizer

（2）原子化器

① Function Transforming the sample into gaseous ground state atoms and absorbing the characteristic spectrum emitted by the absorption radiation.

① 作用　将试样转化为气态的基态原子，并吸收光源发出的特征光谱。

Atomizers must meet the following requirements: a. High efficiency in atomization; b. Good stability and reproducibility; c. Simple operation and low interference.

原子化器应满足如下要求：a. 原子化效率高；b. 稳定性和重现性好；c. 操作简单和干扰小。

② Atomization method

② 原子化方式

a. Flame atomization Flames supply energy to atomize the elements of the interest. The most commonly used flame atomizer is premixing atomizer, including nebulizer, spray chamber and burning chamber.

a. 火焰原子化法　由化学火焰提供能量，使被测元素原子化。常用的是预混合型原子化器，包括雾化器、雾化室和燃烧室三部分。

(a) Nebulizer (fig.3-9, fig.3-10) is used to evaporize the sample solution. It is required that the nebulization

（a）雾化器（图 3-9，图 3-10）。它的作用是将试液雾化，要求雾化效率

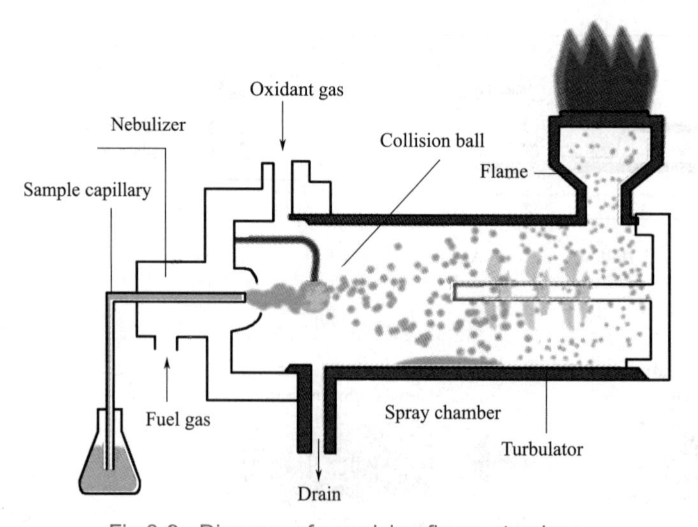

Fig.3-9 Diagram of premixing flame atomizer

图 3-9　预混式火焰雾化器示意图

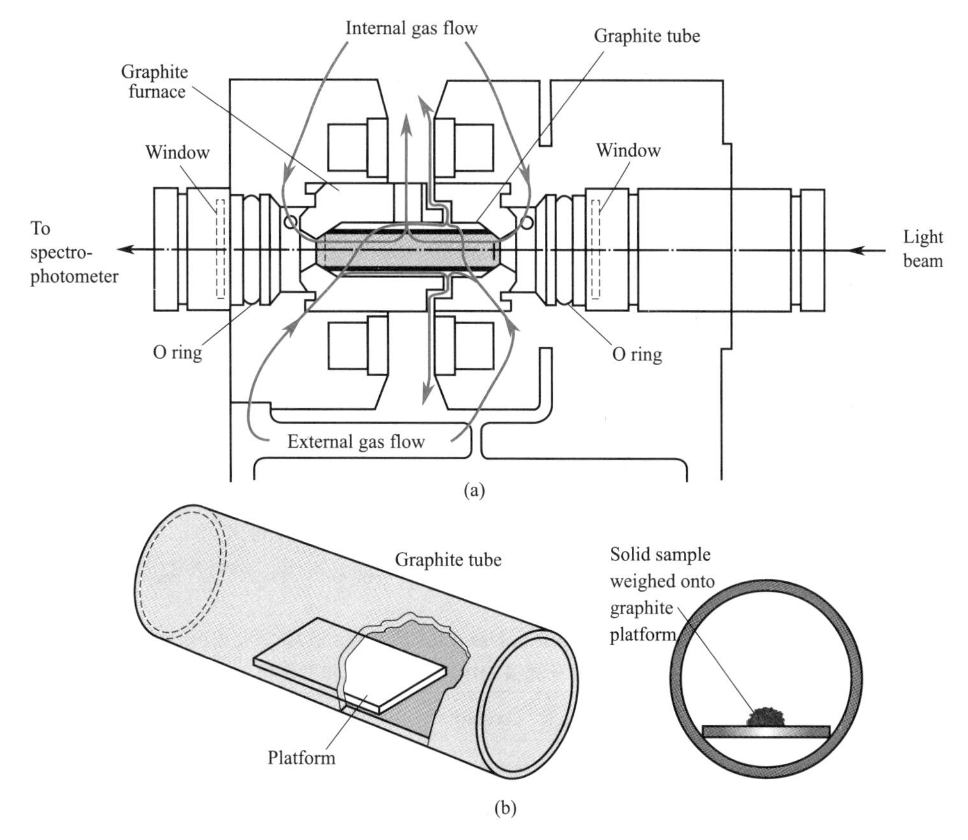

Fig.3-10 Diagram of graphite tube atomizer

图 3-10 石墨管雾化器示意图

efficiency is high (10%-12%), aerosol droplets are fine and spraying is stable.

高（一般为 10% ～ 12%），雾滴细，喷雾稳定。

(b) Spray chamber is used to make droplets further fine so that they can mix intensively with the flame gases. It also has buffering function so as to provide stable droplets to the burner. At the upstream of sprayer, there is a collision ball for the purpose of making the droplets finer. Large droplets are crushed to make aerosol droplets.

（b）雾化室。作用是进一步细化雾滴，使雾滴和燃气得到充分混合的同时起缓冲作用，使供给燃烧器的雾滴平稳。为了细化雾滴，可在雾化器前加碰撞球，使直径较大的雾滴进一步破碎，成为气溶胶。

(c) Burner is used to transform sample vapor to ground state atoms by flames. The atomization efficiency is about 10%.

（c）燃烧器。该装置是利用火焰将试样蒸气转化为基态原子。原子化效率约为 10%。

Use of air compressor pressure reducing valve
空气压缩机与减压阀的使用

Operation of air high pressure cylinder and pressure reducing valve
空气高压钢瓶与减压阀的操作

Operation of acetylene cylinder and pressure reducing valve
乙炔钢瓶与减压阀的操作

(d) Flame and its property: As long as the analyte element dissociated into free ground state atom, the temperature of flame for the atom absorption is sufficient. If the temperature exceeds the necessary temperature, the number of excited atoms and ionization increases, but the number of atoms decreases. The composition of flames affects the sensitivity, stability and interference of the measurement. Common Fuels and Oxidants are shown in table 3-1.

（d）火焰及其性质：原子吸收所使用的火焰，只要其温度能使待测元素离解成自由的基态原子即可。如超过所需温度，则激发态原子增加，电离度增大，基态原子减少，这对原子吸收是很不利的。火焰的组成关系到测定的灵敏度、稳定性和干扰等常用燃料和助燃气见表 3-1。

Burning gas: acetylene, hydrogen, propane.

燃料：乙炔、氢气、丙烷等。

Supporting burning gas: air, nitrous oxide, oxygen.

助燃气：空气、N_2O、氧气等。

Most commonly used flame: air-acetylene, nitrous oxide-acetylene, air-hydrogen, etc.

常用的火焰：空气 - 乙炔、氧化亚氮 - 乙炔、空气 - 氢气等多种。

Table 3-1 Common Fuels & Oxidants Used for Flames in Atomic Spectroscopy[①]

表 3-1 火焰原子光谱法中常用的燃料和助燃气

Fuel	Oxidant	Maximum Temperature/K
Propane	Air	2267
Hydrogen	Air	2380
Acetylene	Air	2540
Hydrogen	Oxygen	3080
Propane	Oxygen	3094
Acetylene	N_2O (nitrous oxide)	3150
Acetylene	Oxygen	3342

① This table was obtained from C.T hJ. Alkemade and R. Herrmann, *Fundamentals of Spectroscopy*, Halsted Press, New York, 1979.

b. Non-flame atomization There are many non-flame atomization methods, in which the most commonly used is electrothermal graphite tube atomizer (fig.3-10, fig.3-11). The advantages of graphite tube atomizer are high atomization, little amount of sample used (1-100μL), solid or viscous sample measurable, high sensitivity and low limit of detection. The disadvantages are low precision, slow determination, lengthy operation and complicated devices.

b. 非火焰原子化法 非火焰原子化的方法有很多，目前广泛应用的是高温石墨炉原子化器（图 3-10、图 3-11）。石墨炉原子化法的优点是原子化程度高，试样用量少（1 ～ 100μL），可测固体及黏稠试样，灵敏度高，检测限低。缺点是精密度差，测定速度慢，操作不够简便，装置复杂。

Procedure of graphite tube atomization: drying, ashing (or pyrolysis), atomization and cleaning (removal of

管式石墨炉使试样原子化的程序：干燥、灰化（或分解）、原子化及高温

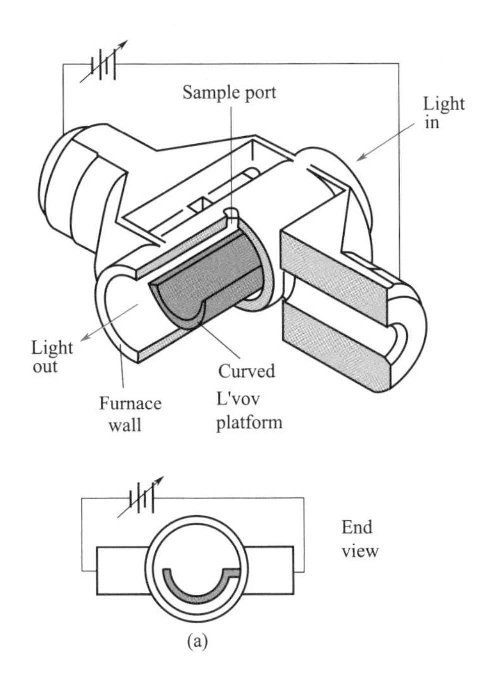

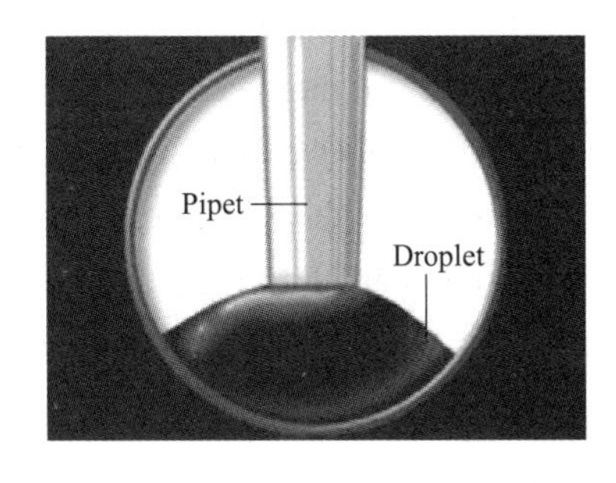

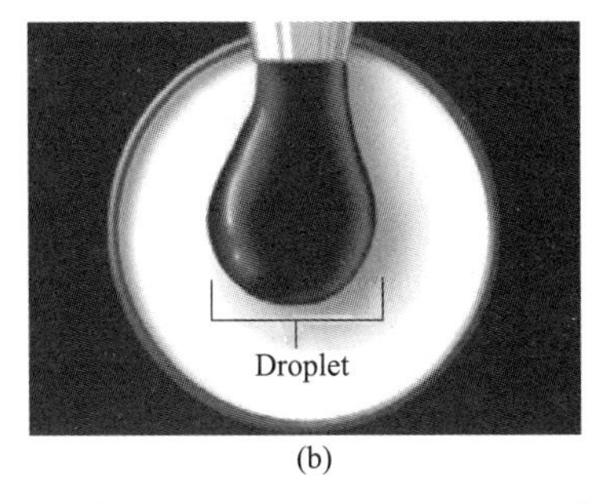

Fig.3-11 (a) Transversely heated graphite furnace maintains nearly constant temperature over its whole length, thereby reducing memory effect from previous runs.(b) Correct position for injecting sample into a graphite furnace deposits the droplet in a small volume on the floor of the furnace. If injection is too high, sample splatters and precision is poor.

图 3-11 （a）横向加热的石墨炉在其整个炉长上保持几乎恒定的温度，以减少之前运行的记忆效应。（b）将样品注入石墨炉的正确位置，使液滴以小体积沉积在炉底。如果进样量过高，样品会飞溅，精度较差。

residues).

(a) Drying: The solvent or the highly-volatile constituents in the sample are evaporated.

(b) Pyrolysis: Low boiling point inorganic and organic substances are removed at high temperature to reduce matrix interference.

(c) Atomization: The analyte element in any form volatizes and dissociates into neutral atoms.

(d) Cleaning: Eventual residues in the graphite tube are removed at high temperature 3000 ℃ to reduce and avoid the influence of memory effect on the next determination.

(3) Optical splitting system Function: seperating the resonance line of the analyte element from the spectral lines in the vicinity. Components: exit slit, entrance slit, reflection mirror and dispersion element.

净化（除残）四个步骤。

（a）干燥。目的是蒸发除去溶剂或样品中挥发性较大的组分。

（b）灰化。在较高温度下除去低沸点无机物及有机物，减少基体干扰。

（c）原子化。使以各种形式存在的分析物挥发并离解为中性原子。

（d）高温净化。升至更高的温度3000℃，除去石墨管中的残留分析物，以减少和避免记忆效应，影响下一次测定。

（3）分光系统　分光系统的作用是将待测元素的共振线与邻近谱线分开。由于采用了锐线光源，谱线比较简单，无需用色散率很高的单色器，最常用的

The monochromator should also collect light to separate the resonance line from the spectral lines in the vicinity. In the test, the width of exit slit has to be adjusted. The selection of slit width is expressed by *band pass*. Because of using sharp line light source, the spectral line is relatively simple Monochromators with high dispersion rates are not needed.

色散元件是光栅，其组件主要包括出射狭缝、入射狭缝、反射镜和色散元件。为了使共振线和邻近的非吸收线分开，也要使单色器有一定的集光本领，实验中必须调整出射狭缝的宽度。狭缝宽度的选择要用光谱通带来表示。

Band pass (*W*): the wavelength range of the light beam passing through the exit slit of the monochromator. It can be shown as the product of slit width (mm) and reciprocal dispersive power of grating (nm · mm^{-1}), that is:

光谱通带（*W*）指通过单色器出射狭缝的光束波长间的范围。可表示为狭缝的宽度（mm）与光栅倒色散率（nm · mm^{-1}）的乘积。

$$W=DS \tag{3.4}$$

D and *S* is reciprocal dispersive power (nm · mm^{-1}) and slit width (mm) respectively.

即光谱通带（nm）＝光栅倒色散率（nm · mm^{-1}）× 狭缝宽度（mm）。

(4) Detection system Function: converting the low-level light signals that have been absorbed in atom vapor and splitted by monochromator to electric signals, which is then amplified by an amplifier so as to be read on the indicator.

Components: detector (photomultiplier), amplifier, readout and recording system.

（4）检测系统　检测系统的作用是将经过原子蒸气吸收和单色器分光后的微弱光信号转换为电信号，再经过放大器放大后，便可在读数装置上显示出来。其主要由检测器（光电倍增管）、放大器、读数和记录系统等组成。

2. Type of Spectrometer

2. 原子吸收分光光度计的类型

(1) Single beam spectrometer [fig.3-12(a)] It is composed of a monochromator and a detector, simple in structure and low in price. But it is easily influenced by the intensity of light source, the preheating takes a long time and the analysis is slow. It meet the requirements of common analysis. But the background cannot be corrected.

（1）单光束原子吸收分光光度计［图 3-12（a）］　由一个单色器和一个检测器组成。结构简单、价廉；但易受光源强度变化影响，灯预热时间长，分析速度慢。能满足一般分析的要求，但背景无法进行精确校正。

(2) Dual beam spectrometer[fig.3-12(b)] It is composed of a monochromator and a detector, with a beam of light passing through flame while another beam passing through monochromator only. This type of spectrometer can offset the factors of the change

（2）双光束原子吸收分光光度计［图 3-12（b）］　由一个单色器和一个检测器组成。一束光通过火焰，一束光不通过火焰，直接经单色器。此类仪器可消除光源强度变化和检测器灵敏度变动

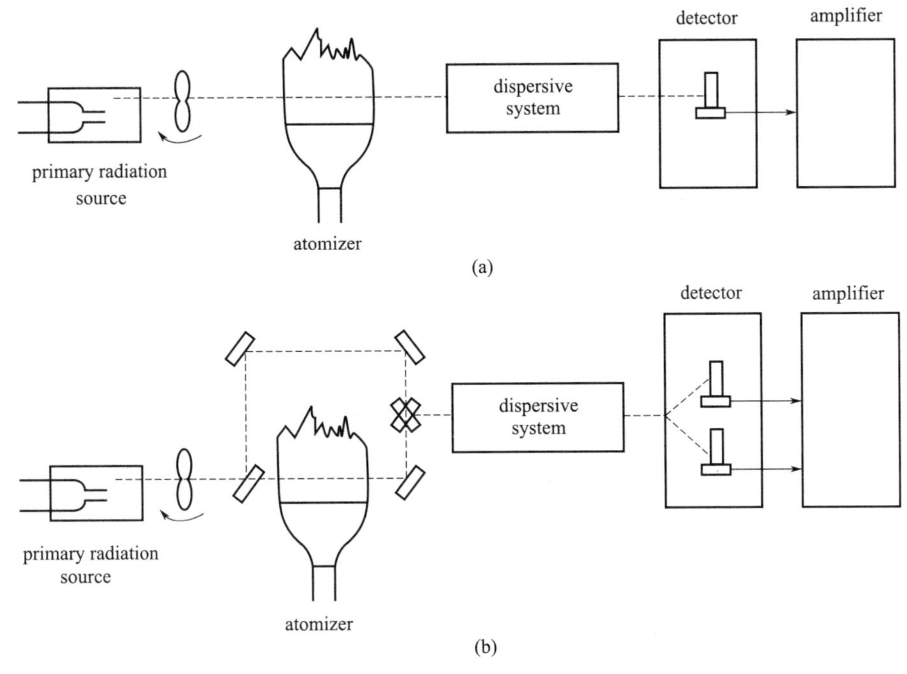

Fig.3-12 Principle of optical system in AAS(a) single beam; (b) dual beam

图 3-12 AAS 光学系统原理（a）单光束；（b）双光束

of light source intensity and detector sensitivity. However, because the reference beam of light does not pass through flame, the flame interference and the background absorption cannot be eliminated.

产生的影响。但由于参比光束不通过火焰，无法消除火焰扰动和背景吸收产生的影响。

(3) Dual channel or multichannel spectrometer There are two light sources, two monochromators and two detectors. This type of spectrometer enjoys high precision and can determine more than two elements at the same time by using internal standard method. But it has complicated structure and high cost. Nowadays, the most commonly used spectrometers are single channel dual beam and dual channel dual beam spectrometers.

（3）双波道或多波道原子吸收分光光度计　由两个光源、两个单色器和两个检测器组成。此类仪器准确度高，可采用内标法，并可同时测定两种以上元素。但装置复杂，仪器价格昂贵。现在使用较多的是单道双光束和双道双光束的原子吸收光谱仪。

3. Maintenance of instruments

3. 仪器日常维护与保养

(1) Constant voltage power supply should be used with power above 2000 W. The main body, graphite tube and air compressor should use different phase. In addition, the power must be earthed.

（1）要使用稳压电源，功率 2000W 以上。实验室电源要进行分相，主机、石墨炉、空气压缩机不要在同一相上。另外，一定要有地线。

(2) High purity (≥ 99%) acetylene is used.

（2）乙炔气体一定要使用高纯乙炔（≥ 99%）。

(3) Caution should be taken on the water drain of air compressor to prevent the blocking of internal tubes by water.

(4) Clean nebulizer and burner head. If the high-efficiency glass nebulizer is blocked, do not clean out with things like metal wire to avoid the damage of nebulizer. The carbon deposition blocking the burner head should be cleaned out with a piece of hard paper, film or bamboo slice instead of a knife blade or other metal sheets. Every time when the measurement is complete, spray distilled water for 5 min to flush the acid, alkali and salt in the burner to prevent the explosion caused by copper acetylide.

(5) Routinely check the leakage of gas and water.

(6) If devices are not in use for a long time, switch them on for 30 min one or two times each week.

(7) The laboratory should be kept clean without dust, large magnetic field, direct sun exposure and corrosive gas. The exhausting fan should be in good condition. The laboratory should be dehumidified to keep the relative humidity of air below 70%. Air conditioner should be generally equipped to maintain the room temperature at 15-30 ℃.

（3）空气压缩机要注意排水，防止水进入仪器内部堵塞管路。

（4）雾化器和燃烧头的清洁。玻璃高效雾化器堵塞时，不能用金属丝之类的物品疏通，否则会损坏雾化器。燃烧头缝口积炭堵塞只能用薄的硬纸片、胶片或竹片疏通，不能用刀片或其他金属片，以免损伤燃烧头缝口。每次测量操作完成后，应喷雾蒸馏水 5min 左右以冲洗燃烧器内酸、碱和盐类物质。若喷雾了高浓度铜溶液时，要用水彻底冲洗燃烧器系统，避免生成乙炔铜化合物引起爆炸危险。

（5）气路和水路要注意经常检漏。

（6）仪器若较长时间不使用，至少应保证每周 1 ～ 2 次打开仪器电源开关通电 30min 左右。

（7）实验室要保持清洁，尽可能做到无尘，无大磁场、电场，无阳光直射和强光照射，无腐蚀性气体，仪器抽风设备良好，经常去湿以保证室内干燥，室内空气的相对湿度应 <70%。一般应装备空调，以保持室温 15 ～ 30℃。

Section 3 Selection of Instrument Measuring Conditions

第 3 节 仪器测量条件的选择

Basic operation of TAS 990 flame atomic absorption spectrometer

TAS 990 火焰原子吸收光谱仪基本操作

1. Selection of Absorption Line

In general, the resonance line is used as absorption line for analysis. For example, the sensitivity of the

1. 吸收线的选择

一般采用共振线作为吸收线进行分析。例如 Cu 的共振线 324.8nm 灵敏度

copper resonance line at 324.8 nm is higher than that of other non-resonance lines. But this is not true to all the elements.

就比其他非共振线测定的灵敏度高。但不是所有元素都是这种情况。

① The sensitivities of the resonance lines of many transition elements are lower than that of other non-resonance lines. For example, chromium 359.35 nm has the highest sensitivity, but it is not a resonance line. The sensitivity of aluminum 309.3 nm is higher than its resonance line, aluminum 308.2 nm.

① 很多过渡元素的共振线灵敏度反而比非共振线低。例如 Cr 359.35nm 灵敏度最高，但却不是共振线。Al 309.3nm 的灵敏度比它的共振线 Al 308.2nm 高。

② The resonance lines of some elements are in the far-UV region, making it difficult for determination. Hence only non-resonance lines can be selected. For example, Hg 185 nm has to be determined in far-UV region, so Hg 253.7 nm is chosen.

② 有的元素的共振线在远紫外区，测定有困难，只能选非共振线。如 Hg 185nm 需在远紫外区测定，只能选 Hg 253.7nm。

③ The resonance lines and non-resonance lines of some elements are very close to each other. For instance, the resonance line of Ni 232.0 nm cannot be separated from the non-resonance line of Ni 231.98 nm even with 0.1 nm band pass. In this case, the non-resonance lines of little interference can be selected as long as the sensitivity is ensured.

③ 有的元素的共振线与非共振线靠得很近。如 Ni 232.0nm 共振线，即使用 0.1nm 通带也不能与 Ni 231.98nm 非共振线分开。这种情况，只要保证有一定的灵敏度，可以选取干扰较小的非共振线。

④ In order to determine the element with high concentration, non-resonance line should be adopted to prevent the dilution of analyte and reduction of sensitivity.

④ 如果是测定高含量元素，为避免稀释，必须降低测定的灵敏度，也应该采用非共振线。

2. Selection of Lamp Current

2. 灯电流的选择

The lamp current of hollow cathode lamp has direct influence on the lifetime of lamp and the sensitivity of measurement. When the lamp current is high, Doppler broadening and self-absorption broadening increase but the sensitivity decreases and the duration of stable state is long. When the lamp current is low, Doppler broadening and natural broadening decrease but the sensitivity increases and the linearity is improved. However the stability of signals is bad. Therefore, in practice the sensitivity and stability have to be considered and confirmed in the selection of lamp current. Hollow cathode lamp should be preheated for

空心阴极灯灯电流的大小直接影响灯的寿命和测定的灵敏度。灯电流较大，多普勒变宽和自吸变宽增大，灵敏度下降，并且达到稳定的时间较长。灯电流较小，多普勒变宽和自吸变宽减小，灵敏度提高，线性也得到改善，但信号稳定性较差。实际分析中灯电流的选择必须兼顾灵敏度和测量稳定性，并通过实验来确定。空心阴极灯一般要预热 20 ～ 30min 才稳定，具有塞曼效应的原子吸收光谱仪，基线漂移减少，但预热时间还是需要的。若吸收

20-30 min before it reaches stability. The spectrometer with Zeeman Effect has little baseline drift, but it still needs preheating. When the absorption value is stable, it shows that the preheating time is sufficient. The most appropriate working current can be chosen according to the measurement of absorption value along with the change of lamp current.

值稳定，表明预热时间已经足够。通过测定吸收值随灯电流的变化而选定最适宜的工作电流。

3. Selection of Band Pass

A proper band pass will ensure high sensitivity, good linear relationship and signal-noise ratio. For elements with simple spectra (e.g. base metal and earth metal), the enlargement of slits can improve signal-noise ratio and stability. But for elements with complicated spectra and many adjacent lines (e.g. Fe, Co, Ni), narrow slit has to be applied. The proper slit width will also be defined in the test.

3. 光谱通带的选择

选择适当的光谱通带，使测定有较高的灵敏度，同时有较好的线性关系和信噪比。对于谱线较简单的元素（如碱金属和碱土金属），增大狭缝可以提高信噪比，增加稳定性。但对于谱线复杂，邻近线多的元素（如 Fe、Co、Ni 等），则必须采用小狭缝。合适的狭缝宽度同样应通过试验确定。

4. Selection of Burner Height

The layouts of ground atoms of different elements vary. The zone of highest atom cloud density is sometimes at the bottom of flame, sometime in the middle and sometimes on the top. Hence, the optimal position of flame must be selected for the measurement by adjusting the burner height, that is the burner position where the maximum absorbance can be obtained when the combustion-supporting ratio is fixed, so that the radiation beam passes through the zone with highest concentration of free atoms in the flame, resulting in the highest sensitivity.

4. 燃烧器高度的选择

基态原子在空间的分布各个元素有所不同。基态原子密度最高的地区有的在火焰下部，有的在中部，有的在上部。因此，必须通过调节燃烧器的高度来选择测定时最佳的火焰位置。即固定一定的燃助比时，能获得最大吸光度的燃烧器位置。即使测量光束从自由原子浓度最大的火焰区通过，以期得到最佳的灵敏度。

5. Selection of Combustion-Supporting Ratio

Flame atmosphere influences measurements a lot. Some elements (e.g. Mg, Zn, Cd, Cu) need oxidizing atmosphere, while some elements (e.g. Ca, Cr, Mn) need reducing atmosphere. Usually, the pressure of supporting gas (air) is fixed to adjust the pressure of combustion gas to obtain the

5. 火焰燃 / 助比的选择

火焰的气氛对测定影响甚大。有的元素（如 Mg、Zn、Cd、Cu 等）需要氧化性气氛，有的元素（如 Ca、Cr、Mn 等）则需还原性气氛。一般是固定助燃气（空气）的压力，调节燃气的压力来获得最佳的燃助比。选定火焰类型

optimal ratio. Experiment has to be carried out to define the proper ratio.

后，应通过试验进一步确定燃气与助燃气流量的合适比例。

6. Interferences and Their Elimination

6. 干扰及其抑制

Sharp line source is employed in AAS, so the interferences are low. But in some circumstances, these interferences cannot be neglected, which includes:

原子吸收光谱法由于采用锐线光源，所以干扰较小，但在某些情况下，干扰问题仍不容忽视。主要包括：

(1) Ionization interference The interference that the the number of ground state atoms reduces due to the electroionization of analyte element in the process of high temperature atomization (mainly existing in the flame atomization), resulting in erroneous low calculation. Elimination method: adding ionization buffer and controlling atomization temperature.

（1）电离干扰　待测元素在高温原子化过程中因电离作用而引起基态原子数减少的干扰（主要存在于火焰原子化中），使测定结果偏低。消除方法：加入消电离剂；控制原子化温度。

For example, When measuring potassium (K), cesium (Cs) ionization potential is lower than potassium and can act as an ionization agent, but expensive; a lot of NaCl is often added to inhibit the ionization of potassium.

例如，测 K 时，Cs 电离电位低，可以作为消电离剂，但价格贵；实际操作中常加入大量 NaCl 抑制 K 的电离。

Since the ionization potential of barium (Ba) is 5.21eV and potassium (K) is 4.3eV, excessive KCl is often added when Ba is measured. Owing to K ionization to produce a large number of electrons, and Ba^{+} obtains electrons to generate atoms, the ionization of barium was inhibited.

由于 Ba 的电离电位为 5.21eV，K 的电离电位 4.3eV，故测 Ba 时，常加入过量 KCl，使 K 电离产生大量电子，Ba^{+} 得到电子而生成原子。

(2) Physical interference The effect that the absorbance is decreased due to the change in physical properties of sample in the procedures of transferring, evaporation and atomization. Elimination method: standard addition method.

（2）物理干扰　试样在转移、蒸发、原子化的过程中，由于试样物理性质的变化而引起吸光度下降的效应。消除方法：标准加入法。

(3) Chemical interference The interference that low volatile and low dissociated compounds are produced in the chemical reaction during atomization. Elimination method: adding protectant and releasing agent.

（3）化学干扰　在原子化的过程中，由于发生化学反应而生成难挥发或难离解的化合物而产生的干扰。消除方法：加保护剂、释放剂等。

① Releasing agent The releasing agent and interfering substances form more stable or more difficult volatile compounds to release the elements to be measured, so as to eliminate the interference. For example, when Ca^{2+} is determined by air acetylene flame, calcium

① 释放剂　释放剂与干扰物质生成更稳定或更难挥发的化合物，使待测元素释放出来，从而排除干扰。例如：用空气 - 乙炔火焰测定 Ca^{2+} 时，由于 Ca^{2+} 与 PO_4^{3-} 反应生成难离解、难原

pyrophosphate $Ca_2P_2O_7$ which is difficult to dissociate and atomize is generated due to the reaction between Ca^{2+} and PO_4^{3-}. In the actual determination, $LaCl_3$ or $SrCl_3$ needs to be added to form more stable phosphate, so as to inhibit the chemical interference of PO_4^{3-} on Ca^{2+}, and Ca^{2+} is released. $LaCl_3$ or $SrCl_3$ is the releasing agent of Ca^{2+}.

子化的焦磷酸钙 $Ca_2P_2O_7$。在实际测定中，需要加入 $LaCl_3$ 或 $SrCl_3$，使之生成更稳定的磷酸盐，从而抑制了 PO_4^{3-} 对 Ca^{2+}的化学干扰，Ca^{2+}被释放出来，$LaCl_3$ 或 $SrCl_3$ 就是 Ca^{2+}的释放剂。

② Protectant The protective agent (usually coordination agents) can form more stable and easily atomized compounds with the elements to be measured, so that the elements to be measured can not be combined with interfering elements. For example, when Ca^{2+} is determined by air acetylene flame, calcium pyrophosphate $Ca_2P_2O_7$ which is difficult to dissociate and atomize is generated due to the reaction between Ca^{2+} and PO_4^{3-}. In the actual determination, when excessive EDTA is added, Ca^{2+} reacts with EDTA to form a more stable complex, which is easier to atomize, thus inhibiting the interference of PO_4^{3-} on Ca^{2+}.

② 保护剂 保护剂能与待测元素形成更稳定且更易原子化的化合物，使待测元素不能与干扰元素结合（通常是配位剂）。例如：用空气 - 乙炔火焰测定 Ca^{2+} 时，由于 Ca^{2+} 与 PO_4^{3-} 反应生成难离解、难原子化的焦磷酸钙 $Ca_2P_2O_7$。在实际测定中，加入过量的 EDTA，Ca^{2+} 与 EDTA 反应生成更稳定的配合物，该配合物更易于原子化，从而抑制了 PO_4^{3-} 对 Ca^{2+} 的干扰。

(4) Background absorption The background interference is caused by molecular absorption and radiation scattering, exhibiting artificially high absorption and an improperly high calculation. Elimination methods: ① background correction by dual wavelength method; ② background correction by using deuterium lamp; ③ background correction by self-absorption; ④ Zeeman-effect background correction.

（4）背景吸收 背景干扰主要是由分子吸收和光散射而产生的，表现为增加表观吸光度，使测定结果偏高。消除方法：①用双波长法扣除背景；②用氘灯校正背景；③用自吸收方法校正背景；④用塞曼效应校正背景。

Section 4 Quantitative Analytical Methods of AAS

第 4 节 定量分析方法

1. The Method of Standard Curve (fig.3-13)

1. 标准曲线法（图 3-13）

The standard curve method is applicable to the determination of analyte of simple composition and low interference. Prepare a series of standard sample solutions with different concentrations. The range of concentration usually falls in the range of absorbance

标准曲线法适用于组成简单、干扰较少的试样。定量依据：配制一系列不同浓度的标准试样，一般使所配的浓度范围为吸光度 0 ～ 0.5 范围内，用试剂（或蒸馏水）作空白，调节仪器的零点

0-0.5. Blank with a reagent (or distilled water), adjust the zero point of instrument (A=0), then measure the absorbances in low-to-high order. Plot a standard curve with the measured absorbance A to concentration. Obtain the absorbance A of sample under same conditions, then the concentration of analyte can be found on the standard curve. Or calculate the concentration by introducing absorbance A of sample into the linear equation based on standard sample data. Note that the standard curve bends easily at high concentration due to the influence of pressure broadening.

（$A=0$），由低到高逐个测出吸光度，将获得的吸光度 A 数据对应于浓度作标准曲线。在相同条件下测定待测试样的吸光度 A 数据，在标准曲线上查出对应的浓度值。或由标准试样数据获得线性方程，将测定试样的吸光度 A 数据带入计算。注意在高浓度时，标准曲线易发生弯曲，是压力变宽影响所致。

Determination of magnesium in hard water（the method of standard curve）

硬水中镁含量的测定（标准曲线法）

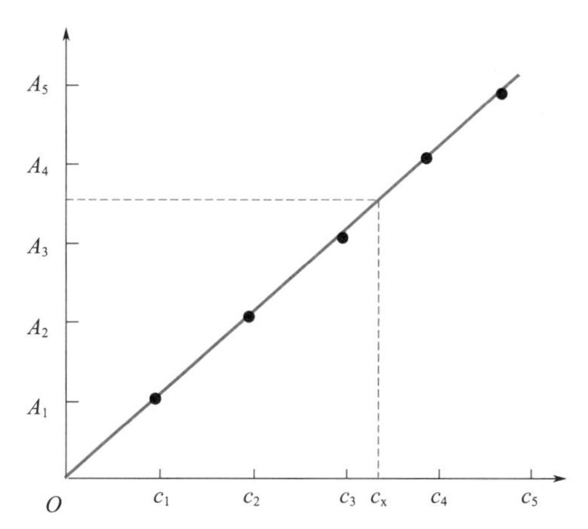

Fig.3-13　Diagram of standard curve method

图 3-13　标准曲线法示意图

2. The Method of Standard Addition

Standard addition method is employed when the sample matrix has great influence, and there isn't pure matrix blank, or trace elements in pure substances are determined. Measure the absorbance A_x of a certain volume c_x of sample solution, add into the sample solution an amount of standard solution whose concentration c_s is close to that of sample solution, then

2. 标准加入法

当试样基体影响较大，且又没有纯净的基体空白，或测定纯物质中极微量的元素时采用。先测定一定体积试液（c_x）的吸光度 A_x，然后在该试液中加入一定量的与未知试液浓度相近的标准溶液，其浓度为 c_s，测得的吸光度为 A，则

Determination of copper in industrial wastewater (the method of standard addition)

工业废水中铜含量的测定（标准加入法）

$$A_x = Kc_x$$
$$A = K(c_x + c_s) \tag{3.5}$$
$$c_x = \frac{A_x}{A - A_x} \cdot c_s$$

Where A is the measured absorbance.

In practice, extrapolation method is used (fig.3-14): add the standard solution into four or five sample solutions of same volume at different ratio of analyte element, and dilute them to same volume, then measure the absorbance A respectively. Take the standard quantity of analyte element as x-axis and the corresponding

其中 A 是测得的吸光度。

实际测定时，通常采用作图外推法（图 3-14）：在 4 份或 5 份相同体积试样中，分别按比例加入不同量待测元素的标准溶液，并稀释至相同体积，然后分别测定吸光度 A。以加入待测元素的标准量为横坐标，相应的吸光度为纵坐

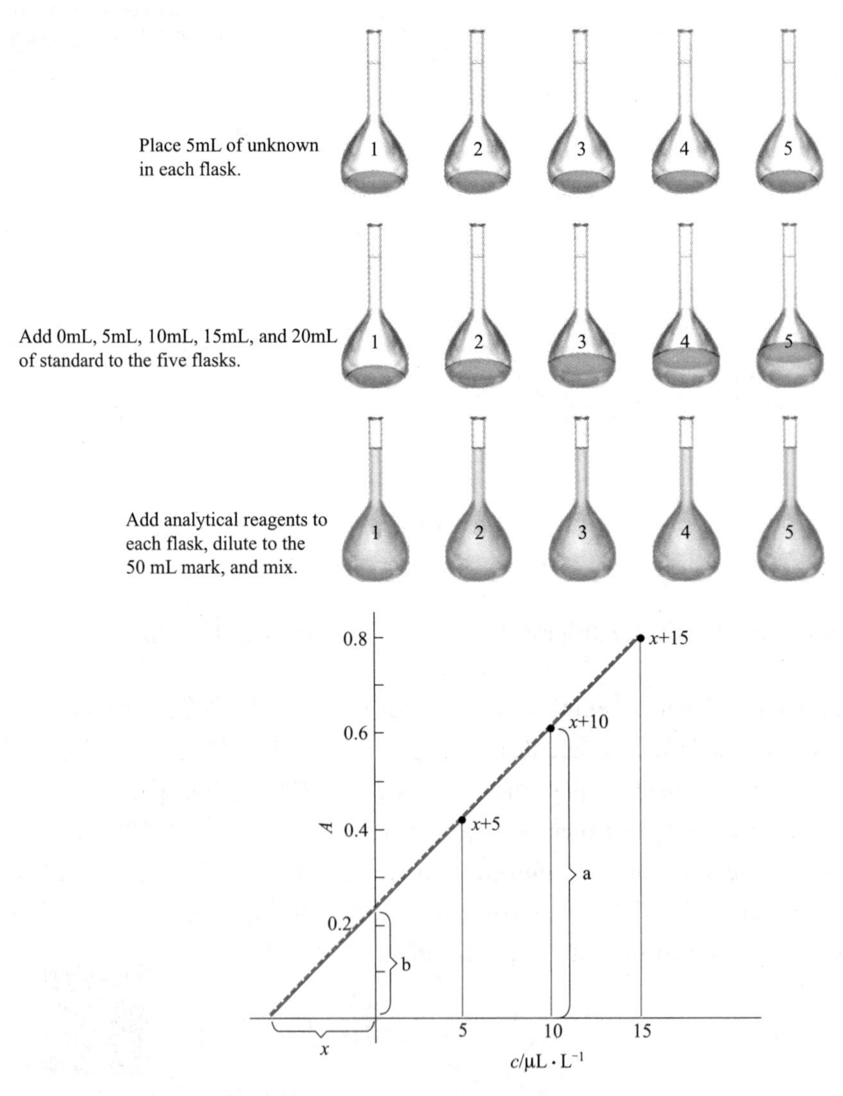

Fig.3-14 Standard addition experiment

图 3-14 标准加入法实验

absorbance as y-axis, a straight line can be obtained. The distance from the crossover point of the extension of this straight line on the x-axis to base point is the quantity of analyte element in the original sample solution.

标作图可得一直线，此直线的延长线在横坐标轴上交点到原点的距离相应的量即为原始试样中待测元素的含量。

Experiment 2 Determination of chromium in sewage by flame atomic absorption spectrometry with N_2O-C_2H_2

1. Purposes of the experiment

(1) Understand the basic structure of atomic absorption spectrometer.

(2) Master basic operation of atomic absorption spectrometer.

(3) Master analytical methods and data processing.

2. Experimental principle

AAS (Atomic Absorption Spectroscopy) is based on absorption of resonance lines by electrons in the outer layer of gaseous ground state atoms of the element to be measured. The atomic number of ground state in gaseous state is proportional to the content of substance, so it can be used for quantitative analysis.

$$A=KLN_0=K'c$$

3. Instruments and reagents

(1) The instrument iCE3000 atomic absorption spectrometer and chromium hollow cathode lamp made by American thermoelectric company; Nitrous oxide (N_2O) gas;Acetylene (C_2H_2).

The working conditions of AAS are shown in the following table.

Working conditions of AAS

Element under test	Element line/nm	Lamp current/mA	The spectral width/nm
Cr	357.9	6	0.5
Burner height/mm	N_2O flow/L・min^{-1}	C_2H_2 flow/L・min^{-1}	Range of measurement/mg・L^{-1}
12	6.0	2.5	1.0 ～ 50

(2) The reagent

① Chromium standard solution: 1.000 g chromium (Cr 99.99% spectral pure) is dissolved in the minimum volume of 1+1 nitric acid and diluted to 1000 mL volumetric flask with distilled water. At this time, the solution contains 1.000 mg・mL^{-1} Chromium (Cr) as a reserve solution.

② Take 50 mL chromium standard storage solution and place it in a 100 mL volumetric flask, dilute it with distilled water to the standard line, and mix well. The solution contains 0.5 mg · mL^{-1} chromium (Cr).

③ 20 g · L^{-1} cesium chloride solution ($CsCl_2$).

④ 40 g · L^{-1} potassium persulfate solution ($K_2S_2O_8$).

⑤ 20 g · L^{-1} diammonium hydrogen citrate solution ($C_6H_{14}O_7N_2$).

⑥ 2 mol · L^{-1} nitric acid solution (HNO_3).

The experimental water shall be deionized water. All reagents used in the experiment are analytically pure or above.

4. Experimental steps

(1) Chromium working curve preparation A certain amount of chromium standard solution (0.5 mg · mL^{-1}) is accurately drawn and placed in a series of 100 mL volumetric bottles (at this time, the concentration of Cr is 0.0 mg · mL^{-1}, 2.0 mg · mL^{-1}, 4.0 mg · mL^{-1}, 6.0 mg · mL^{-1}, 8.0 mg · mL^{-1}, 10.0 mg · mL^{-1}) .5 mL of 20 g · L^{-1} cesium chloride solution, 3 mL of 40 g · L^{-1} potassium persulfate solution, 3 mL of 20 g · L^{-1} hydrogen citrate ii solution and 1 mL of 2 mol · L^{-1} nitric acid solution are added into the volumetric bottles respectively and dilute to the scale with deionized water. The absorption values are measured under the working conditions, and the standard working curves are drawn.

(2) Sample preparation Preparation of sample solution: accurately draw 10 mL of sewage, place it in a 100 mL volumetric bottle (in advance, put 25 mL deionized water in it) , and mix well. Add 5 mL of 20 g · L^{-1} cesium chloride solution, 3 mL of 40 g · L^{-1} potassium persulfate solution, 3 mL of 20 g · L^{-1} diammonium hydrogen citrate solution and 1 mL of 2 mol · L^{-1} nitric acid solutions respectively, dilute to the scale with deionized water. Then measure the absorbance in the same way as in experiment step 1.

5. Data and processing

(1) Fill in the following table with the measurement results

The standard curve

c_{Cr}/mg · mL^{-1}	0.0	2.0	4.0	6.0	8.0	10.0	sample
A							

(2) Drawing and calculation

① Draw the standard curve with chromium concentration as abscissa and absorbance as ordinate, and work out the linear regression equation.

② According to the absorbance of the sample, calculate the content of chromium in the sample.

6. Questions

① Why is N_2O-C_2H_2 flame used in the experiment?

② What is the role of cesium chloride solution?

实验 2　用 $N_2O-C_2H_2$ 火焰原子吸收光谱法测定污水中的铬

1. 实验目的

（1）了解原子吸收光谱仪的基本结构。

（2）掌握原子吸收光谱仪的基本操作。

（3）掌握分析方法的建立和数据处理。

2. 实验原理

原子吸收光谱法是基于待测元素的气态基态原子外层的电子对共振线的吸收。气态的基态原子数与物质的含量成正比，故可用于定量分析。

$$A=KLN_0=K'c$$

3. 仪器与试剂

（1）仪器　美国热电公司制造的 iCE3000 型原子吸收光谱仪；美国热电公司制造的铬空心阴极灯；一氧化二氮 - 乙炔（N_2O）气体；乙炔（C_2H_2）。原子吸收光谱仪的工作条件，见下表。

原子吸收光谱仪的工作条件

待测元素	元素线 /nm	灯电流 /mA	光谱宽度 /nm	燃烧器高度 /mm	N_2O 流量 /$L \cdot min^{-1}$	C_2H_2 流量 /$L \cdot min^{-1}$	测定范围 /$mg \cdot L^{-1}$
Cr	357.9	6	0.5	12	6.0	2.5	1.0 ～ 50

（2）试剂

① 铬标准溶液的制备：称取 1.000g 铬（Cr 99.99% 光谱纯）溶解在最小体积的 1+1 硝酸中，用蒸馏水稀释到 1000mL 容量瓶，此时溶液含 1.000mg·mL^{-1} 铬（Cr），作储备液用。

② 吸取 50mL 铬标准贮备液置于 100mL 容量瓶中，用蒸馏水稀释至标准线混匀。此溶液含铬量为 0.5mg · mL^{-1} 铬（Cr）。

③ 20g · L^{-1} 氯化铯的溶液（$CsCl_2$）。

④ 40g · L^{-1} 过硫酸钾溶液（$K_2S_2O_8$）。

⑤ 20g · L^{-1} 柠檬酸氢二铵溶液（$C_6H_{14}O_7N_2$）。

⑥ 2mol · L^{-1} 硝酸溶液（HNO_3）。

实验用水为去离子水。实验所用试剂均为分析纯或者分析纯以上的试剂。

4. 实验步骤

（1）铬工作曲线制备　准确移取一定量的铬标准溶液（0.5mg · mL^{-1}）置于一系列 100mL 容量瓶中（此时含 Cr 浓度为 0.0mg·mL^{-1}、2.0mg·mL^{-1}、4.0mg·mL^{-1}、6.0mg·mL^{-1}、8.0mg · mL^{-1}、10.0mg · mL^{-1}）。加入 20g · L^{-1} 氯化铯溶液 5mL、40g · L^{-1} 过硫酸钾溶液 3mL、20g · L^{-1} 柠檬酸氢二铵溶液 3mL 和 2mol · L^{-1} 硝酸溶液 1mL，并用去离子水稀释至刻度。在工作条件下进行吸收值测定，并绘制其标准工作曲线。

（2）试样制备　准确移取污水样品 10mL，置于（预先在 100mL 容量瓶中放入 25mL

去离子水）100mL 容量瓶中，混匀。加入 20g · L^{-1} 氯化铯溶液 5mL、40g · L^{-1} 过硫酸钾溶液 3mL、20g · L^{-1} 柠檬酸氢二铵溶液 3mL 和 2mol · L^{-1} 硝酸溶液 1mL，用去离子水稀释至刻度。然后按照实验步骤 1 中同样的方法测量吸光度。

5. 数据及处理

（1）将测量结果填入下表

c_{Cr}/mg · mL^{-1}	0.0	2.0	4.0	6.0	8.0	10.0	试样
A							

（2）绘图及计算

① 以铬的浓度为横坐标，吸光度为纵坐标，绘制测定标准曲线，并求出线性回归方程。

② 根据样品的吸光度，计算样品中铬的含量。

6. 思考题

① 实验所用的气体为什么为 N_2O-C_2H_2 富燃火焰？

② 氯化铯溶液的作用是什么？

Exercises

3-1 In atomic absorption spectrophotometry, when the characteristic spectral line emitted by the light source passes through the sample vapor, it is absorbed by (　　) of the element to be measured in the vapor.

A. excited atomic vapor　　B. ground state atomic vapor

C. molecules in solution　　D. ions in solution

3-2 In the atomic absorption spectrophotometer, the role of the monochromator is (　　).

A. obtaining monochromatic light

B. separating the analysis line of the element to be measured from the adjacent spectral lines

C. getting continuous light

D. None of the above

3-3 There are many factors that broaden the atomic absorption spectrum, and (　　) is the main factor in Non-flame atomic absorption.

A. pressure broadening　　B. Lorentz broadening

C. natural broadening　　D. Doppler broadening

3-4 Atomic absorption spectroscopy belongs to (　　).

A. band spectrum　　B. line spectrum

C. broadband spectrum　　D. molecular spectrum

3-5 Choose different flame types, mainly according to ( ).

A. analysis line wavelength B. lamp current size

C. slit width D. properties of the element to be tested

3-6 The atomic absorption spectrophotometer needs a variety of gases to work. Among the following gases, the one that is not used by AAS is ( ).

A. air B. acetylene C. nitrogen D. oxygen

3-7 The characteristics of atomic absorption spectrophotometry do not include ( ).

A. high sensitivity B. good selectivity

C. good reproducibility D. one lamp for multiple purposes

3-8 If trace magnesium in water is determined, ( ) should be selected.

A. calcium hollow cathode lamp B. magnesium hollow cathode lamp

C. copper hollow cathode lamp D. iron hollow cathode lamp

3-9 A rich flame is a flame whose combustion support ratio is ( ) stoichiometric.

A. greater than B. less than C. equals to D. inner gas pressure

3-10 The atomic absorption spectrometer consists of ( ).

A. light source, atomization system, detection system

B. light source, atomization system, spectroscopic system

C. atomization system, spectroscopic system, detection system

D. light source, atomization system, spectroscopic system, detection system

3-11 For the quantitative analysis method of atomic absorption, the interference that can be eliminated by the standard addition method is ( ).

A. molecular absorption B. background absorption

C. matrix effect D. physical interference

3-12 In atomic absorption spectrophotometer, the commonly-used detector is ( ).

A. Photocell B. Phototube

C. Photomultiplier tube D. Photosensitive plate

3-13 The spectral line broadening caused by the random thermal motion of atoms is called ( ).

A. natural broadening B. Holtsmark broadening

C. Lorentz broadening D. Doppler broadening

3-14 The core component of atomic absorption spectrophotometer is ( ).

A. light source B. atomizer C. spectral system D. detection systems

3-15 In general, when measuring by atomic absorption spectrophotometry, it is always hoped that the light will pass through the part of ( ).

A. highest flame temperature B. lowest flame temperature

C. highest atomic vapor density D. lowest atomic vapor density

3-16 True or false

(1) Atomic absorption spectrophotometry is an analytical method based on the absorption of light of characteristic wavelengths by atoms in the ground state and atoms in the excited state.

(2) In atomic absorption spectrophotometer, the commonly-used light source is hollow

cathode lamp.

(3) Atomic absorption spectrum is a band spectrum, while UV-Vis spectrum is a line spectrum.

(4) In the flame atomization method, the commonly-used gas is air-acetylene.

(5) The atomization device of atomic absorption spectrophotometer is mainly divided into two categories: flame atomizer and non-flame atomizer.

(6) The quantitative analysis of atomic absorption spectrophotometry is based on Beer's law.

(7) Since the transition of electrons from the ground state to the first excited state occurs most easily, the resonance absorption line is the most sensitive line for most elements. Therefore, the resonance line of the element is also called the analytical line.

(8) The light source of the atomic absorption spectrophotometer is a continuous light source.

(9) Compared with the flame atomization method, one of the advantages of the graphite furnace atomization method is the high atomization efficiency.

(10) In the flame atomization method, the sample can be fully decomposed into the atomic vapor state at a sufficient temperature. Therefore, the higher the temperature, the better.

(11) In atomic absorption spectrophotometry, if the concentration of the measured element is very high, or in order to eliminate the interference of adjacent spectral lines, the sub-sensitive line can be selected.

(12) The luminous intensity of the hollow cathode lamp is related to the working current. Increasing the current can increase the luminous intensity. Therefore, the larger the lamp current, the better.

(13) In the analysis of atomic absorption spectrometry, the selection principle of lamp current is to select the lowest working current as far as possible under the condition of ensuring stable discharge and appropriate light intensity output.

(14) The monochromator in the atomic absorption spectrophotometer is placed before the atomization system.

(15) Chemical interference is the main interference factor in atomic absorption spectrometry.

3-17 What is the basic principle of atomic absorption spectrophotometry?

3-18 What is the atomization of the sample? What are the methods for atomization of samples?

3-19 What are the similarities and differences between Atomic Absorption Spectrophotometry and UV-Vis Spectrophotometry?

3-20 How do chemical interferences in atomic absorption spectrometry arise? How to eliminate?

3-21 Briefly describe the names and functions of the main components in the four major systems of the atomic absorption spectrometer.

3-22 Briefly describe the procedure for atomizing the sample in the tubular graphite furnace.

3-23 The inverted line dispersion rate of an atomic absorption spectrometer is 15 $nm \cdot mm^{-1}$. In the determination of Ni, the analytical line of 232.0 nm is used. In order to avoid the interference of the spectral line of 231.6 nm, what is the appropriate width of the slit?

($<$ 0.027 mm)

3-24 The content of Cd in waste liquid of a factory is determined by atomic absorption spectrophotometry. 100.00 mL water sample is accurately taken from the waste liquid outlet, and after appropriate acidification treatment, 10.00 mL methyl isobutyl ketone solution is accurately added for extraction and concentration. The measured element is measured at a wavelength of 228.8 nm, and the measured absorbance is 0.182. Under the same conditions, the standard series of measured absorbances of Cd are shown in the following table. The content of Cd in the waste liquid of the factory is calculated by drawing method (expressed in mg/L).

Mass concentration of Cd/μg · mL^{-1}	0.0	0.1	0.2	0.4	0.6	0.8	1.0
Absorbance	0.000	0.052	0.104	0.208	0.312	0.416	0.520

(0.035 mg · L^{-1})

3-25 The concentration of Mg in a water sample is determined by the standard addition method. 5 samples are taken, and then different amounts of 100 μg · mL^{-1} standard magnesium solution are added to each sample, constant volume to 25 mL. The absorbance is measured as shown in the table below, to obtain the concentration of Mg in the water sample by a graph (the result is expressed in mg · L^{-1}).

Test solution volume/mL	20	20	20	20	20
Volume of Mg standard solution added/mL	0.00	0.25	0.50	0.75	1.00
Absorbance	0.091	0.181	0.282	0.374	0.470

(5.0 mg · L^{-1})

Chapter 4 Electrochemical Methods of Analysis

第4章 电化学分析法

Study Guide 学习指南

Electrochemical Methods of Analysis are a class of techniques in analytical chemistry which is based on the electrochemical properties of substance. Usually a chemical battery is formed by the test substance solution and a suitable electrode system, and then the composition and content of the test substance are determined by measuring the electromotive force of the battery or measuring the changes in physical quantities such as the current and electricity passing through the battery. This chapter shall mainly deal with potentiometric analysis and coulometric analysis. The target of this chapter is to familiarize with the principles, concepts, and industrial applications of potentiometry and coulometry.

电化学分析法是建立在物质的电化学性质基础上的一类分析方法。通常用被测物质溶液与适合的电极系统构成一个化学电池，然后通过测量电池的电动势或测量通过电池的电流、电量等物理量的变化来确定被测物质的组成和含量。本章主要介绍电位分析法、库仑分析法。通过本章的学习，掌握电位分析法、库仑分析法的基本原理及概念，了解这两种方法的工业应用。

Section 1 Type of Electrochemical Method and Potentiometry

第1节 电化学分析法的分类和电位分析法

1. Interfacial Electrochemical Methods

The diversity of interfacial electrochemical methods is evident from the partial family tree shown in figure 4-1. At the first level, interfacial electrochemical methods are divided into static methods and dynamic methods. In static methods no current passes between the electrodes,

1. 表面电化学分析方法

表面电化学分析方法的分类见图 4-1。首先，表面电化学方法分为静态法和动态法。在静态法中，电极之间没有电流通过，电化学电池中的物质浓度保持不变，或者说是静态的。电位分析法就是

and the concentrations of species in the electrochemical cell remain unchanged, or static. Potentiometry, in which the potential of an electrochemical cell is measured under static conditions, is one of the most important quantitative electrochemical methods, shown in table 4-1.

一种在静态条件下测量电化学电池电位的方法，也是最重要的电化学定量方法之一（见表 4-1）。

The largest division of interfacial electrochemical methods is the group of dynamic methods, in which current flows and concentrations change as the result of a redox reaction. Dynamic methods are further subdivided by whether we choose to control the current or the potential. In controlled-current coulometry, we completely oxidize or reduce the analyte by passing a fixed current through

表面电化学分析方法的最大分支是动态分析法，其电流和浓度随着氧化还原反应进行而变化。动态法根据选择控制电流或控制电位可以进一步细分。在控制电流库仑法中，使恒定电流通过分析溶液来完全氧化或还原分析物质。控制电位法进一步细分为控制电位库仑法

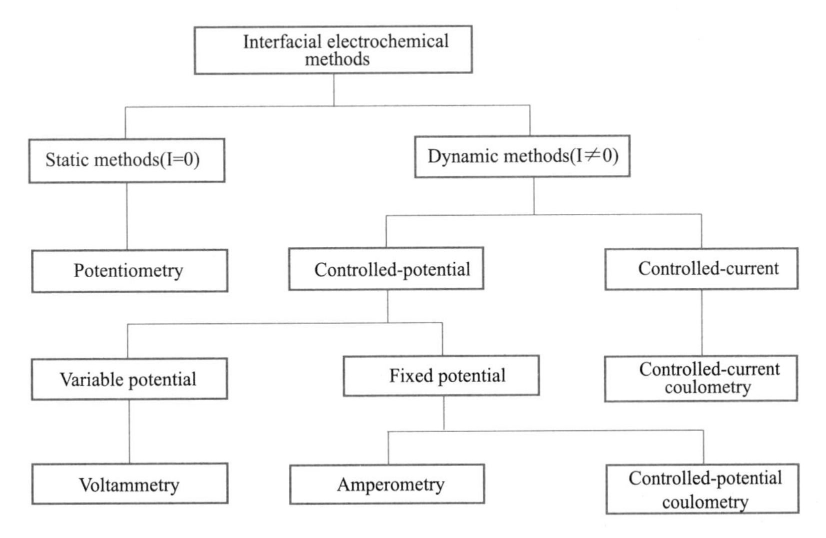

Fig.4-1 Partial family tree for interfacial electrochemical methods of analysis

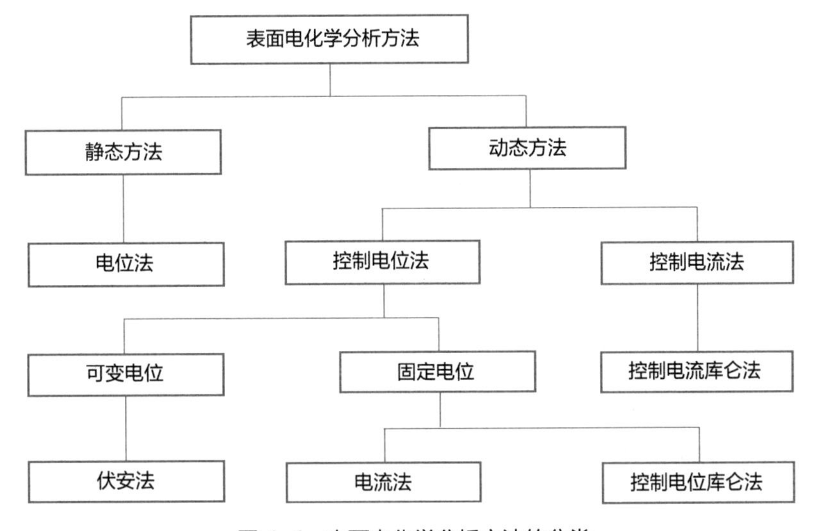

图 4-1 表面电化学分析方法的分类

Table 4-1 Examples of Methods for Electrochemical Analysis

Method	Definition
Potentiometry	Potentiometry is one of the simplest electroanalytical techniques. The principle of potentiometric measuring technique is the measurement of the cell potential, i.e., the potential difference between two electrodes (the indicator and the reference electrodes) in an electrochemical cell under zero current conditions with the aim of obtaining analytical information about the chemical composition of a solution. There are two major types of analysis techniques in potentiometry. When the cell potential is determined and correlated to the activity or the concentration of the individual chemical species, it is called direct potentiometry. And when the variation of the cell potential is monitored as function of the reagent addition to the sample, it is called indirect potentiometry (potentiometric titration)
Amperometry	Amperometry is a voltammetric method, which is based on the measurement of current at a fixed operating potential in stirred (or flowing) solutions or at a rotating working electrode. The current is the result of electrochemical oxidation or reduction of the electroactive compound after applying the potential pulse across the working and the auxiliary electrodes
Voltammetry	Voltammetry comprises microelectrolysis techniques in which the working electrode potential is forced by external instrumentation to follow a known potential-time function and the resultant current-potential and current-time curves which are analyzed to obtain information about the solution composition
Coulometry	Coulometric methods are based on the measurement of the quantity of electricity required for the quantitative electrooxidation or electroreduction of a chemical species. The quantity of electricity is measured in coulombs

表4-1 电化学分析方法举例

方法	定义
电位法	电位法是最简单的电分析技术之一。电位测量技术的原理是测量电池电位，即在零电流条件下测量电化学电池中两个电极（指示电极和参比电极）之间的电位差，目的是获得溶液化学成分的分析信息。电位法有两种主要的分析技术：当测定电池电位并与单个化学物质的活度或浓度相关时，称为直接电位法；当电池电位的变化与添加到样品中的试剂相关时，称为间接电位法（电位滴定法）
电流法	电流法是一种伏安法，它是基于在搅拌（或流动）溶液或在旋转工作电极中测量固定工作电位下的电流的电化学分析方法。电流是在工作电极和辅助电极上施加电位脉冲后，电活性化合物电化学氧化或还原的结果
伏安法	伏安法属于微电解技术，其中工作电极电位通过外部仪器强制遵循已知的电位 - 时间函数，并通过分析所得的电流 - 电位和电流 - 时间曲线，以获得相关溶液成分的信息
库仑法	库仑法是基于对化学物质的定量电氧化或电还原所需电量的测量。电量以库仑为单位

the analytical solution. Controlled-potential methods are subdivided further into controlled-potential coulometry and amperometry, in which a constant potential is applied during the analysis, and voltammetry, in which the potential is systematically varied.

和安培法，二者都是在分析过程中施加恒定电位；还有伏安法，但其电位是系统变化的。

2. Potentiometric Methods of Analysis

In potentiometry the potential of an electrochemical cell is measured under static conditions. Because no current, or only a negligible current, flows while measuring a solution's potential, its composition remains unchanged. For this reason, potentiometry is a useful quantitative method. The first quantitative potentiometric applications appeared soon after the formulation, in 1889, of the Nernst equation relating an electrochemical cell's potential to the concentration of electroactive species in the cell.

When first developed, potentiometry was restricted to redox equilibria at metallic electrodes, limiting its application to a few ions. In 1906, Cremer discovered that a potential difference exists between the two sides of a thin glass membrane when opposite sides of the membrane are in contact with solutions containing different concentrations of H_3O^+. This discovery led to the development of the pH glass electrode in 1909. Other types of membranes also yield useful potentials. Kolthoff and Sanders, for example, showed in 1937 that pellets made from AgCl could be used to determine the concentration of Ag^+. Electrodes based on membrane potentials are called ion-selective electrodes, and their continued development has extended potentiometry to a diverse array of analytes.

(1) Potentiometric electrochemical cells　A schematic diagram of a typical potentiometric electrochemical cell is shown in figure 4-2. Note that the electrochemical cell is divided into two half-cells, each containing an electrode immersed in a solution containing ions whose concentrations determine the electrode's potential. This separation of electrodes is necessary to prevent the redox reaction from occurring spontaneously on the surface of one of the electrodes, short-circuiting the electrochemical cell and making the measurement of cell potential impossible.

We can connect the two halves with a salt bridge.

2. 电位分析法

在电位分析法中，电化学电池的电位是在静态条件下测定的。由于在测定溶液电位时没有电流通过，或者只有可忽略的电流通过，所以溶液的组分保持不变。因此，电位滴定法是一种非常有用的定量方法。1889 年，能斯特方程将电化学电池的电位与活性物质的浓度联系起来，此后不久，就出现了电位分析法在定量分析中的应用实例。

首次应用时，电位分析法仅限于金属电极上的氧化还原平衡，其应用仅限于少数离子。1906 年，Cremer 发现，当玻璃膜两侧与含有不同浓度 H_3O^+ 的溶液接触时，膜两侧间存在电位差。这一发现促进了 1909 年 pH 玻璃电极的诞生。其他类型的膜同样可产生有用的电势，例如，Kolthoff 和 Sanders 在 1937 年研究发现，用氯化银制成的颗粒可以用来测定 Ag^+ 的浓度。基于膜电位的电极称为离子选择电极，其不断发展使电位测定法扩展到各种分析物质。

（1）电化学电池　典型的电化学电池的结构如图 4-2 所示。电化学电池分为两个半电池，在每个含有一定浓度离子的半电池中，插入一个电极，溶液中离子的浓度决定了电极的电位。这种电极的分离是非常必要的，它能防止氧化还原反应自发地发生在某一电极的表面，但同时也造成了电化学电池的断路，使电池电位的测量不能顺利实现。

这时可以用一个盐桥连接这两个半电池。

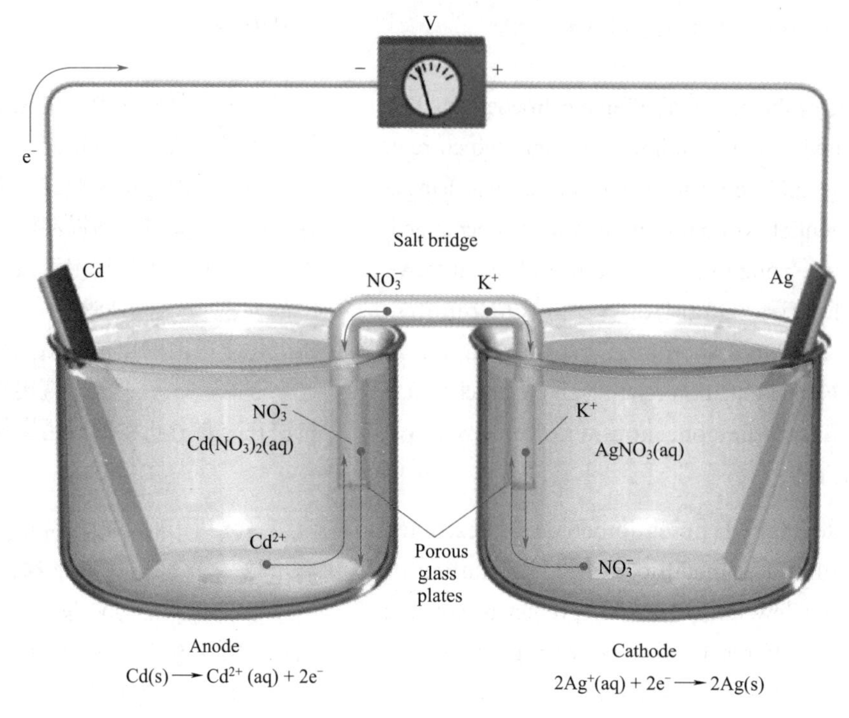

Fig.4-2 A cell that works—thanks to the salt bridge

图 4-2 电池工作的关键在盐桥

The salt bridge is a U-shaped tube filled with a gel containing a high concentration of KNO_3 (or other electrolyte that does not affect the cell reaction). The ends of the bridge are porous glass disks that allow ions to diffuse but minimize mixing of solutions inside and outside the bridge. When the cell is operating, K^+ from the bridge migrates into the cathode compartment and a small amount of NO_3^- migrates from the cathode into the bridge. Ion migration offsets the charge buildup that would otherwise occur as electrons flow into the silver electrode. In the absence of a salt bridge, negligible reaction can occur because of charge buildup. The migration of ions out of the bridge is greater than the migration of ions into the bridge because the salt concentration in the bridge is much higher than the salt concentration in the half-cells. At the left side of the salt bridge, NO_3^- migrates into the anode compartment and a little Cd^{2+} migrates into the bridge to prevent buildup of positive charge. This movement of ions in the salt bridge completes the electric circuit.

盐桥是一个装满了含有高浓度硝酸钾凝胶（或其他不影响电池反应的电解质）的 U 形管，盐桥的两端是多孔玻璃盘，其允许离子扩散，但要尽量减少盐桥内外溶液的混合。当电池工作时，盐桥中的 K^+ 迁移到阴极室，少量的 NO_3^- 从阴极迁移到盐桥中。离子的迁移抵消了当电子流入银电极时可能会发生的电荷积累。在没有盐桥的情况下，由于电荷的积累，会发生可以忽略不计的反应。由于盐桥中的盐浓度远远高于半电池中的盐浓度，造成离子向盐桥外的迁移大于离子向盐桥内的迁移。在盐桥的左侧，NO_3^- 迁移到阳极室，微量的 Cd^{2+} 迁移到盐桥中，以防止正电荷的积累。盐桥中离子的运动使电化学电池形成了通路。

(2) Shorthand notation for electrochemical cells By convention, the electrode on the left is considered to be

（2）电化学电池的简化符号 按照惯例，左边的电极是发生氧化反应

the anode ("Anode" means "a way up" in Greek), where oxidation occurs

的阳极：

$$Zn(s) \rightleftharpoons Zn^{2+}(aq)+2e^-$$

and the electrode on the right is the cathode ("Cathode" means "a way down" in Greek), where reduction occurs The electrochemical cell's potential, therefore, is for the reaction

右边的电极是发生还原反应的阴极：

$$Cu^{2+}(aq)+2e^- \rightleftharpoons Cu(s)$$

Also, by convention, potentiometric electrochemical cells are defined such that the indicator electrode is the cathode (right half-cell) and the reference electrode is the anode (left half-cell).

电化学电池的反应：

$$Zn(s)+Cu^{2+}(aq) \rightleftharpoons Cu(s)+Zn^{2+}(aq)$$

Although figure 4-3 provides a useful picture of an electrochemical cell, it does not provide a convenient representation. A more useful representation is a shorthand, or schematic, notation that uses symbols to indicate the different phases present in the electrochemical cell, as well as the composition of each phase. A vertical slash (|) indicates a phase boundary where a potential develops, and a comma (,) separates species in the same phase, or two phases where no potential develops. Shorthand cell notations begin with the anode and continue to the cathode. The electrochemical cell in figure 4-3, for example, is described in shorthand notation as

虽然图 4-3 提供了一个电化学电池的结构图，但它表示起来并不方便。一个更实用的表示方法就是用符号来表示电化学电池单元中存在的不同相以及每个相的组成。通常用“|”表示相界，“,”表示同一相中的不同物质或没有电位产生的两相。

$$Zn(s)|ZnCl_2(aq,0.01M)||CuSO_4(aq,0.01M)|Cu(s)$$

The double vertical slash (||) indicates the salt bridge, the contents of which are normally not indicated. Note that the double vertical slash implies that there is a potential difference between the salt bridge and each half-cell.

“||”表示盐桥，其意味着盐桥和每个半电池之间存在一个电位差。电池的书写一般从阳极开始，阴极结束。

(3) Liquid junction potentials A liquid junction potential develops at the interface between any two

（3）液接电位　在组成不同且离子迁移率各异的任何两种离子溶液之间的

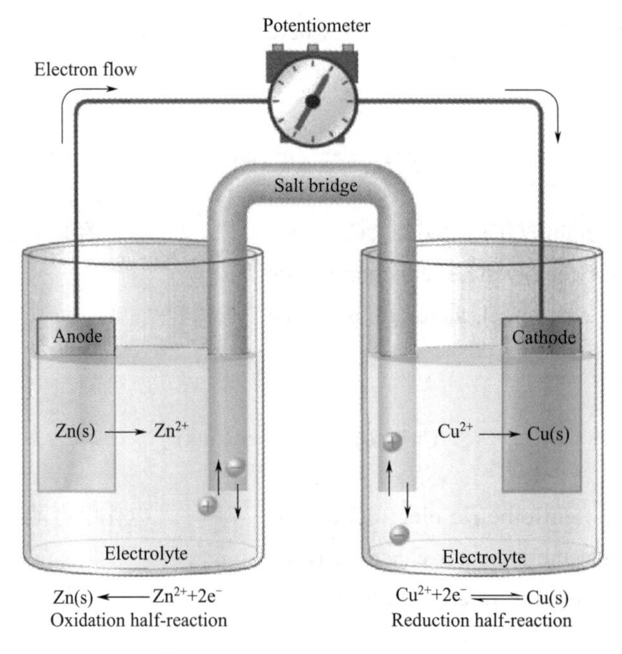

Fig.4-3 The general components of an electrochemical cell for potentiometry.

图 4-3 用于电位测量的电化学电池的一般组成

ionic solutions that differ in composition and for which the mobility of the ions differs. Consider, for example, solutions of 0.1 M HCl and 0.01 M HCl separated by a porous membrane (figure 4-4). Since the concentration of HCl on the left side of the membrane is greater than that on the right side of the membrane, there is a net diffusion of H^+ and Cl^- in the direction of the arrows. The mobility of H^+, however, is greater than that for Cl^-, as shown by the difference in the lengths of their respective arrows. As a result, the solution on the right side of the membrane develops an excess of H^+ and has a positive charge. Simultaneously, the solution on the left side of the membrane develops a negative charge due to the greater concentration of Cl^-. The difference in potential across the membrane is called a liquid junction potential, E_{lj}.

界面处均会形成液接电位。例如，用多孔膜分离 0.1M HCl 和 0.01M HCl 的溶液时，由于膜左侧的 HCl 浓度大于膜右侧的 HCl 浓度，因此在箭头方向存在 H^+ 和 Cl^- 的净扩散（见图 4-4）。然而，H^+ 的迁移率大于 Cl^-，如其各自箭头长度的差异所示。因此，膜右侧的溶液产生过量的 H^+ 并带有正电荷。同时，由于氯离子浓度较高，膜左侧的溶液产生负电荷。膜两侧的电位差称为液接电位 E_{lj}。

The magnitude of the liquid junction potential is determined by the ionic composition of the solutions on the two sides of the interface and may be as large as 30-40 mV. For example, a liquid junction potential of 33.09 mV has been measured at the interface between solutions of 0.1 M HCl and 0.1 M NaCl (figure 4-5).

液接电位的大小由界面两侧溶液的离子组成决定，可达 30 ～ 40mV。例如，在 0.1M 盐酸和 0.1M 氯化钠溶液之间的界面上测量了 33.09mV 的液接电位（见图 4-5）。

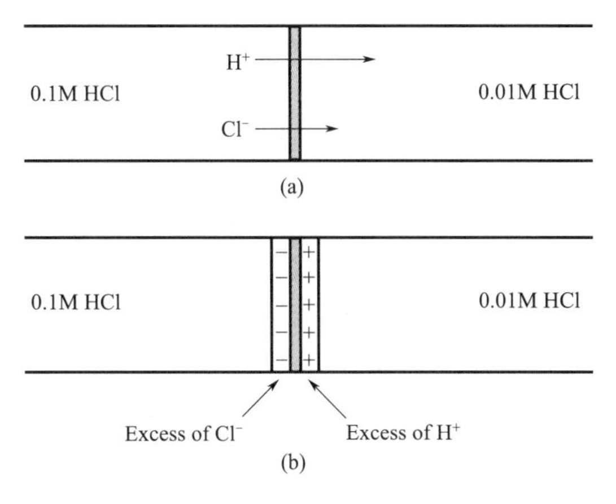

Fig.4-4 Origin of liquid junction potential between solutions of 0.1 M HCl and 0.01 M HCl

图 4-4 0.1M HCl 和 0.01M HCl 溶液之间液接电位的来源

The magnitude of a salt bridge's liquid junction potential is minimized by using a salt, such as KCl, for which the mobilities of the cation and anion are approximately equal. The magnitude of the liquid junction potential also is minimized by incorporating a high concentration of the salt in the salt bridge. For this reason salt bridges are frequently constructed using solutions that are saturated with KCl. Nevertheless, a small liquid junction potential, generally of unknown magnitude, is always present. When the potential of an electrochemical cell is measured, the contribution of the liquid junction potential must be included. Thus, The potential of a potentiometric electrochemical cell is given as

通过加入阳离子和阴离子的迁移率近似相等的盐（如 KCl），可使盐桥的液接电位最小化。通过在盐桥中加入高浓度的盐，液接电位也可最小化，故盐桥常用 KCl 的饱和溶液。由于液接电位的存在，当测量电化学电池的电位时，必须考虑液接电位的影响。因此，电化学电池的电动势可表示为式（4.1）。

$$E_{cell}=E_c-E_a+E_{lj} \tag{4.1}$$

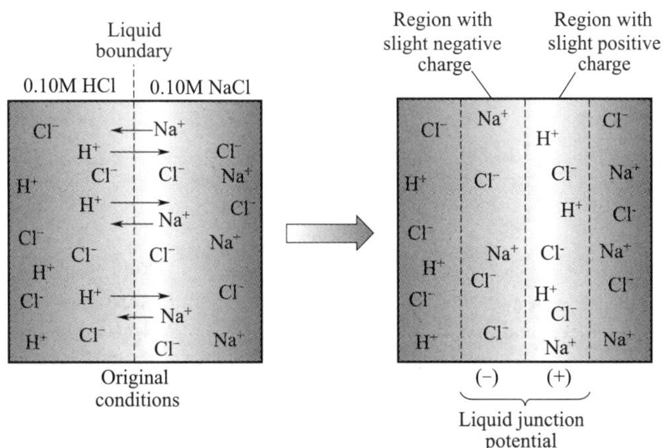

Fig.4-5 An example of the creation of a liquid junction potential.

图 4-5 液接电位产生实例

where E_c and E_a are reduction potentials for the reactions occurring at the cathode and anode.

其中 E_c、E_a 分别为阴极和阳极的电位。

Section 2 Types of Potentiometric Electrodes

第 2 节 电位电极的分类

1. Reference Electrodes

Potentiometric electrochemical cells are constructed such that one of the half-cells provides a known reference potential, and the potential of the other half-cell indicates the analyte's concentration. By convention, the reference electrode is taken to be the anode (shown in figure 4-6); thus, the shorthand notation for a potentiometric electrochemical cell is

Reference||Indicator

and the cell potential is

1. 参比电极

电化学电池是由两个半电池构成，其中一个半电池提供电位已知的参考值，另一个半电池指示待测物的浓度。一般地，参比电极放入阳极，指示电极放入阴极（见图 4-6）。电化学电池的简化符号为：

参比 || 指示

电池的电动势为式（4.2）。

$$E_{cell}=E_{ind}-E_{ref}+E_{lj} \tag{4.2}$$

The ideal reference electrode must provide a stable potential so that any change in E_{cell} is attributed to the indicator electrode, and, therefore, to a change in the analyte's concentration. In addition, the ideal reference

理想的参比电极必须提供稳定的电位，因此电池电动势的任何变化皆因指示电极的电位变化，也即分析物浓度的变化。此外，理想的参比电极还应易于

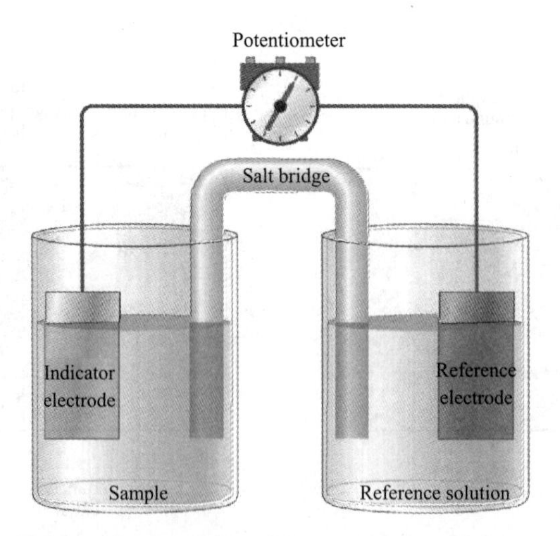

Fig.4-6 Composition of the electrochemical cells

图 4-6 电化学电池的组成

electrode should be easy to make and to use.

(1) Standard hydrogen electrode The standard hydrogen electrode (SHE) is rarely used for routine analytical work, but is important because it is the reference electrode used to establish standard-state potentials for other half-reactions. The SHE consists of a Pt electrode immersed in a solution in which the hydrogen ion activity is 1.00 and in which H_2 gas is bubbled at a pressure of 1 atm (figure 4-7). A conventional salt bridge connects the SHE to the indicator half-cell. The shorthand notation for the standard hydrogen electrode is

制造和使用。

（1）标准氢电极 标准氢电极（SHE）很少用于常规分析工作，但其重要作用是建立其他半反应标准状态电位的参比电极。标准氢电极由浸入氢离子活度为1.00M溶液中的Pt电极组成，且须在Pt电极表面附近溶液中通入1atm的H_2气体（图4-7）。一般通过盐桥将标准氢电极连接到含有指示电极的半电池中。根据定义，在所有温度下，SHE的电极电位均为0.00V。尽管SHE是重要的基本参比电极，但由于其难以制备和使用不便，很少使用。

$$\mathrm{Pt(s),H_2(g,1\ atm)|H^+(aq,}a\mathrm{=1.00)||}$$

and the standard-state potential for the reaction

$$\mathrm{2H^+(aq)+e^-} \rightleftharpoons \mathrm{H_2(g)}$$

is, by definition, 0.00 V for all temperatures. Despite its importance as the fundamental reference electrode against which all other potentials are measured, the SHE is rarely used because it is difficult to prepare and inconvenient to use.

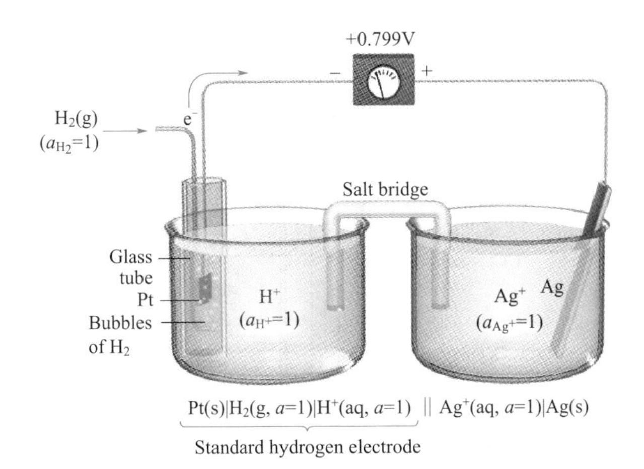

Fig.4-7 The general design of a standard hydrogen electrode

图4-7 标准氢电极组成示意图

(2) Calomel electrodes Calomel reference electrodes are based on the redox couple between Hg_2Cl_2 and Hg (calomel is a common name for Hg_2Cl_2).

（2）甘汞电极 甘汞电极的基础是Hg_2Cl_2和Hg氧化还原电对。

$$Hg_2Cl_2(s)+2e^- \rightleftharpoons 2Hg(l)+2Cl^-(aq)$$

The Nernst equation for the calomel electrode is

其能斯特方程见式（4.3）。

$$E=E^{\ominus}_{Hg_2Cl_2/Hg}-\frac{0.0592}{2}\lg[Cl^-]^2 = +0.2682-\frac{0.0592}{2}\lg[Cl^-]^2 \quad (4.3)$$ [1]

The potential of a calomel electrode, therefore, is determined by the concentration of Cl^-.

The saturated calomel electrode (SCE), which is constructed using an aqueous solution saturated with KCl, has a potential at 25℃ of +0.2444 V. A typical SCE is shown in figure 4-8 and consists of an inner tube, packed with a paste of Hg, Hg_2Cl_2, and saturated KCl, situated within a second tube filled with a saturated solution of KCl. A small hole connects the two tubes, and an asbestos fiber serves as a salt bridge to the solution in which the SCE is immersed. The stopper in the outer tube may be removed when additional saturated KCl is needed. The shorthand notation for this cell is

可见，甘汞电极的电极电位是由Cl^-浓度决定的。饱和甘汞电极（SCE）内充液为饱和的氯化钾溶液，其在25℃时的电极电位为+0.2444V。典型的SCE如图4-8所示，该电极有内外两根玻璃管，内管封接一根铂丝，铂丝插入Hg、Hg_2Cl_2和饱和KCl中；外管装有饱和的KCl溶液（盐桥溶液）。外管下端开有小孔，孔内塞入石棉纤维，当SCE浸入溶液构成电池回路时，起到盐桥作用。

$$Hg(l)|Hg_2Cl_2(sat'd),KCl(aq,sat'd)||$$

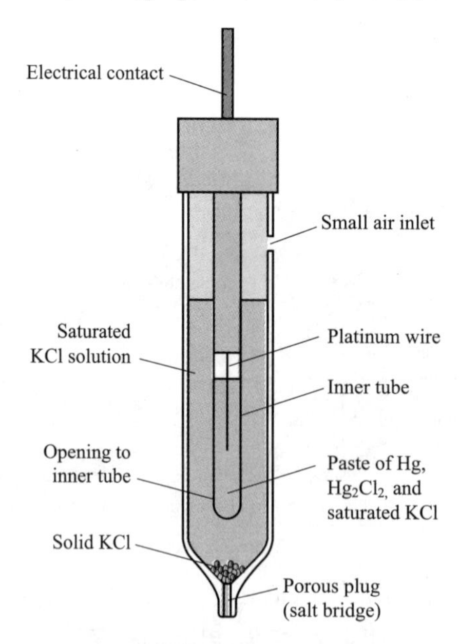

Fig.4-8 The general design of a calomel reference electrode

图4-8 甘汞电极组成示意图

[1] [x] indicates the concentration of x . [x] 表示x的浓度。

The SCE has the advantage that the concentration of Cl^-, and, therefore, the potential of the electrode, remains constant even if the KCl solution partially evaporates. On the other hand, a significant disadvantage of the SCE is that the solubility of KCl is sensitive to a change in temperature. At higher temperatures the concentration of Cl^- increases, and the electrode's potential decreases. For example, the potential of the SCE at 35℃ is +0.2376 V. Electrodes containing unsaturated solutions of KCl have potentials that are less temperature-dependent, but experience a change in potential if the concentration of KCl increases due to evaporation. Another disadvantage to calomel electrodes is that they cannot be used at temperatures above 80 ℃, because Hg begins to decompose above 80 ℃, it affects the stability of the electrode and also harms the health of the user.

SCE的优点是，即使KCl溶液部分蒸发，Cl^-的浓度仍保持恒定，即电极电位保持不变。但是，SCE的一个显著缺点是KCl的溶解度对温度的变化很敏感，在较高的温度下，Cl^-的浓度增加，电极的电位降低。例如，在35℃时的SCE电位为+0.2376V。含有KCl不饱和溶液的电极电位对温度的敏感性较小，但如果KCl的浓度因蒸发而增加，电位就会发生变化。甘汞电极的另一个缺点是其不能在80℃以上的温度下使用，因为80℃以上，Hg开始分解，影响电极使用稳定性，也对使用者身体健康造成危害。

(3) Silver/Silver chloride electrodes Another common reference electrode is the silver/silver chloride electrode, which is based on the redox couple between AgCl and Ag.

（3）银-氯化银电极 另一种常见的参比电极是Ag-AgCl电极，它是基于AgCl和Ag之间的氧化还原电对。

$$AgCl(s)+e^- \rightleftharpoons Ag(s)+Cl^-(aq)$$

As with the saturated calomel electrode, the potential of the Ag/AgCl electrode is determined by the concentration of Cl^- used in its preparation.

与饱和甘汞电极一样，Ag-AgCl电极的电位也是由Cl^-的浓度决定的。

$$E=E^{\ominus}_{AgCl/Ag}-0.0592\lg[Cl^-]=+0.2223-0.0592\lg[Cl^-] \quad (4.4)$$

When prepared using a saturated solution of KCl, the Ag/AgCl electrode has a potential of +0.197 V at 25 ℃. Another common Ag/AgCl electrode uses a solution of 3.5 M KCl and has a potential of +0.205 V at 25 ℃. The Ag/AgCl electrode prepared with saturated KCl, of course, is more temperature-sensitive than one prepared with an unsaturated solution of KCl.

25℃时，饱和KCl溶液的Ag-AgCl电极的电极电位为+0.197V；同一温度下，3.5M KCl溶液的Ag-AgCl电极的电极电位为+0.205V。用饱和KCl制备的Ag-AgCl电极比用不饱和KCl溶液制备的电极对温度更敏感。

A typical Ag/AgCl electrode is shown in figure 4-9 and consists of a silver wire, the end of which is coated with a thin film of AgCl. The wire is immersed in a solution that contains the desired concentration of KCl and that is saturated with AgCl. A porous plug serves as the salt bridge. The shorthand notation for the cell is

一个典型的Ag-AgCl电极如图4-9所示，其是在Ag丝上镀一层AgCl，并浸在KCl、AgCl的饱和溶液中，管下多孔塞起到盐桥的作用。该电池可写为：

$$Ag(s)|AgCl(sat'd), KCl(x)||$$

where x is the concentration of KCl.

其中，x 是氯化钾的浓度。

In comparison to the SCE the Ag/AgCl electrode has the advantage of being useful at higher temperatures. On the other hand, the Ag/AgCl electrode is more prone to reacting with solutions to form insoluble silver complexes that may plug the salt bridge between the electrode and the solution.

与 SCE 相比，Ag-AgCl 电极更适应较高的温度，但 Ag-ACl 电极更易与溶液反应，生成可能堵塞盐桥的不溶性银盐络合物。

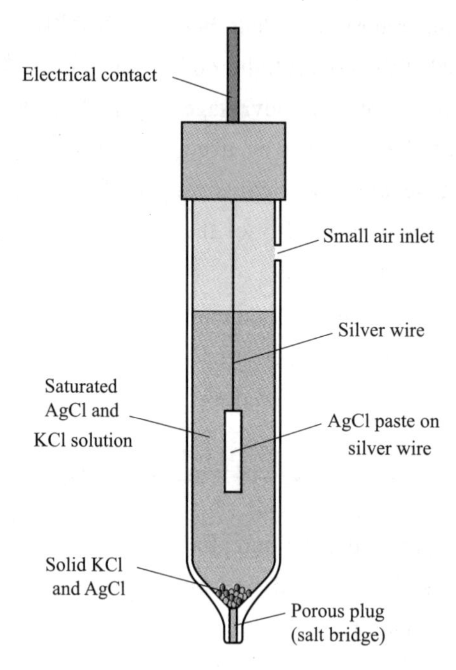

Fig.4-9　The general design of a silver/silver chloride electrode

图 4-9　银－氯化银电极组成示意图

2. Indicator Electrodes

2. 指示电极

The potential of the indicator electrode in a potentiometric electrochemical cell varies with changes in analyte concentration, meet the Nernst equation. Two classes of indicator electrodes are used in potentiometry: metallic electrodes and ion-selective electrodes.

电化学电池中指示电极的电位随被分析物浓度的变化而变化，符合能斯特方程。在电位测定中常用两类指示电极：金属指示电极和离子选择性电极。

(1) Metallic indicator electrodes　The potential of a metallic electrode is determined by the position of a redox reaction at the electrode–solution interface. Three types of metallic electrodes are commonly used in potentiometry, each of which is considered in the following discussion.

（1）金属指示电极　金属电极的电位由电极－溶液界面上的氧化还原反应决定。在电位测量中，常用的金属电极有三类。

① Electrodes of the first kind　When a copper electrode is immersed in a solution containing Cu^{2+}, the potential of the electrode due to the reaction

① 第一类电极　当铜电极浸没在含有 Cu^{2+} 的溶液中时

$$Cu^{2+}(aq)+2e^{-} \rightleftharpoons Cu(s)$$

is determined by the concentration of copper ion.

电极电位由 Cu^{2+} 的浓度决定，其能斯特方程为式（4.5）。

$$E=E^{\ominus}_{Cu^{2+}/Cu}-\frac{0.0592}{2}\lg\frac{1}{[Cu^{2+}]}=+0.3419-\frac{0.0592}{2}\lg\frac{1}{[Cu^{2+}]} \quad (4.5)$$

If the copper electrode is the indicator electrode in a potentiometric electrochemical cell that also includes a saturated calomel reference electrode,

如果以铜电极作指示电极，饱和甘汞电极作参比电极，

$$SCE||Cu^{2+}(unknown)|Cu(s)$$

then the cell potential can be used to determine an unknown concentration of Cu^{2+} in the indicator half-cell.

则可以通过测定电池电动势确定Cu^{2+}的未知浓度，见式（4.6）。

$$\begin{aligned}E_{cell}&=E_{ind}-E_{ref}+E_{lj}\\&=+0.3419-\frac{0.0592}{2}\lg\frac{1}{[Cu^{2+}]}-+0.2444+E_{lj}\end{aligned} \quad (4.6)$$

Metallic indicator electrodes in which a metal is in contact with a solution containing its ion are called **electrodes of the first kind**. In general, for a metal M, in a solution of M^{n+}, the cell potential is given as

金属与含有其离子的溶液组成的金属指示电极称为第一类电极。一般来说，对于金属 M，在 M^{n+} 的溶液中，电池电势为式（4.7）。

$$E_{cell}=K-\frac{0.0592}{n}\lg\frac{1}{[M^{n+}]}=K+\frac{0.0592}{n}\lg[M^{n+}] \quad (4.7)$$

where K is a constant that includes the standard-state potential for the M^{n+}/M redox couple, the potential of the reference electrode, and the junction potential. For a variety of reasons, including slow kinetics for electron transfer, the existence of surface oxides and interfering reactions, electrodes of the first kind are limited to Ag, Bi, Cd, Cu, Hg, Pb, Sn, Tl, and Zn. Many of these electrodes, such as Zn, cannot be used in acidic solutions where they are easily oxidized by H^{+}.

式中 K 是一个常数，受 M^{n+}/M 氧化还原电对的标准电势、参比电极的电势和液接电位的影响。由于各种原因，如电子转移速度慢、表面氧化物的存在及干扰反应等，第一类电极仅限于 Ag、Bi、Cd、Cu、Hg、Pb、Sn、Tl 和 Zn。这类电极（如锌）不能用于酸性溶液，因其很容易被 H^{+} 氧化。

② Electrodes of the second kind　When the potential of an electrode of the first kind responds to the potential of

② 第二类电极　当第一类电极的电位对另一种能与 M^{n+} 平衡的离子电

another ion that is in equilibrium with M^{n+}, it is called an **electrode of the second kind**. Two common electrodes of the second kind are the calomel and silver/silver chloride electrodes. When wrapped with saturated KCl solution, used as reference electrode, when immersed in Cl^- solution of unknown concentration, used as indicator electrode, indicates the concentration of Cl^- and Ag^+. Electrodes of the second kind also can be based on complexation reactions. For example, an electrode for EDTA is constructed by coupling a Hg^{2+}/Hg electrode of the first kind to EDTA by taking advantage of its formation of a stable complex with Hg^{2+} (figure 4-10).

位有响应时，称为**第二类电极**。常见的第二类电极是甘汞电极和银 - 氯化银电极，当裹有饱和 KCl 溶液时，用作参比电极；当浸入未知浓度的 Cl^- 溶液中，可作指示电极，指示 Cl^- 和 Ag^+ 的浓度。第二类电极也可以建立在配位反应的基础上。例如，汞电极就是由金属 Hg 浸入含少量 Hg-EDTA 配合物及待测金属离子的溶液中组成的，其可用作 EDTA 滴定 M^{n+} 的指示电极（图 4-10）。

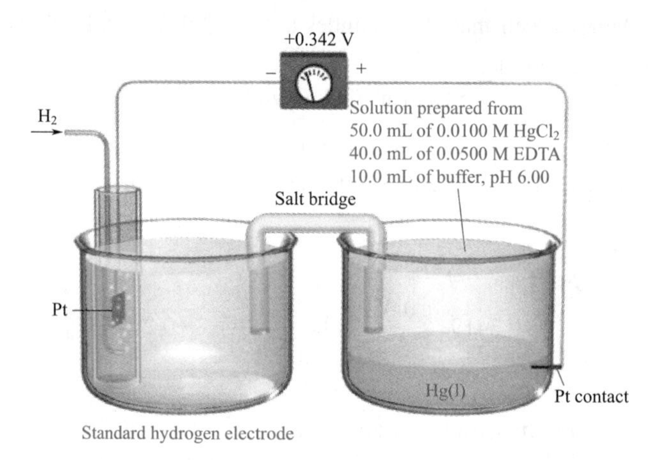

Fig.4-10 A galvanic cell that can be used to measure the formation constant for $Hg(EDTA)^{2-}$

图 4-10 可用于测量 $Hg(EDTA)^{2-}$ 生成常数的原电池

If the solution is saturated with AgI, then the solubility reaction

例如，碘化银的饱和溶液中

$$AgI(s) \rightleftharpoons Ag^+(aq)+I^-(aq)$$

determines the concentration of Ag^+, thus

Ag^+ 浓度的计算公式为式（4.8）。

$$[Ag^+]=\frac{K_{sp,AgI}}{[I^-]} \tag{4.8}$$

where $K_{sp,\,AgI}$ is the solubility product for AgI. Thus

其中 $K_{sp,\,AgI}$ 为 AgI 的溶度积，则银电极的电位是 $[I^-]$ 的函数。

$$E=E^{\ominus}_{Ag^+/Ag}-0.0592\lg\frac{1}{[Ag^+]}=+0.7996-0.0592\lg\frac{1}{[Ag^+]} \tag{4.9}$$

$$E=+0.7996-0.0592\lg\frac{[I^-]}{K_{sp,AgI}} \tag{4.10}$$

shows that the potential of the silver electrode is a function of the concentration of I^-. When this electrode is incorporated into a potentiometric electrochemical cell.

当银电极插入待测液，与参比电极组成电化学电池时，电池电动势为式(4.11)。

$$REF||AgI(sat'd),I^-(unk)|Ag(s)$$

the cell potential is

$$E_{cell}=K-0.0592\lg[I^-] \tag{4.11}$$

where K is a constant that includes the standard-state potential for the Ag^+/Ag redox couple, the solubility product for AgI, the potential of the reference electrode, and the junction potential.

其中 K 是一个常数，包括 Ag^+/Ag 氧化还原电对的标准电势、AgI 的溶度积、参比电极电位和液接电位。

③ Redox electrodes　Electrodes of the first and second kind develop a potential as the result of a redox reaction in which the metallic electrode undergoes a change in its oxidation state. Metallic electrodes also can serve simply as a source of, or a sink for, electrons in other redox reactions. Such electrodes are called **redox electrodes**, also called inert metal electrode. Note that the potential of a redox electrode generally responds to the concentration of more than one ion, limiting their usefulness for direct potentiometry. For example, electrode composed of platinum electrode immersed in $Fe^{3+}$and Fe^{2+}solution, $Pt|Fe^{3+}$, Fe^{3+}, the electrode reaction is

③ 氧化还原电极　第一类和第二类电极都是由于金属电极发生氧化还原反应而产生电位。金属电极也可以简单地作为其他氧化还原反应中电子的来源，这种电极称为**氧化还原电极**，也称惰性金属电极。值得注意的是，氧化还原电极的电位通常响应多个离子的浓度，这就限制了其在直接电位法中的应用。例如，铂电极浸入 Fe^{3+} 和 Fe^{2} 溶液组成的电极 $Pt|Fe^{3+}$，Fe^{2+}。

$$Fe^{3+}+e^- \rightleftharpoons Fe^{2+}$$

The electrode potential at 298 K is

$$E_{Fe^{3+}/Fe^{2+}}=E^{\ominus}_{Fe^{3+}/Fe^{2+}}+0.0592\lg(a_{Fe^{3+}}/a_{Fe^{2+}}) \tag{4.12}$$

(2) Membrane electrodes

(2) 膜电极

① Glass ion-selective electrodes　The glass electrode used to measure pH is the most common ion-selective electrode. A typical pH combination electrode, incorporating both glass and reference electrodes in one body, is shown in figure 4-11. A line diagram of this cell

① 玻璃电极　用于测量 pH 的玻璃电极是最常见的离子选择性电极。图 4-11 是一个典型的 pH 复合电极，其由敏感玻璃膜和参比电极组成。电极的 pH 敏感部分是图 4-11 和图 4-12

can be written as follows:

The pH-sensitive part of the electrode is the thin glass bulb or cone at the bottom of the electrodes in figures 4-11 and 4-12. The reference electrode at the left in the line diagram is the coiled Ag|AgCl electrode in the combination electrode in figure 4-11. The reference electrode at the right side of the line diagram is the straight Ag|AgCl electrode at the center of the electrode in figure 4-11. The two reference electrodes measure the electric potential difference across the glass membrane. The salt bridge in the line diagram is the porous plug at the bottom right side of the combination electrode in figure 4-11.

中电极底部的薄玻璃球或锥体。电池简图左侧的参比电极如图 4-11 中组合电极中卷绕的 Ag|AgCl 电极。电池简图右侧的参比电极是图 4-11 中电极中心竖直的 Ag|AgCl 电极。通过这两个参比电极可以测定玻璃膜产生的膜电位。电池简图中的盐桥是图 4-11 中组合电极右下角的多孔塞。

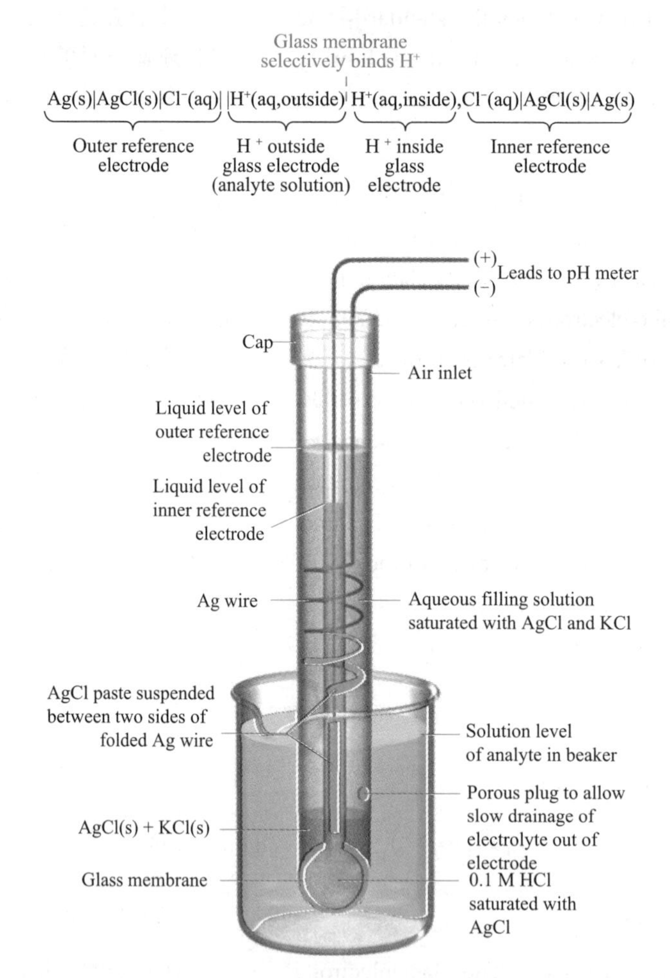

pH composite electrode
pH 复合电极

Fig.4-11 Diagram of a glass combination electrode with a silver-silver chloride reference electrode (The glass electrode is immersed in a solution of unknown pH so that the porous plug on the lower right is below the surface of the liquid. The two Ag|AgCl electrodes measure the voltage across the glass membrane.)

图 4-11 具有银 - 氯化银参比电极的玻璃复合电极示意图（玻璃复合电极浸在未知 pH 溶液中，将其右下方的多孔塞浸没在液面以下。两个 Ag|AgCl 电极测量玻璃膜的产生的电位。）

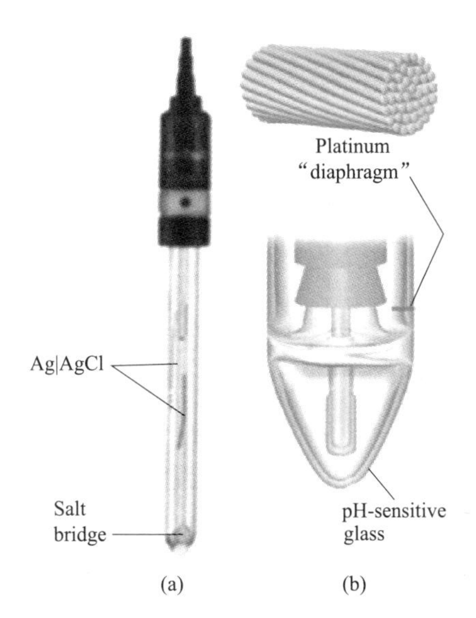

Fig.4-12 (a) Glass-body combination electrode with pH-sensitive glass bulb at the bottom. The porous ceramic plug (the salt bridge) connects analyte solution to the reference electrode. Two silver wires coated with AgCl are visible inside the electrode; (b) A pH electrode with a platinum diaphragm (a bundle of Pt wires), which is said to be less prone to clogging than a ceramic plug

图 4-12 （a）底部带有 pH 敏感玻璃泡的复合电极（在电极内安装两根涂有氯化银的银线，多孔陶瓷塞（盐桥）将分析物溶液连接到参比电极）；（b）带有铂膜片（一束铂丝）的 pH 电极（其比陶瓷塞更不易堵塞）

Figure 4-13 (a) depicts the irregular structure of silicate glass, which is in the bulb of a glass pH electrode. Negatively charged oxygen atoms in glass can bind cations of suitable size. Monovalent cations, particularly Na^+, can move sluggishly through the glass. Figure4-13(b) is an atomic resolution micrograph showing crystalline and amorphous regions of pure silica glass (pure SiO_2). Figure 4-13(c) is a micrograph of an amorphous region of silica glass. Crystalline regions have a repeating array of rings with six SiO_4 units. Amorphous regions have a mixture of ring sizes with irregular orientations. Amorphous silica glass is an approximation of the structure of silicate glass.

图 4-13（a）是 pH 玻璃电极球部中硅酸盐玻璃的不规则结构。玻璃中带负电荷的氧原子可以结合合适大小的阳离子。单价阳离子，特别是 Na^+，可以在玻璃中缓慢移动。图 4-13（b）是玻璃膜的透射电子显微图，显示了纯硅酸盐玻璃（纯二氧化硅）的晶体和非晶形区域，图 4-13（c）是硅酸盐玻璃非晶形区域的显微图。晶体区域是具有六个 SiO_4 单元的重复阵列，非晶状区域是具有不规则取向环的混合物。非晶型硅玻璃与硅酸盐玻璃结构相似。

A schematic cross section of the glass membrane of a pH electrode is shown in figure 4-14. The two surfaces swell as they absorb water. Metal ions in these hydrated gel regions of the membrane diffuse out of the glass and into solution. H^+ can diffuse into the membrane to replace metal ions. The reaction in which H^+ replaces cations in the glass is an ion-exchange equilibrium. A pH electrode responds selectively to H^+ because H^+ is

pH 电极玻璃膜的横截面如图 4-14 所示。玻璃膜电极在使用前必须在水中浸泡一定时间，使玻璃膜表面形成一层水合凝胶层，玻璃膜内侧表面由于长期与内部溶液接触也形成了水合凝胶层。金属离子从水合凝胶层扩散到溶液中，H^+ 则扩散到凝胶层中取代金属离子，H^+ 取代阳离子的反应实质是一种离子交换

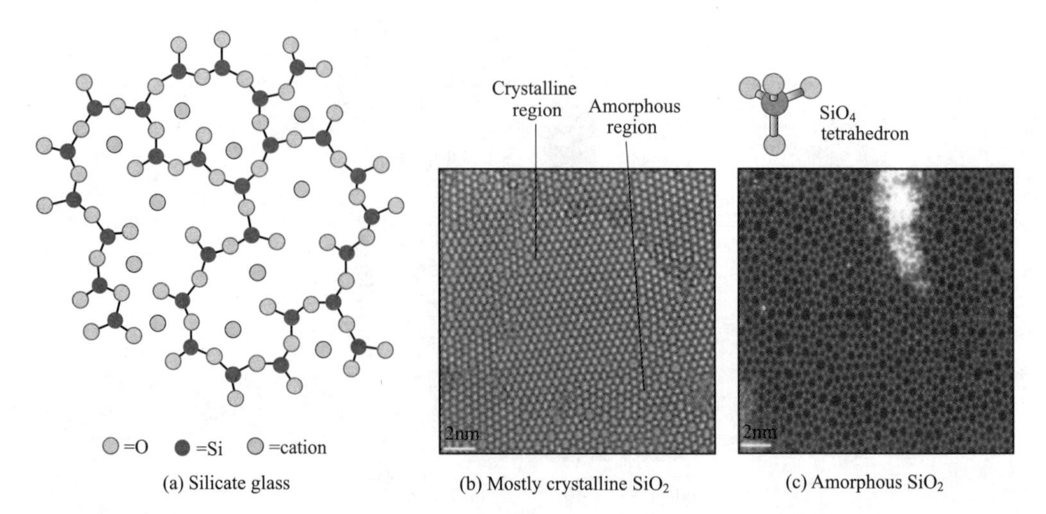

Fig.4-13 (a) Schematic structure of silicate glass, which consists of an irregular network of SiO_4 tetrahedra connected through oxygen atoms. Cations such as Li^+, Na^+, K^+, and Ca^{2+} are coordinated to oxygen atoms. The silicate network is not planar. The diagram is a projection of each tetrahedron onto the plane of the page.(b) and (c) Transmission electron micrographs of thin layer of SiO_2 deposited on graphene. Each vertex of each polygon is looking down the axis of an SiO_4 tetrahedron connected to adjacent SiO_4 tetrahedra through oxygen atoms.

图 4-13 （a）硅酸盐玻璃的结构示意图（其由通过氧原子连接的不规则硅酸盐四面体组成。Li^+、Na^+、K^+ 和 Ca^{2+} 等阳离子与氧原子配位。硅酸盐网状物不是平面的。该图是每个四面体在平面上的投影。)；（b）（c）沉积在石墨烯表面的二氧化硅薄层的透射电镜图（多边形的每个顶点向下看作是一个通过氧原子连接到相邻硅酸盐四面体的轴。）

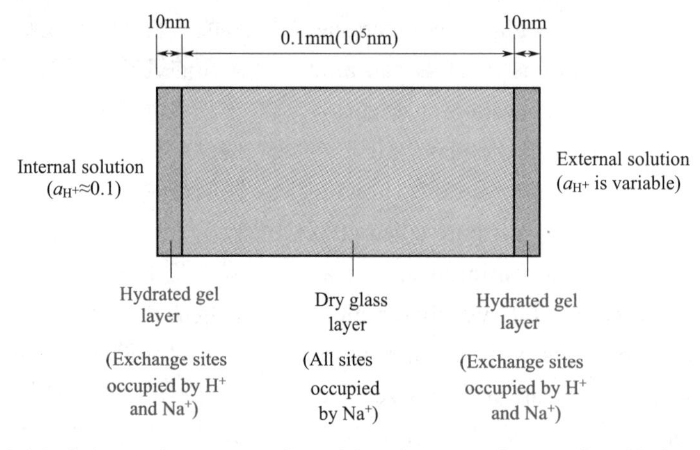

Fig.4-14 Schematic cross section of the glass membrane of a pH electrode

图 4-14 pH 电极玻璃膜的横截面示意图

the main ion that binds significantly to the hydrated gel layer. However, when pH <1 or pH>9, measuring errors may occur. The measuring potential is higher in high acid solution, the error is referred to as the **acid error**; the measuring potential is lower in high alkali solution, it is referred to as the **sodium error** and **alkaline error**, respectively.

平衡。由于 H^+ 是与水合凝胶层显著结合的主要离子，pH 电极会有选择性地响应 H^+。但是当 pH<1 或 pH>9 时，会产生测量误差。在高酸度溶液中，测得的电位值偏高，这种误差称为“**酸差**”；在高碱度溶液中，测得的电位值偏低，这种误差称为“**碱差**”或“**钠差**”。

② Crystalline solid-state ion-selective electrodes A solid-state ion-selective electrode based on an inorganic crystal is shown in figure 4-15. A common electrode of this type is the fluoride electrode, employing a crystal of LaF_3 doped with Eu^{2+}. Doping means adding a small amount of Eu^{2+} in place of La^{3+}. The filling solution contains 0.1 M NaF and 0.1 M NaCl. Fluoride electrodes are used to monitor the fluoridation of municipal water supplies.

② 固态晶体膜离子选择电极 基于无机晶体的固态离子选择性电极如图 4-15 所示。常见的这种类型电极是氟离子电极，其敏化膜为掺杂 Eu^{2+} 的 LaF_3 晶体，即在膜中添加少量的 Eu^{2+} 来代替 La^{3+}。内参比溶液为 0.1M NaF 和 0.1M NaCl。氟离子电极常用于监测城市供水系统中 F^- 的含量。

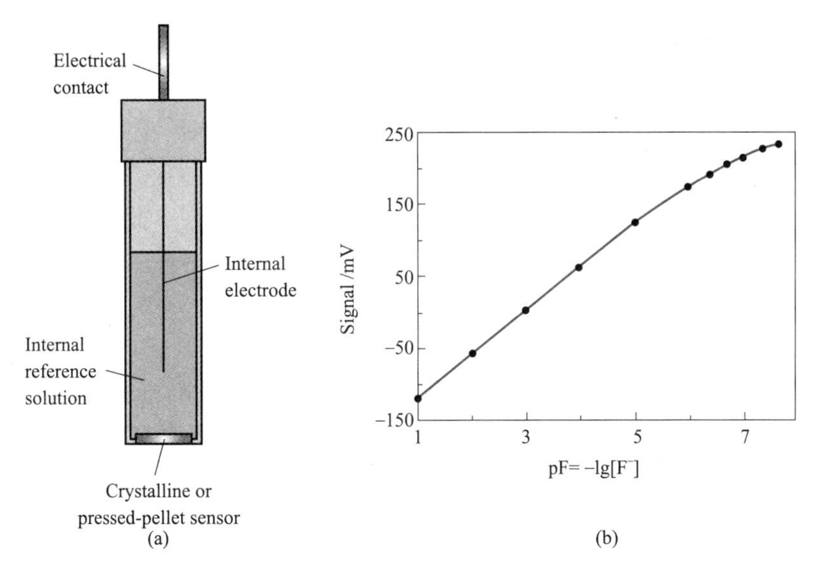

Fig.4-15 (a) The general design of a solid-state ion-selective electrode(b) An example of the response for a fluoride ion-selective electrode that is based on this design

图 4-15 （a）固态离子选择性电极的总体设计（b）基于该设计的氟离子选择性电极的响应示例

To conduct a tiny electric current, F^- migrates across the LaF_3 crystal, as shown in figure 4-16. Anion vacancies are created within the crystal when we dope LaF_3 with EuF_2. An adjacent fluoride ion can jump into the vacancy, leaving a new vacancy behind. In this manner, F^- diffuses from one side to the other.

At 25 ℃,

当用 EuF_2 掺杂 LaF_3 时，LaF_3 的晶格产生空穴，在晶格上的 F^- 迁移到阴离子空穴而导电，如图 4-16 所示。当氟离子电极插入含氟溶液中时，F^- 在电极表面进行交换并扩散。

25℃时，有

$$E=K-0.0592\lg a_{F^-}=K+0.0592\text{pF} \quad (4.13)$$

At low pH, F^- is converted to HF (pK_a=3.17), to which the electrode is insensitive. A routine procedure for measuring F^- is to dilute the unknown in a total ionic strength adjustment buffer (TISAB) containing acetic acid, sodium citrate, NaCl, and NaOH to adjust the pH to 5.5. The buffer keeps all standards and unknowns

在低 pH 值下，F^- 转化为 HF（pK_a=3.17），此时电极不敏感。测量 F^- 时，常用含有醋酸、柠檬酸钠、氯化钠和氢氧化钠的总离子强度调节缓冲溶液（TISAB）调节溶液中离子强度，将 pH 调整到 5.5。TISAB 的加入，使所有标

at a constant ionic strength, so the fluoride activity coefficient is constant in all solutions (and can therefore be ignored). At pH 5.5, there is no interference by OH^- and little conversion of F^- to HF. Citrate complexes Fe^{3+} and Al^{3+}, which would otherwise bind F^- and interfere with the analysis.

准物和未知物保持恒定的离子强度。在 pH=5.5 时，氢氧化物没有干扰，F^- 向 HF 的转化也很小，柠檬酸钠与 Fe^{3+} 和 Al^{3+} 发生络合反应，减少了 Fe^{3+} 和 Al^{3+} 对 F^- 的干扰。

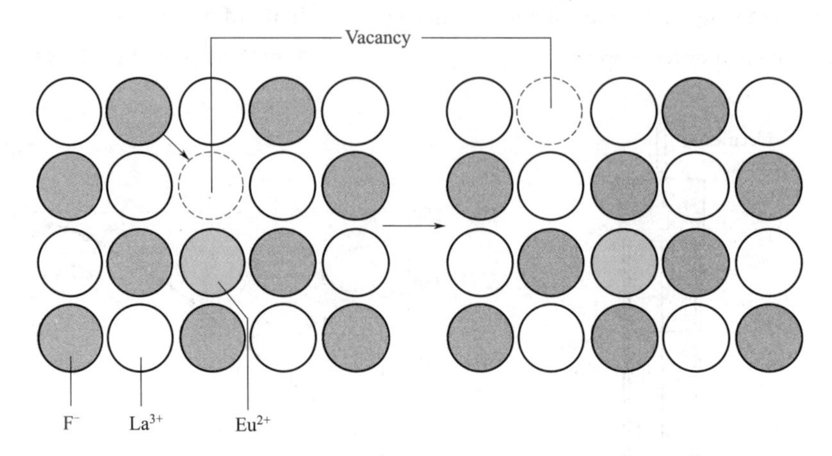

Fig.4-16 Migration of F^- through LaF_3 doped with EuF_2 (Because Eu^{2+} has less charge than La^{3+}, an anion vacancy occurs for every Eu^{2+}. A neighboring F^- can jump into the vacancy, thereby moving the vacancy to another site. Repetition of this process moves F^- through the lattice.)

图 4-16 F^- 通过掺杂 EuF_2 的 LaF_3 实现迁移（由于 Eu^{2+} 的电荷小于 La^{3+}，所以每个 Eu^{2+} 都会出现一个阴离子空穴。相邻的 F^- 迁移到阴离子空穴，从而使空穴移位，重复此过程最终使 F^- 穿过晶格。）

③ Liquid membrane electrodes Electrodes with liquid ion-exchange membranes are typified by a calcium-sensitive electrode (fig.4-17). The membrane-liquid consists of the calcium form of a dialkyl phosphoric acid,$[(RO)_2POO]_2Ca$, which is prepared by repeated treatment of the acid with a calcium salt. The internal solution is calcium chloride and the membrane potential, which is determined by the extent of ion-exchange reactions at the interfaces between the membrane and the internal and sample solutions, is given by

③ 液膜电极 Ca^{2+} 电极是液态膜电极的典型代表（见图 4-17）。电极内装有两种溶液：一种是内参比溶液（0.1M 的 $CaCl_2$ 水溶液），其中插入内参比电极 Ag-AgCl 电极；另一种是液体离子交换剂——0.1M 的二癸基磷酸钙。膜电位由膜内溶液和样品溶液界面上的离子交换反应程度决定。

$$E=K+\frac{0.0592}{2}\lg a_{Ca^{2+}} \qquad (4.14)$$

The electrode responds to Ca^{2+} down to 10^{-6} M and is independent of pH in the range 6 to 10. Other electrodes of this type include ($Ca^{2+}+Mg^{2+}$), K^+, ClO_4^-, NO_3^- and BF_4^-.

该电极对 Ca^{2+} 的响应浓度为 10^{-6}M，适用的 pH 范围为 6 ～ 10。这类电极主要包括（$Ca^{2+}+Mg^{2+}$）、K^+、ClO_4^-、NO_3^- 和 BF_4^-。

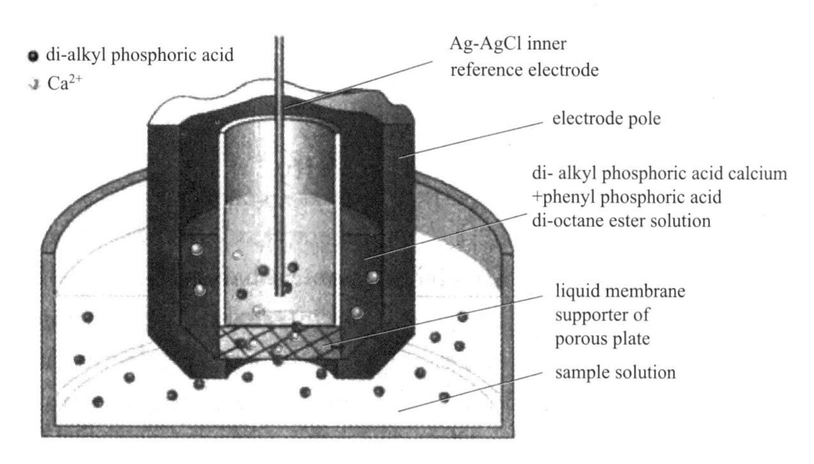

Fig.4-17 Calcium-selective ion-exchange membrane electrode.

图 4-17 钙离子选择性膜电极

④ Gas-sensing electrodes Devices like the pH electrode are not limited to the detection of solution-phase chemicals, but can also be modified for used in other types of measurements. One group of compound electrodes are those that have been modified for the analysis of certain gases. The result is known as a gas-sensing electrode. An ammonia gas-sensing electrode (illustrated in figure 4-18) is both an electrode that can sense a gas and detect a molecular species. This device is a pH electrode covered with a membrane that allows passage of only low molecular-weight gases. This membrane is typically made of a very thin

④ 气敏电极 pH 电极不仅用于液相化学品的测定，经修饰后也可以用于其他相态的测定。其中被修饰后的某复合电极就可以用于气体的分析，这种电极称为气敏电极。图 4-18 表示的是氨电极，其既可检测气体又能检测分子组分，该电极是在 pH 电极敏感膜上覆盖一层透气膜，只允许低分子量的气体通过。这种膜通常是由一块非常薄的特四氟乙烯或聚乙烯制成。在覆盖膜和 pH 敏感膜之间装有体积很小、NH_4^+ 浓度恒定的中介液（0.1mol · L^{-1} KCl）。当

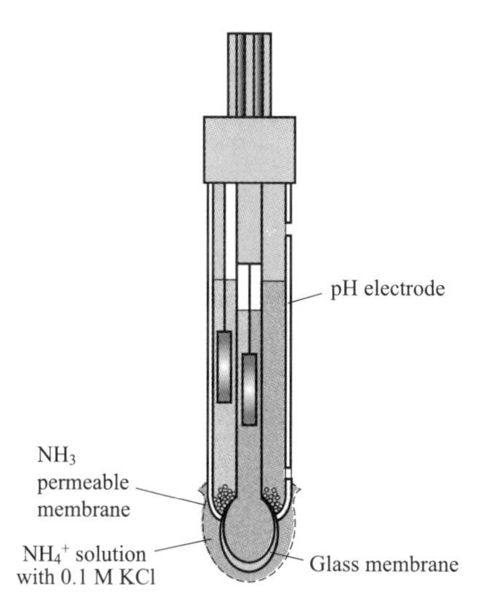

Fig.4-18 Design of a gas-sensing electrode for ammonia, based on the use of a glass-membrane pH electrode for detection

图 4-18 基于玻璃膜 pH 电极设计的一种氨气敏电极

piece of Teflon or polyethylene. Between the covering membrane and the pH-sensitive glass is a small volume of an internal electrolyte solution (0.1 mol•L^{-1} KCl) that has an essentially fixed concentration of NH_4^+. When dissolved ammonia enters this solution through the membrane, the ratio of $[NH_4^+]$ to $[NH_3]$ is changed and the pH is increased. This change creates a response by the pH electrode that is related to the activity of ammonia that was in the sample. Similar electrodes can be made to respond to other basic or acidic gases such as CO_2, SO_2, and NO_2.

氨电极浸入强碱性试液中时，试液中的 NH_4^+ 生成 NH_3 分子，穿过透气膜进入中介液，使中介液中的 pH 值发生变化，并由 pH 玻璃电极测出。类似的电极还可以响应其他碱性或酸性气体，如 CO_2、SO_2 和 NO_2 等。

Section 3 Application of Potentiometric Methods

第 3 节 电位分析法的应用

1. Measurement of pH

With the availability of inexpensive glass pH electrodes and pH meters, the determination of pH has become one of the most frequent quantitative analytical measurements. The potentiometric determination of pH, however, is not without complications, several of which are discussed in this section.

One complication is the meaning of pH. The conventional definition of pH as presented in most introductory texts is

1. pH 值的测定

随着 pH 玻璃电极和 pH 计的出现，pH 值的测定已成为最常见的定量分析方法之一。然而，pH 的测定是非常复杂的。

第一个问题是 pH 的定义。多数文献中，pH 的定义见式（4.15）。

$$pH=-\lg[H^+] \tag{4.15}$$

The pH of a solution, however, is defined by the response of an electrode to the H^+ ion and, therefore, is a measure of its activity.

然而，对特定的溶液而言，其 pH 值是电极对 H^+ 的响应值，测量的是其活度。

$$pH=-\lg(a_{H^+}) \tag{4.16}$$

Calculating the pH of a solution using equation (4.15) only approximates the true pH. Thus, a solution of 0.1 mol · L^{-1} HCl has a calculated pH of 1.00 using

虽然用式（4.15）计算的 pH 值近似于真正的 pH，如计算 0.1mol · L^{-1} 盐酸的 pH 为 1.00，但根据方程（4.16）

equation (4.15) but an actual pH of 1.1 as defined by equation (4.16). The difference between the two values occurs because the activity coefficient for H^+ is not unity in a matrix of 0.1 mol · L^{-1} HCl. Obviously the true pH of a solution is affected by the composition of its matrix. As an extreme example, the pH of 0.01 mol · L^{-1} HCl in 5 mol · L^{-1} LiCl is 0.8, a value that is more acidic than that of 0.1 mol · L^{-1} HCl.

A second complication in measuring pH results from uncertainties in the relationship between potential and activity. For a glass membrane electrode, the cell potential, E_x, for a solution of unknown pH is given as

PHSJ-3F acidity meter

PHSJ-3F 型酸度计

的定义，其实际pH值为1.1。这两个值之间的差异是因为H^+的活度系数在0.1mol · L^{-1}盐酸基质中不一致。显然，真实pH值大小受其基质成分的影响。举一个极端的例子，0.01mol · L^{-1}盐酸在5 mol · L^{-1}氯化锂中的pH值为0.8，其酸度值大于0.1mol·L^{-1}盐酸的pH值。

第二个难题是在进行pH值测量时，电位和活性之间关系的不确定性。对于玻璃膜电极而言，设未知液的电池电动势为E_x，则pH值的计算式为式（4.17）。

$$E_x=K-\frac{RT}{F}\ln\frac{1}{a_{H^+}}=K-\frac{2.303RT}{F}\,pH_x \tag{4.17}$$

where K includes the potential of the reference electrode, the asymmetry potential of the glass membrane and any liquid junction potentials in the electrochemical cell. All the contributions to K are subject to uncertainty and may change from day to day, as well as between electrodes. For this reason a pH electrode must be calibrated using a standard buffer of known pH. The cell potential for the standard, E_s, is

其中K包括参比电极的电位、玻璃膜的不对称电位以及电化学电池中的液接电位，这些都是不确定因素，并经常发生变化，因此，必须使用已知pH值的标准缓冲溶液来校准pH电极。标准缓冲溶液的pH值见式（4.18）。

$$E_s=K-\frac{2.303RT}{F}\,pH_s \tag{4.18}$$

where pH_s is the pH of the standard. Subtracting equation (4.18) from equation (4.17) and solving for pH gives

两式相减，可得式（4.19），该式为国际纯粹与应用化学联合会采用的pH的实用定义，亦称为pH标度。

$$pH_x=pH_s-\frac{(E_x-E_s)F}{2.303RT} \tag{4.19}$$

which is the operational definition of pH adopted by the International Union of Pure and Applied Chemistry.

Equation (4.17)-equation (4.19) are defined for a potentiometric electrochemical cell in which the pH electrode is the cathode. In this case an increase in

式（4.17）～式（4.19）适用于电化学电池，其中pH电极是阴极。在这种情况下，pH值的增加会降低

pH decreases the cell potential. Many pH meters are designed with the pH electrode as the anode so that an increase in pH increases the cell potential The operational definition of pH then becomes $pH_x=pH_s-\frac{(E_x-E_s)F}{2.303RT}$. This difference, however, does not affect the operation of a pH meter.

电极电位。许多pH计以pH电极为阳极，使pH值随电极电势的增加而增加。pH值的操作定义可表达为$pH_x=pH_s-\frac{(E_x-E_s)F}{2.303RT}$。然而，这并不影响pH仪器的操作。

Calibrating the electrode presents a third complication since a standard with an accurately known activity for H^+ needs to be used. Unfortunately, it is not possible to calculate rigorously the activity of a single ion. For this reason pH electrodes are calibrated using a standard buffer whose composition is chosen such that the defined pH is as close as possible to that given by equation (4.16) Table 4-2 gives pH values for several primary standard buffer solutions.

第三个难题是校准电极需要使用H^+活度准确已知的标准溶液。然而，现实是不可能严格地计算单个离子的活度，因此通常使用基准缓冲溶液校准pH电极。表4-2列出了几种主要基准缓冲溶液的pH值。

A pH electrode is normally standardized using two buffers: one near a pH of 7 and one that is more acidic or basic depending on the sample's expected pH. The pH electrode is immersed in the first buffer, and the "standardize" or "calibrate" control is adjusted until the meter reads the correct pH. The electrode is placed in the second buffer, and the "slope" or "temperature" control is adjusted to the buffer's pH. Some pH meters are equipped with a temperature compensation feature, allowing the pH meter to correct the measured pH for any change in temperature.

pH电极通常用两种基准缓冲溶液进行校准：其中一个是pH接近7的溶液，另一个根据样品的预期pH值选择酸性或碱性基准缓冲溶液，见图4-19。具体操作时，先将pH电极浸在第一个缓冲溶液中，调整“标准化”或“校准”按钮，直到pH计显示正确的pH值；再将电极放置在第二个缓冲溶液中，并将“斜率”或“温度”控制器调整到缓冲溶液的pH值。一些pH计配备了温度补偿功能，允许pH计校准测量pH时的温度变化。

Measurement of pH
溶液pH的测定

Table 4-2 pH Values for Selected NIST Primary Standard Buffers

表4-2 标准缓冲溶液的pH值

Temperature /℃	Saturated (25℃) $KHC_4H_4O_6$ (tartrate)	0.05 M $KH_2C_6H_5O_7$ (citrate)	0.05 M $KHC_8H_4O_4$ (phthalate)	0.025 M KH_2PO_4, 0.025 M Na_2HPO_4	0.008695 M KH_2PO_4, 0.03043 M Na_2HPO_4	0.01 M $Na_4B_4O_7$ (borax)	0.025 M $NaHCO_3$, 0.025 M Na_2CO_3
0	—	3.863	4.003	6.984	7.534	9.464	10.317
5	—	3.840	3.999	6.951	7.500	9.395	10.245
10	—	3.820	3.998	6.923	7.472	9.332	10.179
15	—	3.802	3.999	6.900	7.448	9.276	10.118

续表

Temperature /℃	Saturated (25℃) $KHC_4H_4O_6$ (tartrate)	0.05 M $KH_2C_6H_5O_7$ (citrate)	0.05 M $KHC_8H_4O_4$ (phthalate)	0.025 M KH_2PO_4, 0.025 M Na_2HPO_4	0.008695 M KH_2PO_4, 0.03043 M Na_2HPO_4	0.01 M $Na_4B_4O_7$ (borax)	0.025 M $NaHCO_3$, 0.025 M Na_2CO_3
20	—	3.788	4.002	6.881	7.429	9.225	10.062
25	3.557	3.776	4.008	6.865	7.413	9.180	10.012
30	3.552	3.766	4.015	6.854	7.400	9.139	9.966
35	3.549	3.759	4.024	6.844	7.389	9.102	9.925
40	3.547	3.753	4.035	6.838	7.380	9.068	9.889
45	3.547	3.750	4.047	6.834	7.373	9.038	9.856
50	3.549	3.749	4.060	6.833	7.367	9.011	9.828

注 1. Values taken from Bates, R. G. *Determination of pH: Theory and Practice,* 2nd ed. Wiley：New York, 1973.

2. Concentrations are given in molality (moles solute per kilograms solve).

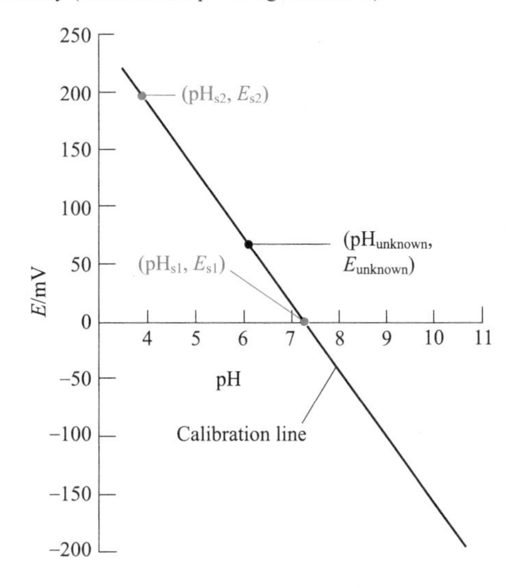

Fig.4-19 Two-point calibration of a pH electrode.

图 4-19 pH 电极的两点校准

2. Pion (pX) Determination

Direct potentiometry offers a way to determine the concentration or the activity of the chemical species for which ion-selective electrodes are available with the help of a calibration graph. The electrode calibration should be carried out for two reasons: ① to check the properties of the potentiometric cell; and ② to serve as a basis for activity or concentration evaluation. As known in practice, the response of ion-selective electrodes deviates from the ideal Nernstian behavior, and can be approximated by the following equation:

2. 离子活度（浓度）的测定

直接电位法提供了一种用离子选择性电极测定化学物质的浓度或活性的方法。进行电极校准有两个原因：①检查电位电池的特性；②作为活性或浓度评价的基础。实践可知，离子选择性电极的响应偏离了理想的能斯特方程，但可以用方程（4.20）近似表达。

$$E=K \pm S\lg a_i \qquad (4.20)$$

Where E is the cell voltage, K is the constant term, whose value corresponds to a solution where $\lg a_i=0$, S is the slope of the electrode response, theoretically=59.2 mV/pI$^-$ on unit for a monovalent ion at 25 ℃ (positive for cations, negative for anions) and a_i is the activity of the measured ion, i.

式中，E 是电池的电动势，K 是常数项，其值对应于 $\lg a_i=0$ 时的 E 值，S 是电极响应的斜率，理论上对于 25℃ 的单价离子，S=59.2mV/pI$^-$，a_i 是 i 离子的活度。

According to equation (4.20) E vs $\lg a_i$ provides, over a wide activity range, a straight-line calibration graph, which can be constructed on the basis of two-point calibration (fig.4-20), using at least two standard solutions. The ion activity was determined by the standard curve method and the standard addition method (fig.4-21).

绘制 E-$\lg a_i$ 曲线，即为电极校准曲线（图 4-20）。离子活度的测定方法有标准曲线法和标准加入法（图 4-21）。

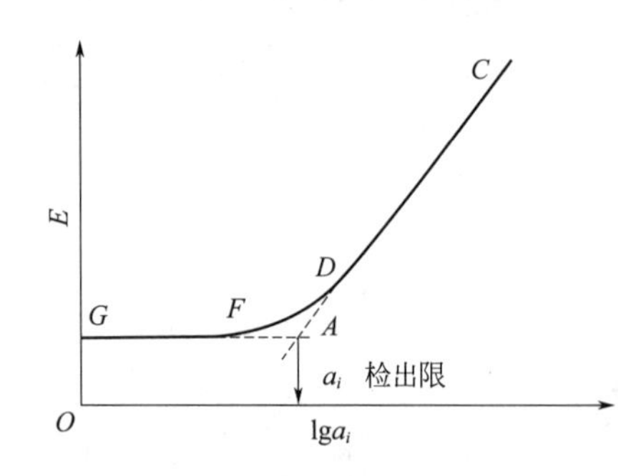

Fig.4-20 The calibration curve of the electrodes

图 4-20 电极的校准曲线

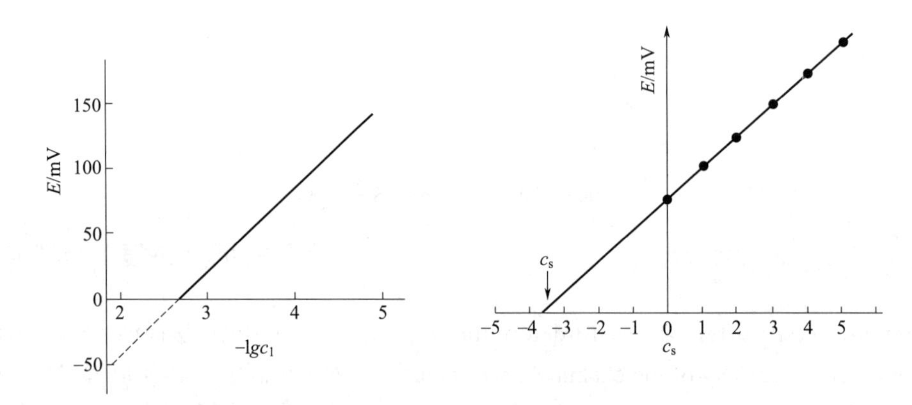

Fig.4-21 Standard curve method and standard addition graph for ion-selective electrode

图 4-21 离子选择性电极的标准曲线法和标准加入法

3. Potentiometric titrations

3. 电位滴定

(1) Principle and experimental installation

① Principle The potentiometric titration is a quantitative analytical method by measuring the potential change of

（1）原理及实验装置

① 原理 利用指示电极电位突变指示滴定终点的容量分析方法（滴定至

indicator electrode to determine the end point of titration. Automatic potentiometer is available now.

终点，浓度突变，从而导致电位突跃）。现已有自动滴定仪。

② Experimental installation The common instruments and devices used for potentiometric titration are shown in figure 4-22. It comprises a buret, a titration pools, an indicator electrode, a reference electrode, a stirrer and an instrument for measuring potential. The cell potential can be measured by potentiometer or DC millivoltmeter. During potentiometric titration, an indicator electrode and a reference electrode are inserted into the solution to be measured to form electrochemical cells. With addition of titrant, the concentration of the ion to be measured will continue to change due to the chemical reaction, so the potential of the indicator electrode will also change accordingly. Near the stoichiometric point, the ion concentration will change suddenly, causing the potential to change suddenly. Therefore, the titration end-point can be determined by measuring change of the potential of the electrochemical cells.

② 实验装置 电位滴定所用的基本仪器装置如图 4-22 所示。它包括滴定管、滴定池、指示电极、参比电极、搅拌器和测量电动势的仪器。测量电动势可用电位计，也可以用直流毫伏计。进行电位滴定时，在待测溶液中插入一支指示电极和一支参比电极组成工作电池。随着滴定剂的加入，由于发生化学反应，待测离子的浓度将不断发生变化，因而指示电极的电位也发生相应的变化，在化学计量点附近，离子浓度发生突变，引起电位的突变，因此通过测量工作电池电动势的变化，就能确定滴定终点。

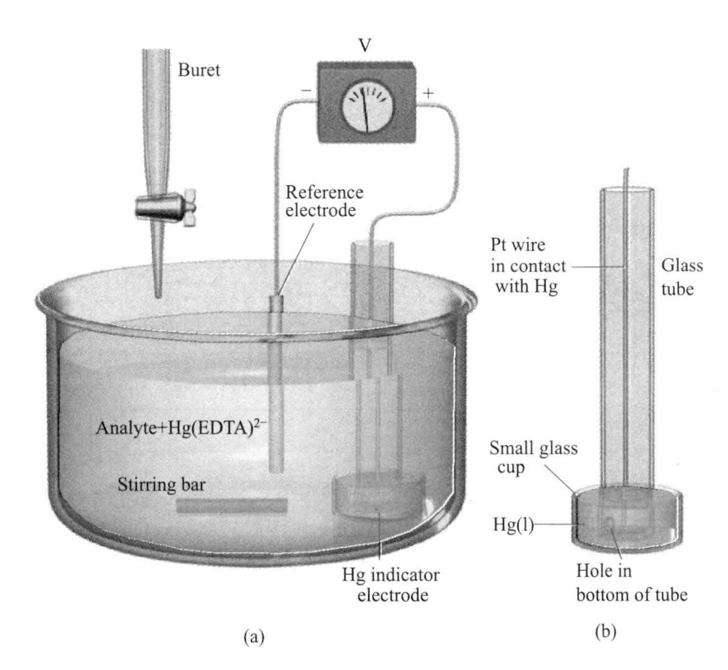

Fig.4-22 Diagram of potentiometric titration (pH meter or millivoltmeter, burette, indicator electrode, reference electrode, agitator)

图 4-22 电位滴定法原理图（pH 计或毫伏计、滴定管、指示电极、参比电极、搅拌器）

Automatic potentiometer is shown in fig.4-23.

自动电位滴定仪见图 4-23。

(2) Determination of the titration endpoint Table 4-3

（2）滴定终点的确定 表 4-3 为使

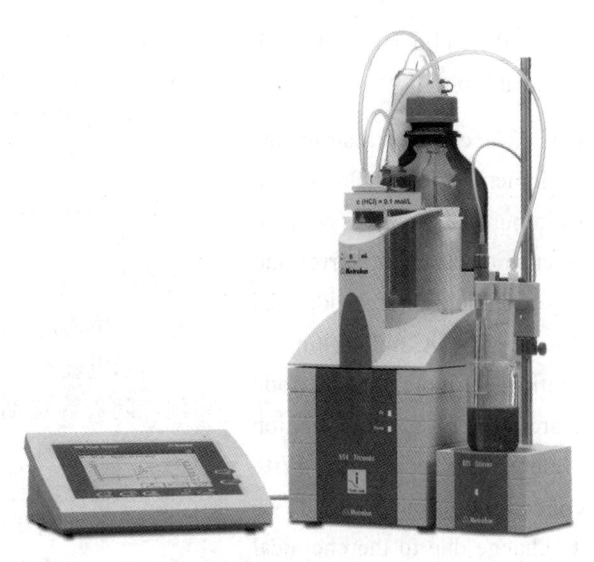

Fig.4-23 Automatic potentiometer

图 4-23 自动电位滴定仪

gives experimental data for the 20.00 mL NaCl solution titrated by 0.1000 mol・L^{-1} $AgNO_3$ standard solution.

用 0.1000mol・L^{-1} $AgNO_3$ 标准溶液滴定 20.00mLNaCl 溶液的实验数据。

Table 4-3 The experimental data of NaCl titrated by 0.1000 mol・L^{-1} $AgNO_3$

表4-3 0.1000 mol・L^{-1} $AgNO_3$ 滴定 NaCl 的实验数据

$V(AgNO_3)$/mL	E/mV	$\frac{\Delta E}{\Delta V}$ /mV・mL^{-1}	$\frac{\Delta^2 E}{\Delta V^2}$ /mV・mL^{-2}
5.00	62	2	
15.00	85	4	
20.00	107	8	
22.00	123	15	
23.00	138	16	
23.50	146	50	
23.80	161	65	
24.00	174	90	
24.10	183	110	
24.20	194	390	2800
24.30	233	830	4400
24.40	316	240	−5900
24.50	340	110	−1300
24.60	351	70	−400
24.70	358	50	
25.00	373	24	
25.50	385		

By monitoring the change of E_{cell} during the course of a titration, where the indicator electrode responds to one of the reactants or products, the stoichiometric or equivalence point can be located. A plot of E_{cell} against volume of titrant added gives a characteristic 'S-shaped'curve [fig.4-24(a)]owing to the logarithmic relation between E_{cell} and activity. If the reactants are in a 1∶1 mole ratio, the curve is symmetrical as shown and the equivalence point is the mid-point of the inflection. This is true for all acid-base, silver-halide and many other titrations. Fig.4-24(b) and fig.4-24(c) show the first and second derivative curves of a normal symmetrical titration curve. The highest point on the first derivative curve is the titrated end-point volume. The point of second derivative equal to 0 on the curve is the titrated end-point volume.

若指示电极对其中一种反应物或产物有响应，可以通过测定滴定过程中电池电动势的变化，找到化学计量点或滴定终点。E_{cell}与滴定剂体积的曲线见图 4-24（a），呈典型的“S 形”曲线，E_{cell}与活度呈对数关系。如果反应物的摩尔比为 1 ∶ 1，则曲线为对称性曲线，化学计量点为曲线拐点的中点。所有的酸碱、卤化银和其他滴定法都是如此。图 4-24（b）和图 4-24（c）为正态对称滴定曲线的一阶和二阶微商曲线。一阶微商曲线上最高点对应的横坐标即为滴定终点体积，曲线上二阶微商等于 0 的点对应的体积即为滴定终点体积。

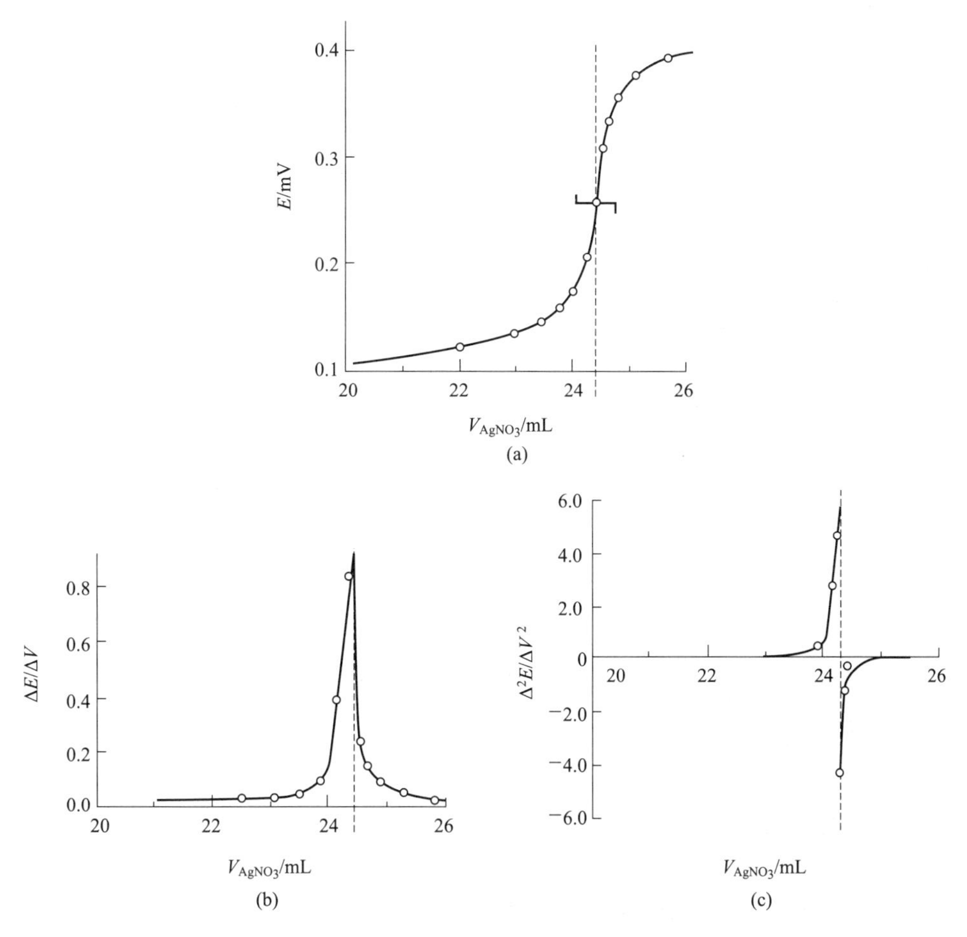

Fig.4-24 Potentiometric titration curves (a) Normal curve; (b) First derivative curve; (c) Second derivative curve

图 4-24 电位滴定曲线（a）正常曲线；（b）一阶微商曲线；（c）二阶微商曲线

Section 4 Coulometric Methods of Analysis

第4节 库仑分析法

1. Principle of Coulometric Analysis

(1) Electrolytic analysis Electrolytic analysis(fig.4-25) is a method used to determine the concentration of the analyte ion by calculating the mass of metal or metallic oxide of the analyte ion deposited on the electrode via electrolytic reaction. Also known as electrogravimetry (fig.4-26). Electrogravimetry is an old analytical technique, time consuming and rarely used. Nowadays, coulometric analysis is widely used.

1. 库仑分析法基本原理

（1）电解分析法 通过电解反应（图4-25），使被测离子以金属或金属氧化物沉积在电极上，根据电极增加的质量计算出被测离子的含量，又称电重量法（图4-26）。电重量法是比较古老的分析技术，费时、使用较少；目前，库仑分析法应用广泛。

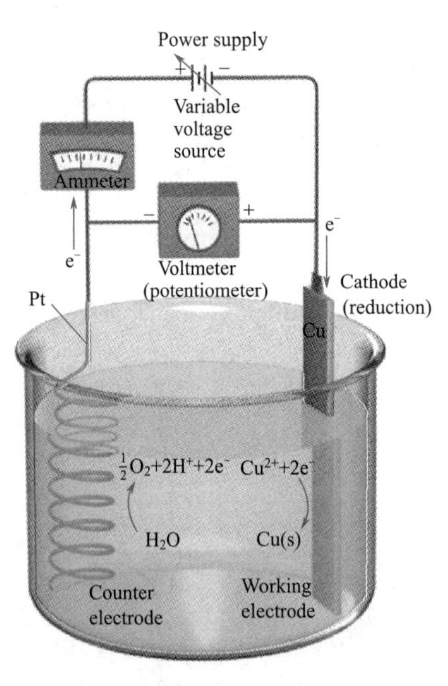

Fig.4-25 Electrolysis experiment

图4-25 电解实验装置

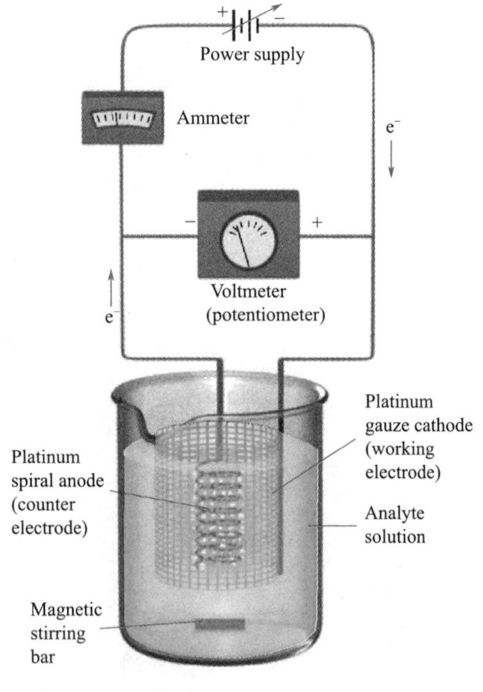

Fig.4-26 Diagram of Electrogravimetry

图4-26 电重量法原理

(2) Coulometric analysis The coulometric analysis is also called electricity quantity analytical method. On the basis that the electrolytic reaction goes on quantitatively,

（2）库仑分析法 库仑分析法也称电量分析法，在电解反应定量进行的基础上，记录电解过程中某电极反应消耗

recording the electricity quantity consumed in the electrode reaction, and then calculate the amount of electrode reaction products according to Faraday's law, so as to measure the content of a certain component in the sample.

的电量，根据法拉第（Faraday）定律计算出电极反应产物的量，从而测得试样中某成分的含量。

(3) Faraday's law Faraday's law of electrolytic refers to the relationship between the mass of the matter precipitated on the electrode and the amount of charge passing through the electrolytic cell, expressed as

（3）法拉第定律 法拉第电解定律是指在电解过程中电极上所析出的物质的质量与通过电解池的电荷量的关系，其表达式见式（4.21）。

$$m=\frac{M}{nF}Q \tag{4.21}$$

Where m is mass of deposited substance on working electrode during electrolysis, Q is electricity consumed by the electrolysis process, M is molar mass of deposited substance, n is electron number transferred in electrode reaction, F is Faraday coefficiency (96487 C/mol).

For a constant current, i, the charge is given as

式中，m 指电解时工作电极上析出物质的质量，Q 为电解过程消耗的电量，M 为析出物质的摩尔质量，n 为电极反应转移的电子数，F 为法拉第常数（96487 C/mol）。

若通过电解池的电流是恒定的，则电荷量为式（4.22）。

$$Q=it_e \tag{4.22}$$

where t_e is the electrolysis time, i is current intensity. If current varies with time, as it does in controlled potential coulometry, then the total charge is given by

其中 t_e 为电解时间。若通过电解池的电流随时间而变，即控制电位库仑法，其总电荷量为式（4.23）。

$$Q=\int_{t=0}^{t=t_e} i(t)\mathrm{d}t \tag{4.23}$$

(4) Forms of coulometry According to the different electrolytical methods, there are three forms of coulometry: controlled-potential coulometry, in which a constant potential is applied to the electrochemical cell, and controlled-current coulometry, in which a constant current is passed through the electrochemical cell, and microcoulometry analysis method with electronic technology for automatic adjustment according to the change of analyte concentration.

（4）库仑法的类型 根据电解方式不同，库仑分析法分为控制电位库仑法和恒电流库仑法，以及根据被测物浓度变化，应用电子技术进行自动调节的微库仑分析法。

2. Controlled-Potential Coulometry

2. 控制电位库仑法

The easiest method for ensuring 100% current efficiency

确保 100% 电流效率的最简单方法

is to maintain the working electrode at a constant potential that allows for the analyte's quantitative oxidation or reduction, without simultaneously oxidizing or reducing an interfering species. The current flowing through an electrochemical cell under a constant potential is proportional to the analyte's concentration. As electrolysis progresses the analyte's concentration decreases, as does the current. When the current tends to be zero, the electrolysis is completed. The resulting current-versus-time profile for controlled-potential coulometry, which also is known as potentiostatic coulometry, is shown in figure 4-27. Integrating the area under the curve [equation (4.23)], from t=0 until t=t_e, gives the total charge.

是将工作电极保持在恒定电位，该电位只允许分析物定量氧化或还原，而不同时氧化或还原干扰物质。恒电位下流过电化学电池的电流与分析物的浓度成正比。随着电解的进行，分析物的浓度降低，电流也随之降低，当电流趋于0时，电解即完成。恒电位库仑法的电流 - 时间曲线如图4-27所示。根据积分曲线下面积［式（4.23）］即可计算出总电荷。

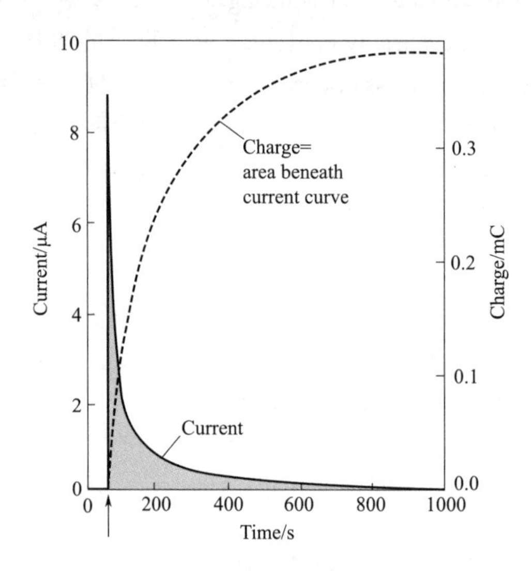

Fig.4-27 Current-time curve for controlled-potential coulometry

图4-27 控制电位库仑法的电流 - 时间曲线

3. Controlled-Current Coulometry

A second approach to coulometry is to use a constant current in place of a constant potential. Controlled-current coulometry, also known as amperostatic coulometry or coulometrictitrimetry, has two advantages over controlled-potential coulometry (figure 4-28). First, using a constant current makes for a more rapid analysis since the current does not decrease over time. Thus, a typical analysis time for controlledcurrent coulometry is less than 10 min, as opposed to approximately 30–60

3. 控制电流库仑法

第二种库仑法是使用恒定电流代替恒定电位（图4-28）。控制电流库仑法，也称为恒电流库仑法或库仑滴定法。与控制电位库仑法相比有两个优点，首先，使用恒定电流有助于提高分析效率，因为电流不会随时间的推移而减少，一般分析时间小于10min，而控制电位库仑法的分析时间约为30～60min。其次，在恒定电流

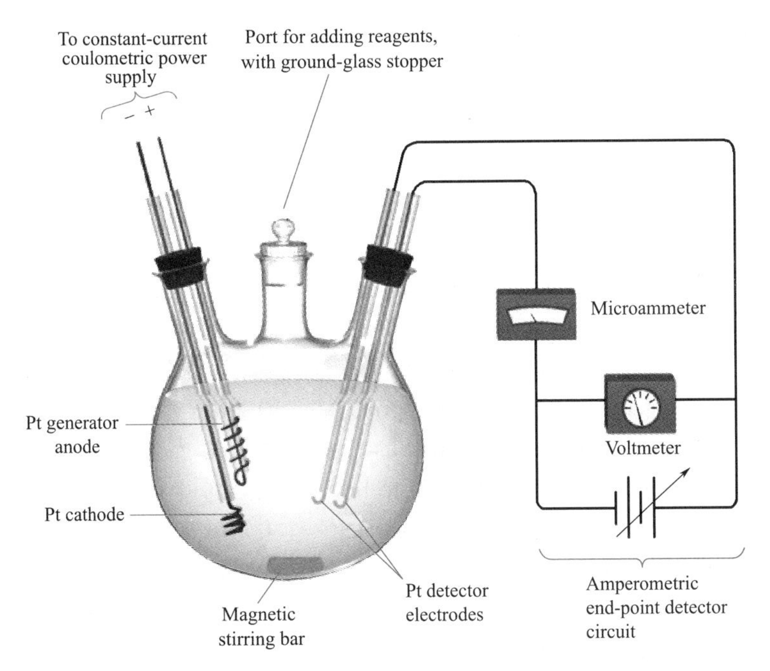

Fig.4-28 Apparatus for coulometrictitration of cyclohexene with Br_2(The reaction is carried out at a constant current. Br_2 generated at the Pt anode at the left reacts with cyclohexene. When cyclohexene is consumed, the concentration of Br_2 suddenly rises, signaling the end of the reaction.)

图 4-28 用溴测定环己烯的库仑法实验装置（反应在恒定电流下进行，在左侧的 Pt 阳极上产生的溴与环己烯发生反应。当环己烯被消耗时，溴的浓度突然上升，表明反应结束。）

min for controlled-potential coulometry. Second, with a constant current the total charge is simply the product of current and time [equation (4.22)]. Just multiply the accurately recorded electrolysis time by the constant current to calculate the total charge. A method for integrating the current–time curve, therefore, is not necessary. Using a constant current does present two important experimental problems that must be solved if accurate results are to be obtained. First, as electrolysis occurs, the current due to its oxidation or reduction steadily decreases. To maintain a constant current the cell potential must change until another oxidation or reduction reaction can occur at the working electrode. Unless the system is carefully designed, these secondary reactions will produce a current efficiency of less than 100%. The second problem is the need for a method of determining when the analyte has been exhaustively electrolyzed. In controlled-current coulometry, a constant current continues to flow even when the analyte has been completely oxidized or reduced. A suitable

下，总电量只是电流和时间的乘积［式(4.22)］。因此，只需用准确记录的电解时间乘以恒定电流即可计算出总电量，无须使用电流 - 时间积分曲线来计算。对于控制电流库仑法，要获得准确的实验结果，必须解决两个重要的实验问题。首先，当电解发生时，因氧化或还原反应而产生的电流稳步降低，为了保持恒定电流，电池电位必须改变，直到在工作电极上发生另一个氧化或还原反应。除非精心设计反应系统，否则这些二次反应将产生低于 100% 的电流效率。第二个问题是需要一种确定分析物何时被完全电解的方法。在受控电流库仑法中，即使分析物已完全氧化或还原，但恒定电流仍继续流动，这就需要一种确定反应终点 t_e 的合适方法。

means of determining the end-point of the reaction, t_e, is needed.

4. Microcoulometry Analysis

The Karl Fischer titration, which measures traces of water in solvents, foods, polymers, and other substances. The titration is usually performed by delivering titrant from an automated buret or by coulometric generation of titrant. The volumetric procedure tends to be appropriate for larger amounts of water (but can go as low as ,1 mg H_2O), and the coulometric procedure tends to be appropriate for smaller amounts of water. We illustrate the coulometric procedure in figure 4-29, in which the main compartment contains anode solution plus unknown. The smaller compartment at the left has an internal Pt electrode immersed in cathode solution and an external Pt electrode immersed in the anode solution of the main compartment. The two compartments are separated by an ion-permeable membrane. Two Pt electrodes are used for end-point detection.

4. 微库仑分析法

卡尔·费休滴定法可测量溶剂、食品、聚合物和其他物质中的微量水。滴定通常通过自动滴定管输送滴定液或通过库仑法生成滴定液来进行。容量法一般适用于大量水的测定（但可低至 1mg H_2O），库仑法则适用于微量水的测定。图 4-29 中，主反应室包含阳极溶液和未知溶液，左侧较小的反应室有一个浸在阴极溶液中的内部铂电极和一个浸在主反应室阳极溶液中的外部铂电极，两个反应室由离子渗透膜隔开，右侧两个 Pt 电极用于终点检测。

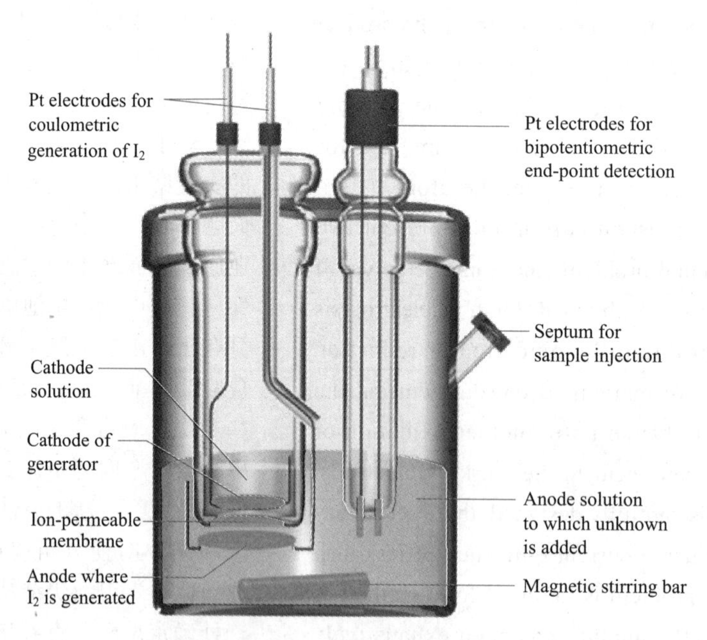

Fig.4-29 Apparatus for coulometric Karl Fischer titration

图 4-29 卡尔·费休滴定法装置

Experiment 3 Determination of fluoride ion in tap water by ion-selective electrode

1. Purpose of the experiment

(1) Understand the principle and method of determining fluoride ion content in water with ion-selective electrode.

(2) Master standard curve method, standard addition method and pX direct determination method.

(3) Understand the structure, characteristics and application conditions of fluoride ion selective electrode.

2. Experimental principle

The fluoride ion selective electrode is composed of lanthanum trifluoride (LaF_3) single crystal sensitive film, Ag-AgCl internal reference electrode and 0.001 mol · L^{-1} NaF-0.1 mol · L^{-1} NaCl internal reference solution. Under the condition that the temperature, ionic strength and pH value of solution remain unchanged, electrode membrane can transform activity of fluoride ions in solution into potential signal. When fluorine electrode is inserted into the solution, its sensitive membrane responds to F^- ion and generates a certain membrane potential between membrane and solution:

$$E_M = K - 2.303\frac{RT}{F}\lg a_{F^-}$$

Under certain conditions, membrane potential is linearly related to the $\lg a_{F^-}$. When fluorine electrode and saturated calomel electrode are inserted into the solution to form a galvanic cell, the electromotive force E of the cell also has a linear relationship with the $\lg a_{F^-}$ under certain conditions:

$$E = K' - 2.303\frac{RT}{F}\lg a_{F^-}$$

In the formula, K' value is the constant including potential of internal and external reference electrodes and liquid connection potential, etc. Activity of fluoride ions can be determined by measuring electromotive force of the cell. When the total ionic strength of solution remains unchanged and ionic activity coefficient is a certain value, so,

$$E = K' - 2.303\frac{RT}{F}\lg c_{F^-} = K' + 2.303\frac{RT}{F}\text{pF}$$

E is linearly related to log value of fluorine ion concentration. Therefore, in order to determine concentration of fluoride ions, an equal sufficient amount of total ion strength regulating buffer is often added to standard solution and sample solution. The total ion strength adjustment buffer can also be called TISAB, which is composed of inert electrolyte, buffer solution and masking agent (usually citric acid, DCTA, EDTA, Sulfosalicylic acid and phosphate, etc.). Addition of TISAB can make their total ionic strength be the same, and pH is stable within the normal working range of fluorine electrode (5～7), while masking the response of interfering ions (such as Al, Fe, Zr, Th,

Ca, Mg, Li and rare earth elements). When concentration of fluoride ion is in the range of 1-10^{-6} mol · L^{-1}, potential of fluoride electrode is in a linear relationship with pF, which can be measured by standard curve method or standard addition method.

3. Apparatus and reagents

(1) Apparatus　pH meter (PHS-25) or Other models acidity meter, fluorine electrode (PBF4-1-01), calomel electrode (232), electromagnetic mixer, volumetric flask (50 mL), volumetric flask (100 mL), beaker (100 mL), pipette (10 mL), pipette (20 mL).

(2) Reagents

① 0.1 mol · L^{-1} NaF standard solution　Weigh out 3.4000 g analytical pure sodium fluoride dried at 120℃ for 2 hours, after dissolving in deionized water, pour it into 1000 mL volumetric bottle, dilute to 1000 mL, mix well, and move it to plastic bottle for storage.

② Total ionic strength adjustment buffer solution (TISAB)　Dissolve 58.8 g sodium citrate and 20.2 g potassium nitrate in appropriate amount of water, add about 800 mL water, adjust pH of the solution to 6.8 with 1: 2 hydrochloric acid or 2% sodium hydroxide, dilute to 1000 mL, shake well for use.

4. Experimental steps

(1) Preparation and notes of fluorine electrode

① Before using, fluoride electrode is immersed in a beaker containing 0.001 mol · L^{-1} sodium fluoride solution for one or two hours for activation treatment.

② The electrode is washed with distilled water to the blank potential, that is, potential of the battery composed of fluoride electrode and calomel electrode in deionized water is about −300 mV, and two measured values are similar before using.

③ During measurement, the electrode is washed with distilled water and dried with filter paper for testing. The sample and standard solution should be at the same temperature, and the stirring speed of sample and standard solution should be the same.

④ When the electrode is immersed in the solution, it is necessary to prevent bubbles attached to the outer side of the lanthanum trifluoride wafer, and there are also no bubbles in the inner side of wafer and the internal reference solution, otherwise it will affect the determination.

(2) Adjusting instrument　Turn on the power and preheat for 10-20 minutes, press the keys "−mV" and "reading" of PHS-25 pH meter, then adjust the knob position to 0 mV. If ion meter is used, after preheating, adjust short-circuit the input terminal to 0 mV. Then connect electrode connector.

(3) Standard curve method

① Preparation of solution　Accurately draw 0.100 mol · L^{-1} F^{-}standard solution 10.00 mL, into a 100 mL volumetric flask, add 10.00 mL TISAB, dilute, to the mark and shake well to obtain a solution with pF of 2.00. Draw 10.00 mL solution with pF of 2.00 into a100 mL volumetric flask, add 9.00 mL TISAB, dilute, and shake well to obtain a solution with pF of 3.00.

Follow the above steps to configure solutions with pF of 4.00, pF of 5.00, and pF of 6.00.

② Pour each solution mentioned above into a dry 100 mL beaker separately, put the stir bar into the solution, use digital ion meter or PHS-25 pH meter to measure electromotive force corresponding to the solutions of different concentration in order from low concentration to high concentration, then record the data. The electrodes must be thoroughly cleaned for different concentrations of solution, moreover, electrode should not be immersed in the concentrated solution for a long time. The standard curve is drawn on a basis of the measured EMF value E (mV) as the ordinate and pF as the abscissa.

③ Determination of fluoride ion concentration in water samples   Pour 20.00 mL water sample into a 100 mL volumetric flask, dilute with TISAB and shake well, then measure the electromotive force (EMF). Check pF value on the standard curve according to the EMF value, and calculate the fluoride ion concentration in the water sample (expressed as $g \cdot L^{-1}$) based on pF value. Then clean the electrode and various glass apparatus.

(4) One-time standard addition method

① Accurately draw 20.00 mL tap water into a 100 mL volumetric flask, add 10 mL TISAB solution, dilute with deionized water, shake it well and pour it all into a dry plastic beaker to measure the electromotive force E_x (mV measurement function of the instrument).

② Accurately add 1.00 mL F^- standard solution with concentration of 100 $\mu g \cdot mL^{-1}$ to the above-mentioned test solution, stir evenly, and measure its electromotive force value E_{x+s}.

③ Pour all the blank solution (Number zero in the standard series) into the solution measured E_{x+s} above, stir well, and determine its electromotive force value E_0.

④ Record data and process

E_x=_________mV; E_{x+s}=_________mV; E_0=_________mV; c_s=100 $\mu g \cdot mL^{-1}$; V_s=1.00mL; V_x=100.0mL.

The fluorine content in water sample test solution can be calculated by the following formula.

$$c_F=\frac{\Delta c}{10^{|E_{x+s}-E_x|/S}-1}$$

In the formula, Δc is the increased concentration of F^- after addition of F^- standard solution,$\mu g \cdot mL^{-1}$, it can be calculated by the following formula.

$$\Delta c=\frac{c_s V_s}{V_s+V_x}\approx\frac{c_s V_s}{V_x}$$

In the formula, S is the response slope of electrode,

$$S=\frac{2.303RT}{nF}$$

The theoretical value of S is often different from the actual one. In order to avoid introducing errors, the actual value needs to be found, which can be obtained by calculating the slope of standard curve, or by doubling the dilution method. The dilution method is to add blank solution of the same volume to the solution after measuring E_x and E_{x+s}, and then measure its electromotive force value E_0, so the actual slope is:

$$S=\frac{E_0-E_{x+s}}{\lg 2}=\frac{E_0-E_{x+s}}{0.301}$$

The fluorine content of water sample can be calculated by the following formula.

$$c_{F^-}^0=\frac{c_{F^-}\times 100.00}{20.00}$$

5. Questions

① What is the principle of fluorine ion concentration determination with fluorine electrode?

② How does the acidity of the solution affect the determination?

③ What components should be included in total ion strength regulating buffer solution (TISAB)? What role does each play?

④ When determining the potential value of the standard solution, why is it determined in order from low concentration to high concentration?

⑤ Try to compare the advantages and disadvantages of standard addition method and standard curve method through experiments.

⑥ How to ensure the accuracy of standard addition method in experiments?

实验 3　离子选择性电极测定自来水中氟离子的含量

1. 实验目的

（1）了解用离子选择性电极测定水中氟离子含量的原理和方法。

（2）掌握标准曲线法、标准加入法、pX 直接测定法。

（3）了解氟离子选择性电极的结构、特性和使用条件。

2. 实验原理

氟离子选择性电极由三氟化镧（LaF_3）单晶敏感膜、Ag-AgCl 内参比电极和 0.001mol·L^{-1} NaF-0.1mol · L^{-1}NaCl 内参比溶液构成。在保持溶液的温度、离子强度和 pH 值不变的条件下，电极膜可将溶液中氟离子活度转变成电位信号。当氟电极插入溶液时其敏感膜对 F^-产生响应，在膜和溶液间产生一定的膜电位：

$$E_M=K-2.303\frac{RT}{F}\lg a_{F^-}$$

在一定条件下，膜电位与 $\lg a_{F^-}$ 值成直线关系。当氟电极与饱和甘汞电极插入溶液中组成原电池时，电池的电动势 E 在一定条件下也与 $\lg a_{F^-}$ 值成直线关系：

$$E=K'-2.303\frac{RT}{F}\lg a_{F^-}$$

式中 K' 值为包括内、外参比电极的电位、液接电位等的常数。通过测量电池的电动势可以测定氟离子的活度。当溶液的总离子强度不变，离子的活度系数为一定值时，则有：

$$E=K'-2.303\frac{RT}{F}\lg c_{F^-}=K'+2.303\frac{RT}{F}\text{pF}$$

E 与氟离子浓度的对数值成直线关系。因此，为了测定氟离子浓度，常在标准溶液和试样溶液中同时加入相等的足够量的由惰性电解质、缓冲溶液和掩蔽剂（通常用柠檬酸，DCTA，EDTA，磺基水杨酸及磷酸盐等）组成的总离子强度调节缓冲溶液（TISAB），使它们的总离子强度相同，且 pH 稳定在氟电极正常工作的范围内（5 ～ 7），同时掩蔽干扰离子（如 Al、Fe、Zr、Th、Ca、Mg、Li 及稀土元素等）的响应。当氟离子浓度在 1 ～ 10^{-6}mol·L^{-1} 范围内时，氟电极电位与 pF 成直线关系，可用标准曲线法或标准加入法进行测定。

3. 仪器和试剂

（1）仪器　PHS-25 型酸度计或其他型号酸度计，PBF4-1-01 型氟电极，232 型甘汞电极，电磁搅拌器，50mL 容量瓶，100mL 容量瓶，100mL 烧杯，10mL 吸量管，20mL 移液管。

（2）试剂

① 0.1 mol•L^{-1} NaF 标准溶液：称取在 120℃下干燥 2h 的分析纯氟化钠 3.4000g，用去离子水溶解后转移至 1000mL 容量瓶中，稀释至 1000m，混匀，转移至塑料瓶中保存。

② 总离子强度调节缓冲溶液（TISAB）：溶解 58.8g 柠檬酸钠和 20.2g 硝酸钾于适量水中，再加入约 800mL 水，以 1 ∶ 2 盐酸或 2% 氢氧化钠调节溶液的 pH 值为 6.8，稀释至 1000mL，摇匀备用。

4. 实验步骤

（1）氟电极的准备和注意事项

① 氟电极在使用前放在装有 0.001mol · L^{-1} 氟化钠溶液的烧杯中浸泡 1 ～ 2h，进行活化处理。

② 用蒸馏水淌洗电极到空白电位，即清洗到氟电极和甘汞电极组成电池在去离子水中的电位约为 −300mV，并且两次测定值相近方可使用。

③ 在测量时，电极用蒸馏水冲洗后，应用滤纸擦干后进行测试，试样和标准溶液应在同一温度，试样和标准溶液的搅拌速度应相等。

④ 电极浸入溶液时要防止三氟化镧晶片外侧附有气泡，同样晶片内侧及内参比溶液中也不得有气泡，否则影响测定。

（2）仪器调节　打开电源，预热 10 ～ 20min，按下 pHS-25 型酸度计的“−mV”及“读数”键，调节“定位”旋钮至 0mV。若用离子计，则预热后，短接输入端，调至 0mV。连接电极接头。

（3）标准曲线法

① 溶液的配制　准确吸取 0.100mol · L^{-1} F^- 标准溶液 10.00mL，置于 100mL 容量瓶中，加入 TISAB 10.00mL，用水稀释至刻度，摇匀，得 pF=2.00 的溶液。吸取 pF=2.00 溶液 10.00mL，置于 100mL 容量瓶中，加入 TISAB 9.00mL，用水稀释至刻度，摇匀，得 pF=3.00 的溶液。

依照上述步骤，配置 pF=4.00、pF=5.00 和 pF=6.00 溶液。

② 将上述溶液分别倒入烘干的 100mL 烧杯中，放入搅拌棒，用数字离子计或 pHS-25 型酸度计，按由稀到浓的次序，测出对应不同浓度溶液的电动势值，记下数据。在不同浓

度溶液的测定之间必须充分清洗电极，电极不宜在浓溶液中长时间的浸泡。以测得的电动势值 E（mV）为纵坐标，以 pF 值为横坐标，绘制标准曲线。

③ 水样中氟离子浓度的测定　移取水样 20.00mL 于 100mL 容量瓶中，用 TISAB 稀释至刻度，摇匀，测定电动势值。根据电动势值在标准曲线上查得 pF 值，并由 pF 值计算水样中的氟离子浓度（以 $g \cdot L^{-1}$ 表示）。清洗电极及各种玻璃仪器。

（4）一次标准加入法

① 准确吸取 20.00mL 自来水于 100mL 容量瓶中，加入 TISAB 溶液 10mL，用去离子水稀释至刻度，摇匀后全部转入干燥的塑料烧杯中，测定电动势值 E_x(用仪器的 mV 测量功能)。

② 向上述待测试液中准确加入 1.00mL 浓度为 $100\mu g \cdot mL^{-1}$ 的 F^-标准溶液，搅拌均匀，测定其电动势值 E_{x+s}。

③ 将空白溶液（即标准系列中的 0 号）全部加到上面测过 E_{x+s} 的溶液中，搅拌均匀，测定其电动势值 E_0。

④ 数据记录及处理

E_x=______mV; E_{x+s}=______mV; E_0=______mV; $c_s=100\mu g \cdot mL^{-1}$; V_s=1.00mL; V_x=100.0mL。

水样试液中氟含量可由下式计算

$$c_F=\frac{\Delta c}{10^{|E_{x+s}-E_x|/S}-1}$$

式中 Δc 为加入 F^-标准溶液后增加的 F^-浓度，$\mu g \cdot mL^{-1}$，可由下式计算得到：

$$\Delta c=\frac{c_s V_s}{V_s+V_x}\approx\frac{c_s V_s}{V_x}$$

式中，S 为电极的响应斜率，$S=\frac{2.303RT}{nF}$。

S 的理论值和实际值常有出入，为避免引入误差，需找到实际值，可由计算标准曲线的斜率求得，也可采用稀释一倍的方法求得。稀释法是在测出 E_x 和 E_{x+s} 后的溶液中加入同体积的空白溶液，然后测定其电动势值 E_0，则实际斜率为：

$$S=\frac{E_0-E_{x+s}}{\lg 2}=\frac{E_0-E_{x+s}}{0.301}$$

水样氟含量可由下式计算：$c^0_{F^-}=\frac{c_{F^-}\times 100.00}{20.00}$。

5. 思考题

① 用氟电极测定氟离子浓度的原理是什么？

② 溶液的酸度对测定有何影响？

③ 总离子强度调节缓冲溶液（TISAB）应包含哪些组分？各起什么作用？

④ 测定标准溶液的电位值时，为什么按由稀到浓的次序测定？

⑤ 试通过实验比较标准加入法和标准曲线法的优缺点。

⑥ 实验中如何保证标准加入法的准确度？

Exercises

4-1 Known: $Fe^{3+}+e^- \longrightarrow Fe^{2+}$ $E^\ominus= 0.77V$; $Cu^{2+}+2e^- \longrightarrow Cu$ $E^\ominus=0.34$ V;

$Fe^{2+}+2e^- \longrightarrow Fe$ $E^\ominus=-0.44$ V; $Al^{3+}+3e^- \longrightarrow Al$ $E^\ominus=-1.66$ V.

The strongest reducing agent is ( ).

A. Al^{3+} B. Fe^{2+} C. Fe D. Al

4-2 When pH=10, the electrode potential of the hydrogen electrode is ( ).

A. −0.59 V B. −0.30 V C. 0.30 V D. 0.59 V

4-3 Which of the following electrode potentials has nothing to do with pH? ( )

A. MnO_4^-/Mn^{2+} B. H_2O_2/H_2O C. O_2/H_2O_2 D. $S_2O_8^{2-}/SO_4^{2-}$

4-4 A primary battery: Pt|Fe^{3+} (1 mol · dm^{-3}), Fe^{2+} (1 mol · dm^{-3}) || Ce^{4+} (1 mol · dm^{-3}), Ce^{3+} (1 mol · dm^{-3})|Pt, The battery reaction of this battery is ( ).

A. $Ce^{3+}+Fe^{3+} \longrightarrow Ce^{4+}+Fe^{2+}$ B. $Ce^{4+}+Fe^{2+} \longrightarrow Ce^{3+}+Fe^{3+}$

C. $Ce^{3+}+Fe^{2+} \longrightarrow Ce^{4+}+Fe$ D. $Ce^{4+}+Fe^{3+} \longrightarrow Ce^{3+}+Fe^{2+}$

4-5 As indicator electrode in potentiometry, the relationship between its potential and the activity (concentration) of the measured ion is ( ).

A. irrelevant B. proportional

C. that it is proportional to the logarithm of the measured ion activity (concentration)

D. fit for the Nernst equation

4-6 Commonly-used reference electrodes is ( ).

A. glass electrode B. gas sensing electrode

C. saturated calomel electrode D. silver-silver chloride electrode

4-7 About the causes of pH glass electrode membrane potential, which of the following statements is true? ( )

A. The reduction of hydrogen ions on the glass surface transfers electrons.

B. Sodium ions move through the glass membrane.

C. Hydrogen ions penetrate the glass membrane, resulting in a difference in the concentration of hydrogen ions inside and outside the membrane.

D. It is the result of ion exchange and diffusion of hydrogen ions on the surface of glass membrane.

4-8 The asymmetric potential generated by the pH glass electrode is derived from ( ).

A. Difference of the surface properties of the inner and outer glass films.

B. Difference of the H^+ concentration in the inner and outer solutions.

C. Difference of the H^+ activity coefficients in the inner and outer solutions.

D. Difference of internal and external reference electrodes.

4-9 The selectivity coefficient of the ion selective electrode can be used for ( ).

A. estimating the detection limit of the electrode

B. estimating the level of interference from coexisting ions

C. correcting method error

D. estimating the linear response range of the electrode

4-10 When measuring with an ion-selective electrode, a magnetic stirrer is used to stir the solution in order to (   ).

A. reduce concentration polarization　　B. accelerate response speed

C. keep electrode surfaces clean　　D. reduce electrode resistance

4-11 When measuring the pH of a solution with a glass electrode, which quantitative analysis method is used? (   )

A. Standard Curve Method　　B. Direct Comparison

C. One-time Standard Accession　　D. Incremental Method

4-12 When using a silver chloride crystal membrane ion-selective electrode to measure chloride ions, if a saturated calomel electrode is used as a reference electrode, the salt bridge is (   ).

A. KNO_3　　B. KCl　　C. KBr　　D. KI

4-13 Which ionic strength adjusting buffer should be used when measuring fluoride ions in water (Contains small amount of Fe^{3+}, Al^{3+}，Ca^{2+}，Cl^-) with a fluoride ion selective electrode? (   )

A. 0.1 mol · $L^{-1}KNO_3$　　B. 0.1 mol · L^{-1}NaOH

C. 0.05 mol · $L^{-1}C_6H_5Na_3O_7$ (pH adjusted to 5 to 6)

D. 0.1 mol · L^{-1}NaAc (pH adjusted to 5 to 6)

4-14 When conducting quantitative analysis by the standard curve method with an ion-selective electrode，the requirement is that (   ).

A. The ionic strength of the test solution is consistent with that of the standard series of solutions.

B. The ionic strength of the test solution and the standard series solution is more than 1.

C. The ion activity to be measured in the test solution is consistent with that in the standard series of solutions.

D. The ionic strength to be measured in the test solution is consistent with that in the standard series of solutions.

4-15 In potentiometric titration，plot the $E-V$ titration curve，the end point of the titration is (   ).

A. maximum slope point of the curve　　B. minimum slope point of the curve

C. the most positive point of E　　D. the most negative point of E

4-16 In potentiometric titration，plot the $\Delta E/\Delta V-\overline{V}$ titration curve，the end point of the titration is (   ).

A. the turning point of the curve jump　　B. maximum slope point of the curve

C. minimum slope point of the curve　　D. the point where the slope of the curve is zero

4-17 In potentiometric titration，plot the $\Delta^2E/\Delta V^2-\overline{V}$ titration curve，the end point of the titration is (   ).

A. the most positive point of $\Delta^2E/\Delta V^2$　　B. the most negative point of $\Delta^2E/\Delta V^2$

C. the point where $\Delta^2E/\Delta V^2$ is zero　　D. the point where the slope of the curve is zero

4-18 Electrolytic acid $CuSO_4$ solution with a current of 2.0A, if 400 mg of copper is deposited, how many seconds are needed? [*Ar*(Cu)= 63.54][1] ( )

A. 22.4 B. 59.0 C. 304 D. 607

4-19 In the $CuSO_4$ solution, energized with a platinum electrode at a current of 0.100 A for 10 min, how many milligrams of copper are deposited on the cathode? [*Ar*(Cu)= 63.54] ( )

A. 60.0 B. 46.7 C. 39.8 D. 19.8

4-20 How many grams of iron can be deposited from $Fe_2(SO_4)_3$ solution with 96484 C electricity? [*Ar*(Fe)=55.85, *Ar*(S)= 32.06, *Ar*(O)= 16.00] ( )

A. 55.85 B. 29.93 C. 18.62 D. 133.3

4-21 Two platinum sheets are used as electrodes to electrolyze a $CuSO_4$ solution containing H_2SO_4. The reaction that occurs on the anode is ( ).

A. OH^- lose electrons B. Cu^{2+} get electrons

C. SO_4^{2-} lose electrons D. H^+ get electrons

4-22 The effect of adding a large amount of irrelevant electrolytes in coulometric titration is to ( ).

A. slow down migration speed B. increase migration current

C. increase current efficiency D. guarantee current efficiency of 100%

4-23 In coulometric analysis, in order to improve the selectivity of the determination, it is generally used for ( ).

A. large working electrode B. large current

C. controlling potential D. control time

4-24 In the process of controlled potential electrolysis, in order to keep the working electrode potential constant, it is necessary to keep ( ).

A. the applied voltage constantly changing B. the applied voltage unchanged

C. auxiliary electrode potential unchanged D. electrolysis current constant

4-25 The theoretical basis of Coulomb analysis is ( ).

A. Electrolysis equation B. Faraday's Law

C. Nernst equation D. Fick's Law

4-26 The $NiSO_4$ solution is electrolyzed with a nickel electrode as the cathode, and the cathode product is ( ).

A. H_2 B. O_2 C. H_2O D. Ni

4-27 The prerequisites for controlling potential coulomb analysis are ( ).

A. 100% current efficiency B. 100% titration efficiency

C. to control electrode potential D. to control current density

4-28 Actual decomposition voltage includes ( ).

A. reversible EMF B. overvoltage

C. sum of reversible electromotive force and overvoltage

D. sum of reversible electromotive force, overvoltage, and IR voltage drop

4-29 The basis for determining whether an electrode is an anode or a cathode is ( ).

[1] *Ar* (x) means the relative atomic mass of x. *Ar* (x) 表示 x 的原子量。

A. the level of electrode potential B. the nature of the electrode reaction

C. properties of Electrode Materials D. degree of electrode polarization

4-30 The basis for determining whether the electrode is positive or negative is ( ).

A. the level of electrode potential B. the nature of the electrode reaction

C. properties of Electrode Materials D. degree of electrode polarization

4-31 A cell is formed with a lead electrode in 0.015 mol · L^{-1} lead acetate and a cadmium electrode in 0.021 mol · L^{-1} cadmium sulfate. The two solutions are joined by a salt bridge containing ammonium nitrate. What is the potential of the cell at 25 ℃? (The activity coefficients in the two solutions can be considered equal.)

(E=0.273 V or E=−0.2727 V)

4-32 A 100.0 mL solution containing 0.1000 mol · L^{-1} NaCl was titrated with 0.1000 mol · L^{-1} $AgNO_3$, and the voltage of the cell is controlled by an Ag electrode (+) and an SCE (−). Calculate the voltage after the addition of 65.00 and 100.00 mL of $AgNO_3$.[$E^{\ominus}(Ag^+/Ag)$=0.799 V, E_{SCE}=0.241 V, K_{sp}(AgCl)=1.8×10^{-10}]

(E=0.0790 V or E=0.339 V)

4-33 When a saturated calomel electrode and a fluoride (F) ion-selective electrode are inserted into a 1.000×10^{-3} mol · L^{-1} standard solution, the voltage reading is −0.1590 V with SCE positive. For a test solution containing fluoride (F^-) the reading becomes −0.2120 V. Calculate the concentration of the fluoride in the test solution.

(1.26×10^{-4}mol · L^{-1})

4-34 At 25 ℃, when a saturated calomel electrode and a Cl^- ion-selective electrode are inserted into a 0.001 mol · L^{-1} standard solution, the voltage reading is 0.159 V with SCE positive. For a test solution containing chloride the reading becomes 0.212 V. Calculate the concentration of the Cl^- in the test solution.

(7.9×10^{-3}mol · L^{-1})

4-35 A cell is set up as follows: silver electrode, unknown solution, salt bridge, saturated KCl solution, Hg_2Cl_2 (s), mercury electrode, (a) Which electrode is the reference and which is the indicator? (b) What is the purpose of the salt bridge, and what electrolyte should it contain? (c) If the potential of the cell is found to be 0.300 V, with the silver electrode more positive than the mercury, what is the concentration of Ag^+ ion in the unknown?

[(a) Satarated Calomel electrode is reference electrode Silverelectrode is indicator electrode.(b) The salt bridge can reduce liquid junction potential. (c) 4.73×10^{-5} mol · L^{-1}]

Chapter 5 Gas Chromatography

第5章 气相色谱分析法

Study Guide 学习指南

Gas chromatography (GC) is a technique for separating compounds according to the tiny differences in partition coefficients of two phases. When the two phases are moving relatively, the analyte will be repeatedly distributed between two phases, magnifying the original tiny difference so that each constituent is separated and analyzed and the physical and chemical constants are measured. This charpter shall mainly focus on the introduction of the principle, basic concepts, commonly used detectors and quantitative analytical methods of GC. The learning objectives of this chapter are to master the basic principles and concepts, to understand the main components and their functions of gas chromatograph as well as to be familiar with the usage of the related work station software.

色谱法是一种分离方法，它利用物质在两相中分配系数的微小差异进行分离。当两相做相对移动时，使被测物质在两相之间进行多次分配，这样原来的微小差异产生了很大的效果，使各组分分离，以达到分离分析及测定一些物理化学常数的目的。本章重点介绍气相色谱法原理、基本概念、常用检测器及定量分析方法等内容。通过本章的学习，掌握气相色谱法的基本原理及概念、了解气相色谱仪的主要部件和作用，熟悉相关工作站软件的使用。

Section 1 An Introduction to Chromatography

第1节 色谱分析法简介

1. What is Chromatography?

In 1903 in Warsaw, the botanist M.T. Swett invented adsorption chromatography to separate plant pigments using a hydrocarbon solvent and inulin powder (a carbohydrate) as stationary phase. The separation of colored bands led to the name chromatography, from the

1. 色谱的定义

1903年，俄国植物学家 M.T. Swett 发明了吸附色谱。他用碳氢化合物和菊根粉（一种碳水化合物，可作为固定相）来分离植物叶绿素，这种色带的分离称为色谱，该词源于希腊语“颜色”。

Greek chromatos (color).

Figure 5-1 shows a solution containing solutes A and B placed on top of a column packed with solid particles and filled with solvent. When the outlet is opened, solutes A and B flow down into the column. Fresh solvent is then applied to the top of the column and the mixture is washed down the column by continuous solvent flow. If solute B is more strongly adsorbed than solute A on the solid particles. Solute B moves down the column more slowly than solute A and emerges at the bottom after solute A.

图 5-1 显示的是把含有溶质 A 和 B 的溶液加入有固体颗粒的填充柱中，打开下端出口后，A 和 B 在填充柱中向下流动。然后将新鲜的溶剂加到填充柱柱头，并通过连续的溶剂淋洗后混合物被洗出柱体。如果溶质 B 比溶质 A 对固体颗粒的吸附能力更强，溶质 B 比溶质 A 在柱中的迁移速度更慢，即溶质 B 出现在 A 流出后的柱子底部。

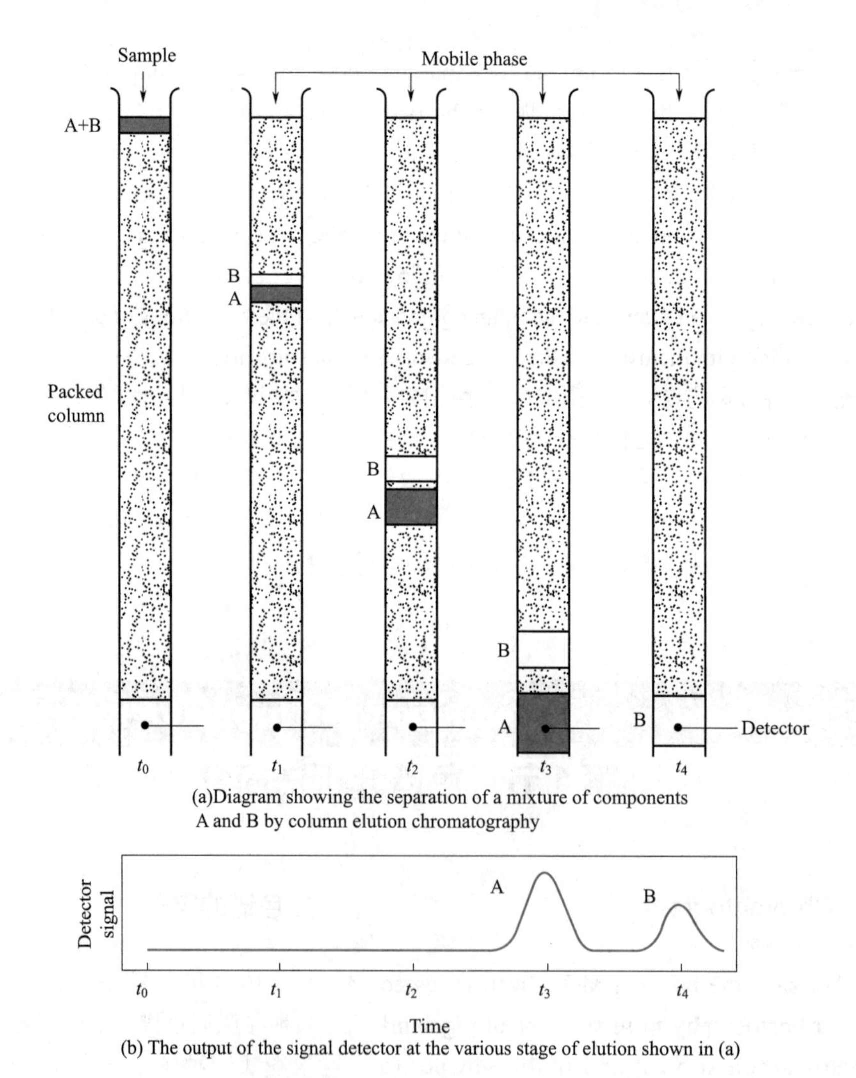

(a)Diagram showing the separation of a mixture of components A and B by column elution chromatography

(b) The output of the signal detector at the various stage of elution shown in (a)

Fig.5-1 The separation of a mixture by column chromatography

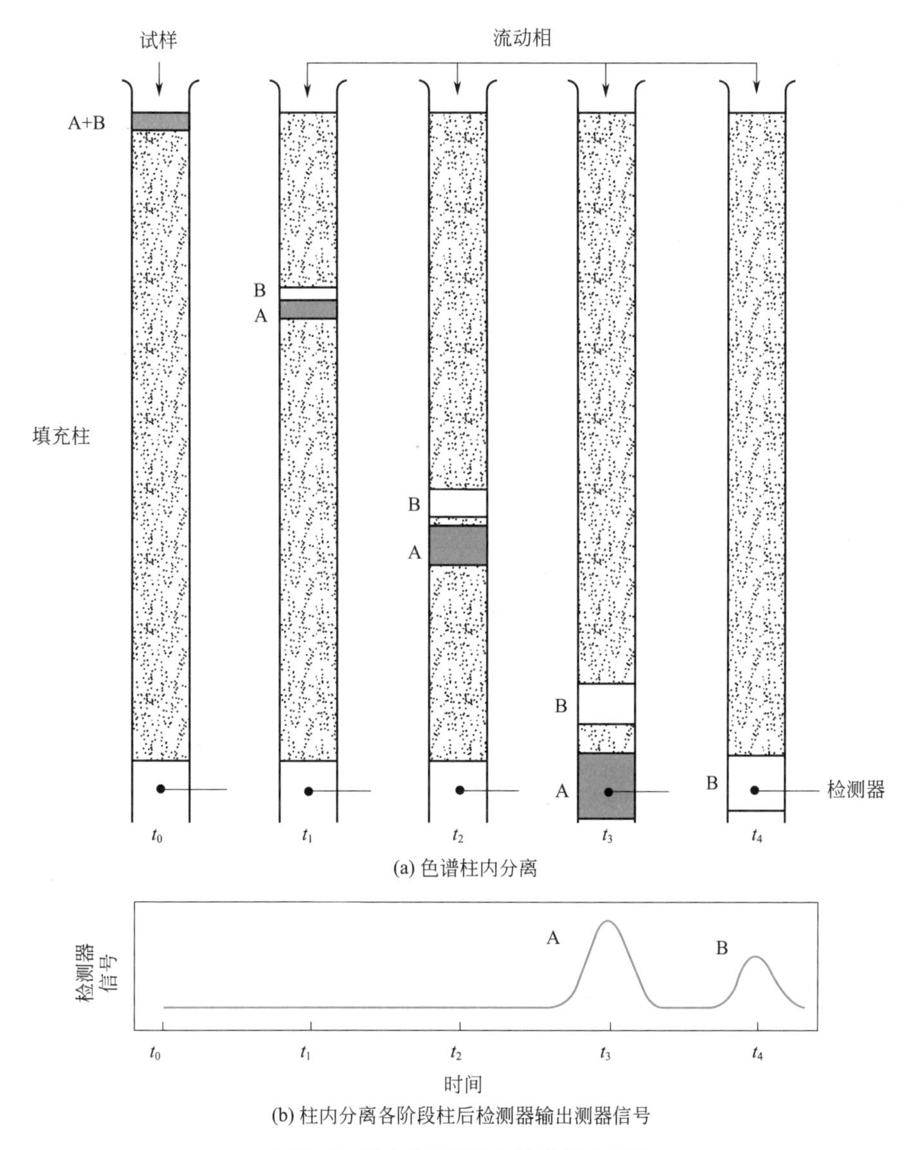

图 5-1　混合物两组分色谱分离示意图

The mobile phase (the solvent moving through the column) in chromatography is either a liquid or a gas. The stationary phase (the one that stays in place inside the column) is most commonly a viscous liquid chemically bonded to the inside of a capillary tube or onto the surface of solid particles packed in the column (figure 5-2). Alternatively, as in figure 5-1, the solid particles themselves may be the stationary phase. In any case, the equilibration of solutes between mobile and stationary phases gives rise to separation.

色谱法中的流动相（通过柱的溶剂）可以是液体，也可以是气体。固定相（装填在柱内的物质）通常是一种黏性液体，其与毛细管内部或填充在柱内的固体颗粒表面发生化学键合，如图 5-2 所示。或者如图 5-1，固体颗粒本身可以作为固定相。在任何情况下，混合物在流动相和固定相之间的分配，最终导致了混合物中各组分的分离。

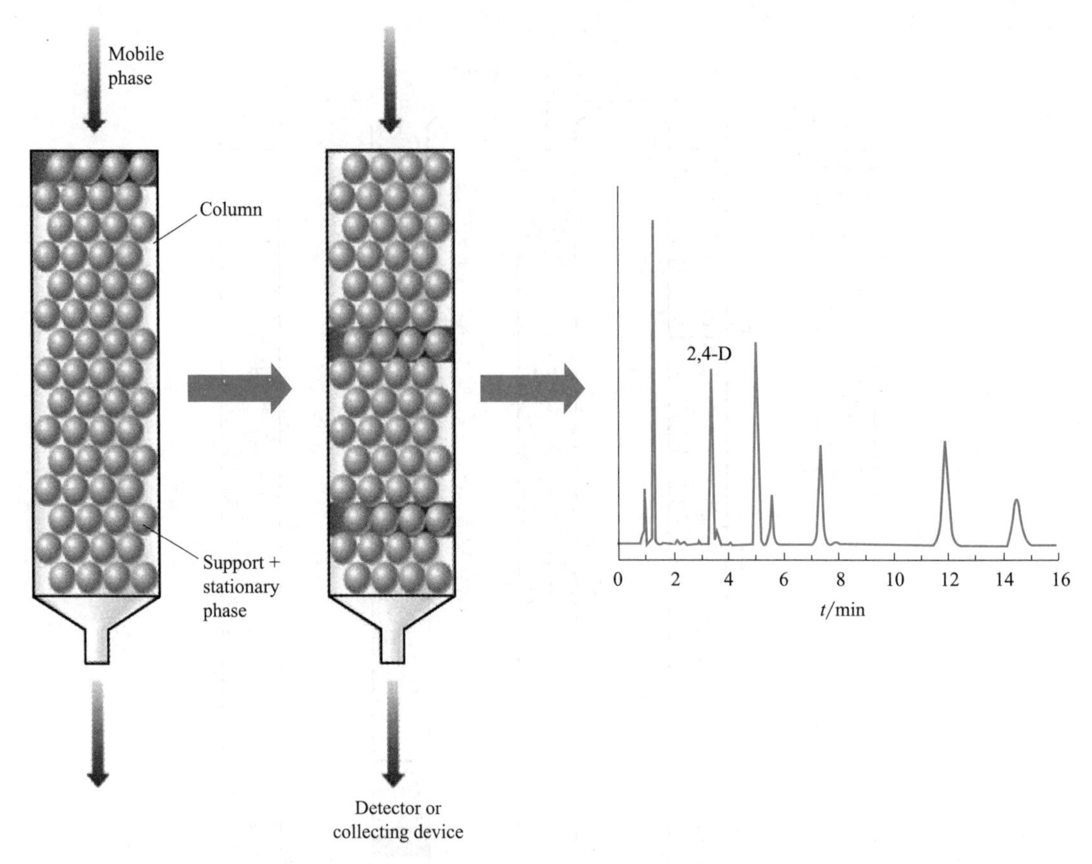

Fig.5-2 Separation of 2,4-D from other sample components by chromatography[The main components of the chromatographic system (the mobile phase, stationary phase, and support, which together make up the" column") are shown on the left. The plot on the right (called a chromatogram) shows the measured amount of each chemical that leaves the column after a given amount of time (or volume of applied mobile phase).]

图 5-2 通过色谱法将 2,4-D 从其他样品组分中分离出来 [色谱系统的主要组成 (流动相、固定相和支持物，即色谱柱) 见左图。右图 (称为色谱图) 显示了在给定时间 (或应用的流动相体积) 后离开色谱柱的每种化学物质的测量信号。]

2. Classification of Chromatography methods

The most widely used classification method is based upon the nature of the mobile phase, as shown in the first column of table 5-1, chromatographic methods fall into three categories: liquid, gas, and supercritical fluid. The second column of the table reveals that there are five types of liquid chromatography and three types of gas chroma-tography, which differ in the nature of the stationary phase and the types of equilibria between phases. Chromatography is divided into categories on the basis of the mechanism of interaction of the solute with the stationary phase, as shown in table 5-2.

2. 色谱分析法的分类

应用最广泛的分类方法是基于流动相的性质，如表 5-1 的第一列所示，色谱法可分为液相、气相和超临界流体三类。该表的第二列显示，有 5 种液相色谱和 3 种气相色谱，它们在固定相的性质和相间平衡的类型上各不相同。根据不同组分与固定相的作用机理，对色谱法进行分类，如表 5-2 所示。

Table 5-1 Classification of column chromatographic methods

general classification	specific method	stationary phase	type of equilibrium
gas chromatography (GC) (mobile phase: gas)	gas-solid	solid	adsorption
	gas-bonded phase	organic species bonded to a solid surface	partition between liquid and bonded surface
	gas-liquid	liquid adsorbed on a solid	partition between gas and liquid
liquid chromatography (LC) (mobile phase: liquid)	liquid-liquid partition	liquid adsorbed on a solid	partition between immiscible liquids
	liquid-bonded phase	organic species bonded to a solid surface	partition between liquid and bonded surface
	liquid-solid adsorption	solid	adsorption
	ion exchange	ion-exchange resin	ion exchange
	size exclusion	liquid in interstices of a polymeric solid	partition/sieving
supercritical-fluid chromatography (SFC) (mobile phase: supercritical-fluid)		organic species bonded to a solid surface	partition between supercritical fluid and bonded surface, separation process of elution chromatography

表5-1 柱层析方法分类

一般分类	具体方法	固定相	平衡类型
气相色谱（流动相：气体）	气固吸附色谱	固体	吸附
	键合（气）相色谱	有机物键合到固体表面	气体和键合相之间的分配
	气液色谱	涂渍在固体上的液体	气液之间
液相色谱（流动相：液体）	液液分配色谱	液体涂渍在固体上	不混溶的液体之间的分配
	键合（液）相色谱	固体表面键合有机物	液体和键合相之间的分配
	液固吸附色谱	固体	吸附
	离子交换色谱	离子交换树脂	离子交换
	尺寸排阻	化学惰性的多孔凝胶	体积排阻作用
超临界流体色谱法（SFC）（流动相：超临界流体）		固体表面键合有机物	超临界流体与键合固定相之间的分配

Table 5-2 Types of chromatography

Types	Introduction	Principle
Adsorption chromatography	A solid stationary phase and a liquid or gaseous mobile phase are used. Solute is adsorbed on the surface of the solid particles. The more strongly a solute is adsorbed, the slower it travels through the column.	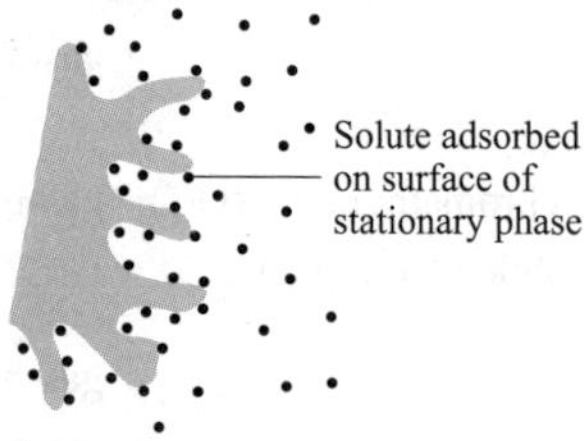 Solute adsorbed on surface of stationary phase
Partition chromatography	A liquid stationary phase is bonded to a solid surface, which is typically the inside of the fused silica (SiO_2) chromatography column in gas chromatography. Solute equilibrates between the stationary liquid and the mobile phase, which is a flowing gas in gas chromatography.	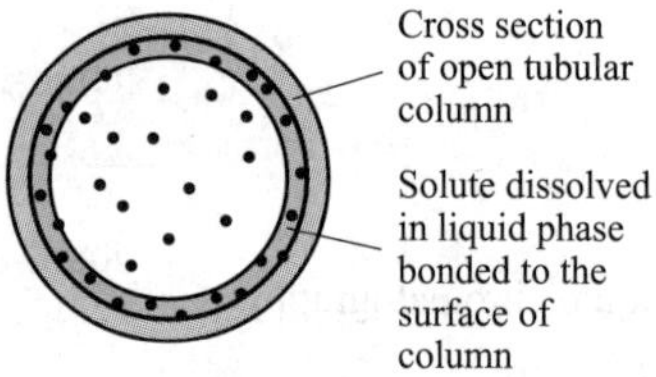 Cross section of open tubular column Solute dissolved in liquid phase bonded to the surface of column
Ion-exchange chromatography	Anions such as $-SO_3^-$ or cations such as $-N(CH_3)_3^+$ are covalently attached to the stationary solid phase, usually a *resin*. Solute ions of the opposite charge are attracted to the stationary phase. The mobile phase is a liquid.	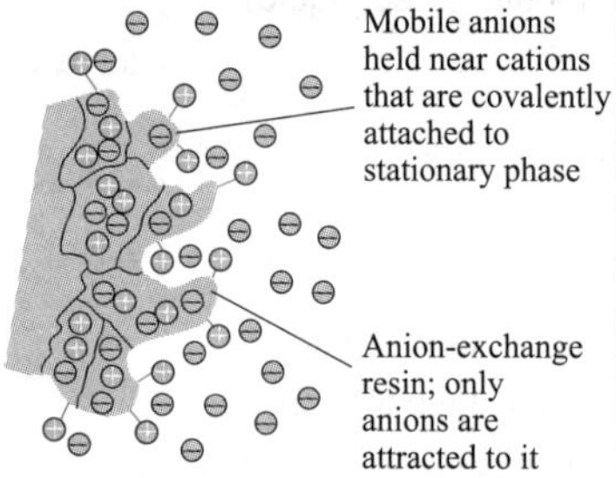 Mobile anions held near cations that are covalently attached to stationary phase Anion-exchange resin; only anions are attracted to it
Molecular exclusion chromatography	Also called size exclusion, gel filtration, or gel permeation chromatography. This technique separates molecules by size, with the larger solutes passing through most quickly. In the ideal case of molecular exclusion, there is no attractive interaction between the stationary phase and the solute. Rather, the liquid or gaseous mobile phase passes through a porous gel. The pores are small enough to exclude large solute molecules but not small ones. Large molecules stream past without entering the pores. Small molecules take longer to pass through the column because they enter the gel and therefore must flow through a larger volume before leaving the column.	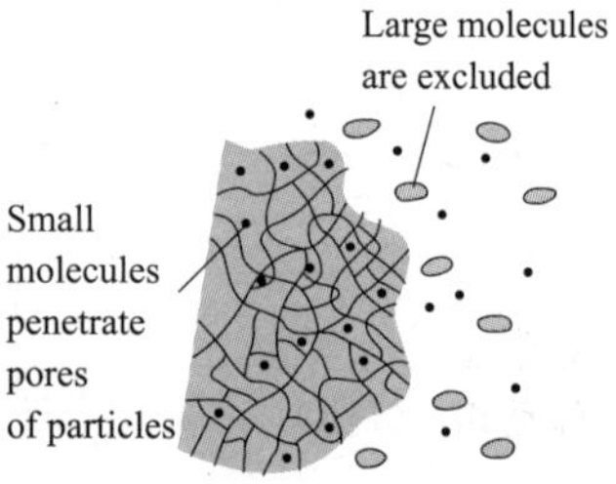 Large molecules are excluded Small molecules penetrate pores of particles
Affinity chromatography.	This most selective kind of chromatography employs specific interactions between one kind of solute molecule and a second molecule that is covalently attached (immobilized) to the stationary phase. For example, the immobilized molecule might be an antibody to a particular protein. When a mixture containing a thousand proteins is passed through the column, only the one protein that interacts with the antibody binds to the column. After all other solutes have been washed from the column, the desired protein is dislodged by changing the pH or ionic strength.	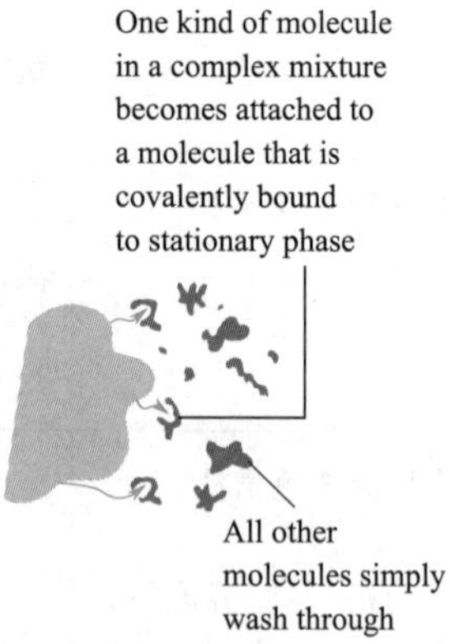 One kind of molecule in a complex mixture becomes attached to a molecule that is covalently bound to stationary phase All other molecules simply wash through

表5-2　色谱类型

类型	介绍	原理
吸附色谱	使用固体固定相和液体或气体流动相。溶质吸附在固体颗粒的表面。溶质被吸附得越强，它通过色谱柱的速度就越慢	利用吸附剂表面对不同组分物理吸附性能的差别而使之分离
分配色谱	固定液键合到固体表面。对于气相色谱而言，柱材料多为熔融石英（SiO_2），流动相为气体，溶质在固定液和流动相间达到平衡	利用固定液对不同组分分配性能的差别而使之分离
离子交换色谱	阴离子如 SO_4^{2-} 或阳离子如 $N(CH_3)_3^+$ 共价连接到固定相（通常是树脂）上，带相反电荷的溶质离子被固定相吸引。流动相是液体	利用被分离组分与固定相之间发生离子交换的能力差异来实现分离
尺寸排阻色谱	也称为尺寸排阻、凝胶过滤或凝胶渗透色谱。该技术按溶质分子大小进行分离。体积大的分子不能渗透到凝胶孔穴中去而被排阻，被较早地淋洗出来。尺寸小的组分渗入凝胶颗粒内，流程长，流动速度慢，后流出色谱柱	利用分子大小顺序进行分离的一种色谱方法
亲和色谱	这种选择性最高的色谱法利用了一种溶质分子和另一种共价键合到固定相的分子之间的特定相互作用。例如，固定相分子可能是针对特定蛋白质的抗体。当含有一千种蛋白质的混合物通过色谱柱时，只有一种与抗体相互作用的蛋白质能与色谱柱结合，其他溶质则从色谱柱中淋洗掉。通过改变 pH 值或离子强度来分离所需的蛋白质	利用相互间具有高度特异亲和性的两种物质之一作为固定相，由于溶质分子与固定相不同程度的亲和性，使组分与杂质分离

Section 2　Elution Profile (Chromatogram) and Terminology

第2节　色谱流出曲线及相关术语

1. Elution Profile

As long as the analyte is taken into the chromatographic column by the mobile phase, the gases or liquids exiting the column are detected by a detector and a profile is plotted to record the change of signal with the retention time. This profile is called the elution profile, shown in figure 5-3. When the analyte exits the column, the detector can obtain the concentrations of its compositions, expressing as peaks in the elution profile. Chromatogram: when a mixture compound is separated by a chromatographic column, various components are collected at the bottom of the column, which are converted into electrical signals by a detector. The

1. 色谱流出曲线

从流动相携带着被测组分进入色谱柱起，用检测器检测流出柱后的气体或液体，并用记录器记录信号随时间变化的曲线，此曲线就叫色谱流出曲线，如图 5-3。当待测组分流出色谱柱时，检测器就可检测到其组分的浓度，在流出曲线上表现为峰状，叫色谱峰。

Chromatographic legal parameters
色谱法定性参数

signals are proportional to the concentration of the components separated and are visually outputted on a profile called chromatogram.

色谱图是指试样中各组分经色谱柱分离后，在柱的末端收集各组分，经检测器转换为电信号，用记录仪将各组分浓度记录下来，即得到色谱图。

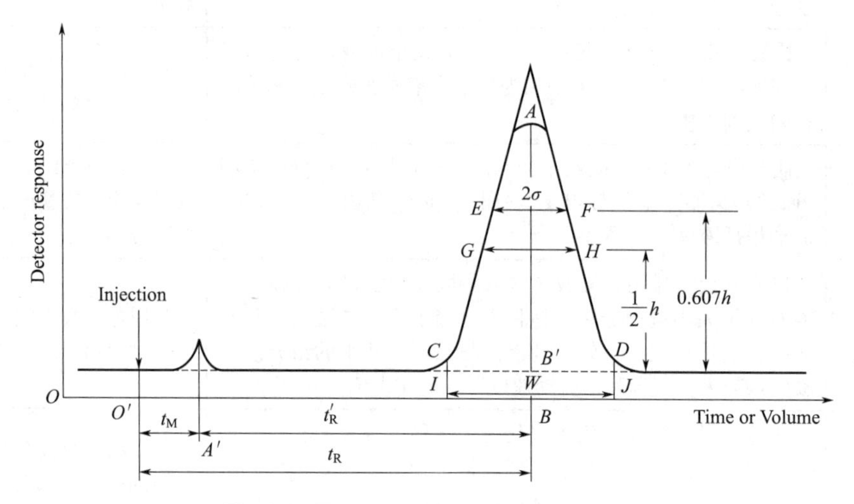

Fig.5-3 Elution profile (chromatogram)

图 5-3 色谱流出曲线（色谱图）

(1) Base line Under experimental conditions, the elution profile is named as base line if only mobile phase enters the detector. The base line should be a horizontal straight line under steady conditions. Its straightness reflects the steadiness of the experiment conditions.

（1）基线 在实验条件下，色谱柱后仅有纯流动相进入检测器时的流出曲线称为基线。基线在稳定的条件下应是一条水平的直线。它的平直与否可反映出实验条件的稳定情况。

(2) Standard deviation The peaks in the elution profile tends to spread into a Gaussian shape with standard deviation σ. It is half the peak width at a peak height of 0.607, which is half of the *EF* shown in figure 5-3.

（2）标准偏差 色谱峰一般扩散成对称的具有标准偏差 σ 的高斯形状，σ 是 0.607 倍峰高处峰宽度的一半，即图 5-3 中 *EF* 的一半。

(3) Peak base width (W) The peak base width is the distance between two intersection points of the base line with the tangent lines of two inflection points at both sides of a peak.

（3）色谱峰底宽（W） 由色谱峰的两边拐点作切线，与基线交点的距离，即图 5-3 中的 *IJ*。

$$W=4\sigma \tag{5.1}$$

(4) Peak width at half height ($W_{1/2}$) It is the peak width at half of the peak height.

（4）半峰高宽度（$W_{1/2}$） 色谱峰高一半处的峰宽，也称为色谱峰半高宽度，即图 5-3 中的 *GH*。

$$W_{1/2}=2\sigma\cdot\sqrt{21n2}=2.354\sigma \tag{5.2}$$

(5) Peak height and peak area　The peak height (h) is the distance between the peak point and the base line. The peak area (A) is the area surrounded by the peak point and the base line.

（5）峰高或峰面积　色谱峰顶点与基线的距离叫峰高，即图 5-3 中的 AB'。色谱峰与峰底基线所围成区域的面积叫峰面积，即图 5-3 中 ACD 内的面积（A）。

$$A=1.065hW_{1/2} \tag{5.3}$$

2. Terminology

2. 相关术语

(1) Retention time

（1）保留时间

① Dead time　The dead time is the time taken by the substance that is not interactive with the stationary phase from the sample introduction to the peak maximum value. It is directly proportional to the void column of the chromatographic column. Because the substance is not interactive with the stationary phase, its flow rate is close to that of the mobile phase. The mean flow rate of the mobile phase can be calculated via t_M.

① 死时间（t_M）　不与固定相作用的物质从进样到出现峰极大值时的时间，它与色谱柱的空隙体积成正比。由于该物质不与固定相作用，因此，其流速与流动相的流速相近。根据 t_M 可求出流动相平均流速。

$$t_M=\frac{L}{\mu} \tag{5.4}$$

Where L is length of the column, μ is average line velocity of the mobile phase.

其中，L 为柱长，μ 为流动相的平均线速度。

② Retention time　The retention time is the time taken from the sample introduction to the appearing of peak maximum value[fig.5-4(a)], including the time of components travelling through the column with the mobile phase (t_0) and the time of components lingering in the stationary phase.

② 保留时间（t_R）　试样从进样到出现峰极大值时的时间［图 5-4（a）］。它包括组分随流动相通过柱子的时间 t_0 和组分在固定相中停留的时间。

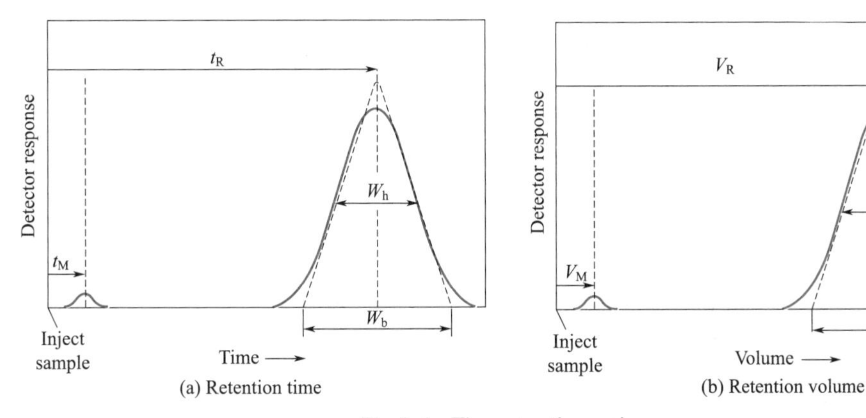

Fig.5-4　The retention value

图 5-4　保留值

③ Adjusted retention time The adjusted retention time is the remaining time of the retention time deducting the dead time of a component, which is the lingering time of the component in the stationary phase, that is

③ 调整保留时间（t_R'） 某组分的保留时间减去死时间后的保留时间，即是组分在固定相中的停留时间。

$$t'_R=t_R-t_M \tag{5.5}$$

(2) Retention volume

（2）保留体积

① Dead volume The volume of the mobile phase flowing through the chromatographic column in the dead time is called the dead volume (V_M), which is equal to the volume of the mobile phase within the chromatographic column.

① 死体积（V_M） 死时间内流经色谱柱的流动相的体积称为死体积 V_M，即等于色谱柱内流动相的体积。

$$V_M=t_MF_c \tag{5.6}$$

Where F_c is the volume flowrate, the average volume flowing through the chromatographic column in unit of time, mL · min^{-1}.

其中 F_c 为体积流量，为单位时间内流经色谱柱的平均体积，单位：mL · min^{-1}。

② Retention volume The volume of the mobile phase flowing through the chromatographic column in the retention time[(fig.5-4(b)] is called the retention volume (V_R).

② 保留体积（V_R） 保留时间内流经色谱柱的流动相的体积［图 5-4(b)］，称为保留体积，以 V_R 表示。

$$V_R=t_RF_c \tag{5.7}$$

③ Adjusted retention volume V_R' refers to the volume of the mobile phase flowing through the chromatographic column in the adjusted retention time.

③ 调整保留体积（V_R'） 调整保留体积（V_R'）是指在调整保留时间内流过色谱柱的流动相的体积。

$$V'_R=t'_RF_c=(t_R-t_M)\cdot F_c=V_R-V_M \tag{5.8}$$

(3) Relative retention time The ratio of the adjusted retention times of component 2 and 1, and it is also called **selectivity factor (α)**.

（3）相对保留时间 相对保留值是指组分 2 的调整保留时间与另一组分 1 的调整保留时间之比，又称为**选择性因子（α）**。

$$\alpha=\frac{t'_{R_2}}{t'_{R_1}}=\frac{V'_{R_2}}{V'_{R_1}} \tag{5.9}$$

Relative retention values can be used to indicate the selectivity of the chromatographic column. The greater the relative retention, the greater the separation between two components. When the relative retention is equal to 1, the two components cannot be separated.

相对保留值可用来表示色谱柱的选择性。相对保留值越大，相邻两组分的调整保留时间差别越大，分离越好；相对保留值等于 1 时，两组分不能被分离。

(4) Retention factor For each peak in the chromatogram, the retention factor (also to be called capacity factor, capacity ratio, or partition ratio), k, is the time required to elute that peak minus the time t_M required for mobile phase to pass through the column, expressed in multiples of t_M.

（4）保留因子　对于色谱图中的每个峰而言，保留因子 k（或称容量因子、容量比、分配比）是指该峰出现所需的时间 t_R 减去流动相通过色谱柱所需的时间 t_M，再除以 t_M，见式（5.10）。

Retention factor:

$$k=\frac{t_R-t_M}{t_M}=\frac{t'_R}{t_M} \tag{5.10}$$

The longer a component is retained by the column, the greater is the retention factor.

可见，保留时间越长，保留因子越大。

(5) Relation between retention time and the partition coefficient The retention factor in equation (5.10) is equivalent to

（5）保留时间和分配系数间的关系　式（5.10）中的保留因子可理解为溶质在固定相中的停留时间与在流动相中的停留时间之比。

$$k=\frac{\text{time solute spends in stationary phase}}{\text{time solute spends in mobile phase}} \tag{5.11}$$

Why this is true? If the solute spends all its time in the mobile phase and none in the stationary phase, it would be eluted in time t_M. Putting $t_R=t_M$ into equation (5.10) gives k=0, because solute spends no time in the stationary phase. Suppose that solute spends equal time in the stationary and mobile phases. The retention time would then be $t_R=2\ t_M$ and $k=(2\ t_M-t_M)/t_M=1$. If solute spends three times as much time in the stationary phase as in the mobile phase, $t_R=4\ t_M$ and $k=(4\ t_M-t_M)/t_M=3$. Then, there will be three times as many moles of solute in the stationary phase as in the mobile phase at any time. The quotient in equation (5.10) is equivalent to

这是为什么呢？如果溶质所有时间内都停留在流动相中，则在固定相中的停留时间为0，即组分将在时间 t_M 时被洗脱。将 $t_R=t_M$ 代入式（5.10）中，得到 k=0。假设溶质在固定相和流动相中停留的时间相等，则保留时间 $t_R=2t_M$，k=1。如果溶质在固定相中停留的时间是在流动相中的三倍，$t_R=4t_M$，k=3。而且在任一时间，固定相中溶质的摩尔数都是流动相的三倍。
故保留因子的求解可转化为式（5.12）。

$$k=\frac{c_S V_S}{c_M V_M} \tag{5.12}$$

Where c_S is the concentration of solute in the stationary phase, V_S is the volume of the stationary phase, c_M is the concentration of solute in the mobile phase, and V_M is the volume of the mobile phase.

The quotient $\frac{c_S}{c_M}$ is the ratio of concentrations of solute

其中 c_S 为固定相中溶质的浓度，V_S 为固定相的体积，c_M 为流动相中溶质的浓度，V_M 为流动相的体积，$\frac{c_S}{c_M}$ 为固定相和流动相中溶质浓度的比值。如果色谱柱中组分在两相中的分配达到平

in the stationary and mobile phases. If the column is run slowly enough to be at equilibrium, the quotient $\frac{c_S}{c_M}$ is the **partition coefficient**, K, introduced in connection with solvent extraction. Therefore, we cast equation (5.13) in the form.

衡时，$\frac{c_S}{c_M}$称为**分配系数**K。保留因子与分配系数的关系见式（5.13）。

$$k=K\frac{V_S}{V_M}=\frac{t_R-t_M}{t_M}=\frac{t'_R}{t_M} \tag{5.13}$$

Which relates retention time to the partition coefficient and the volumes of stationary and mobile phases. Because $t'_R \propto k \propto K$, relative retention can also be expressed as:

可见保留因子与分配系数、溶质在色谱柱固定相和流动相中的体积比有关。由于保留时间正比于保留因子及分配系数，故两种溶质的相对保留值还可表示为：

$$\alpha=\frac{t'_{R_2}}{t'_{R_1}}=\frac{k_2}{k_1}=\frac{K_2}{K_1} \tag{5.14}$$

That is, the relative retention of two solutes is proportional to the ratio of their partition coefficients. This relation is the physical basis of chromatography.

因此，两种溶质的相对保留值与其分配系数的比值成正比。这种关系是色谱法的物理学基础。

Physical basis of chromatography: The greater the ratio of partition coefficients between mobile and stationary phases, the greater the separation between two components of a mixture.

色谱的物理学基础：流动相和固定相之间的分配系数之比越大，混合物的两个组分就越容易分离。

Example 5-1

In a chromatographic analysis of low-molecular-weight acids, butyric acid elutes with a retention time of 7.63 min. The column's dead time is 0.31 min. Calculate the retention factor for butyric acid. The retention time for isobutyric acid is 5.98 min. What is the selectivity factor for isobutyric acid and butyric acid?

Solution

(1) $k=\frac{t_R-t_M}{t_M}=\frac{7.63\text{min}-0.31\text{min}}{0.31\text{min}}=23.6$

(2) First we must calculate the retention factor for isobutyric acid.

$k=\frac{t_R-t_M}{t_M}=\frac{5.98\text{min}-0.31\text{min}}{0.31\text{min}}=18.3$

The selectivity factor, therefore, is $\alpha=\frac{k_2}{k_1}=\frac{23.6}{18.3}=1.29$

Section 3 Fundamental Theory of Chromatography

第3节 色谱分析基本理论

1. Plate Theory

In 1941, Martin and Synge Nobel Prize (1952) proposed the plate theory when studying the chromatographic separation process. The theory takes the chromatographic column assimilate to a fractionating column, that is, the chromatographic column is composed of a series of continuous and identical horizontal plates. The height of each plate is expressed as H, which is called height equivalent to theoretical plate. The plate theory assumes that on each plate, the solute quickly reaches the distribution equilibrium between mobile and stationary phases, and then moves forward with the mobile phase in a plate by plate. For a chromatographic column with length L, number of theoretical plates is N_T, then

1. 塔板理论

1941年马丁（Martin）、辛格（Synge）在研究色谱分离过程时，提出了塔板理论（1952年获得了诺贝尔奖）。该理论是将色谱柱比作为一个精馏塔，即色谱柱是由一系列连续的、相同的水平塔板组成。每一块塔板的高度用H表示，称为理论塔板高度。塔板理论假设：在每一块塔板上，溶质在两相间很快达到分配平衡，然后随着流动相按一个一个塔板的方式向前转移。对一根长为L的色谱柱，柱内的理论塔板数为N_T为柱长与板高之比。

$$N_T=\frac{L}{H} \tag{5.15}$$

$$N_T=5.54\left(\frac{t_R}{W_{1/2}}\right)^2 \tag{5.16}$$

$$N_T=16\left(\frac{t_R}{W}\right)^2 \tag{5.17}$$

By the theory of distillation, number of theoretical plates can be obtained.

The efficiency of chromatographic columns increases as the plate number becomes greater and as the plate height becomes smaller.

根据塔板理论，可利用式（5.16）和式（5.17）求得理论塔板数，可见，当色谱柱长L一定时，H越小，N_T越多，说明组分流经的塔板数目越多，柱效能也就越高。

In practical work, the dead volume V_M (or corresponding to dead time t_M) in the column does not participate in the distribution process. The number of theoretical plates N_T calculated according to equations (5.16) and (5.17) cannot reflect the real situation of component

在实际工作中，由于柱内的死体积V_M（或对应的死时间t_M）并不参与分配过程，所以按式（5.16）和式（5.17）计算出的理论塔板数N_T并不能反映柱内组分分配的真实情况，应扣除V_M或

distribution in the column, then V_M or t_M should be deducted, so number of effective plates N_{eff}:

t_M，故有效塔板数和有效塔板高度的计算见式（5.18）和式（5.19）。

$$N_{eff}=5.54\left(\frac{t'_R}{W_{1/2}}\right)^2=16\left(\frac{t'_R}{W}\right)^2 \tag{5.18}$$

$$H_{eff}=\frac{L}{N_{eff}} \tag{5.19}$$

Typical lengths, plate numbers, and plate heights for chromatographic columns are shown in table 5-3.

色谱柱的柱长、塔板数和塔板高度见表 5-3。

Table 5-3 Typical lengths, plate numbers, and plate heights for chromatographic columns

Type of column	Length, L	Number of theoretical plates, N	Plate height, H
Gas Chromatography			
Classic packed column	2 m	3,600-4,000	0.50-0.55 mm
SCOT capillary column	15 m	15,800-27,300	0.55-0.95 mm
WCOT capillary column	30 m	43,900-480,000	0.06-0.68 mm
Liquid Chromatography			
Packed column (4.6 mm inner diameter)	10-25 cm	6,000-25,500	0.01-0.04 mm
Microbore column (1.0 mm inner diameter)	25-100 cm	18,000-100,000	0.01-0.04 mm

表 5-3 色谱柱的柱长、塔板数和塔板高度

色谱柱类型	柱长 L	理论塔板数 N	板高 H
气相色谱			
经典填充柱	2m	3,600 ～ 4,000	0.50 ～ 0.55mm
SCOT 毛细管柱	15m	15,800 ～ 27,300	0.55 ～ 0.95mm
WCOT 毛细管柱	30m	43,900 ～ 480,000	0.06 ～ 0.68mm
液相色谱			
填充柱（内径 4.6mm）	10 ～ 25cm	6,000 ～ 25,500	0.01 ～ 0.04mm
微孔柱（内径 1mm）	25 ～ 100cm	18,000 ～ 100,000	0.01 ～ 0.04 mm

Example 5-2

A solute with a retention time of 407s has a width at the base of 13.0 s on a column 12.2 m long. Find the number of plates and plate height.

Solution

(1) $N_T=16\left(\frac{t_R}{W}\right)^2=16\times\left(\frac{407s}{13s}\right)^2=1.57\times10^4$

(2) $N_T=\frac{L}{H}=\frac{12.2m}{1.57\times10^4}=0.78mm$

TEST YOURSELF The half-width of the same peak is 7.6 s. Find the plate height.
(Answer: 0.77 mm)

2. Kinetic Theory(Rate Theory)

2. 速率理论

Plate height, H, is proportional to the variance of a chromatographic band: the smaller the plate height, the narrower the band. The Van Deemter equation tells us how the column and flow rate affect plate height:

塔板高度 H 与色谱带的标准差成正比：板高越小，色谱峰越窄。范第姆特方程见式（5.20）。

$$H=A+\frac{B}{u}+Cu \tag{5.20}$$

where u is the linear velocity and A, B, and C are constants for a given column and stationary phase. Changing the column and stationary phase changes A, B, and C. The Van Deemter equation says there are band-broadening mechanisms that are proportional to linear velocity, inversely proportional to linear velocity, and independent of linear velocity（fig.5-5）. At the optimum linear velocity, the plate height H of the column is lowest, so the number of plates N_T is greatest. Below the optimum linear velocity, longitudinal diffusion broadening, B, is most significant. Above the optimum, equilibration time broadening, C, is dominant. At the optimum linear velocity, the B and C terms contribute equally to plate height.

式中，u 是线速度，A、B 和 C 是与色谱柱和固定相有关的常数。改变色谱柱和固定相，会改变 A、B 和 C。范第姆特方程表明了谱带展宽的机理：即板高既受线速度成正比项的影响，又受与线速度成反比项的影响，还受与线速度无关项的影响（图 5-5）。在最佳线速度下，色谱柱板高 H 最小，塔板数 N_T 最大。在最佳线速度以下，纵向扩散是谱带展宽的主要因素。在最佳线速度以上，传质（平衡时间）是引起谱带展宽的主要因素。在最佳线速度下，B 和 C 项对板高的贡献相等。

In packed columns, all three terms contribute to band broadening. For open tubular columns, the multiple path term, A, is 0, so bandwidth decreases and resolution increases. In capillary electrophoresis, both A and C go to 0, thereby reducing plate height to submicron values and providing extraordinary separation powers.

对于填充柱，这三项共同导致谱带展宽；对于开管式色谱柱，多径项 A 为 0，因此带宽减少，分辨率增加。对于毛细管电泳，A 和 C 都为 0，从而将板高降低到亚微米级，并提供超强的分离能力。

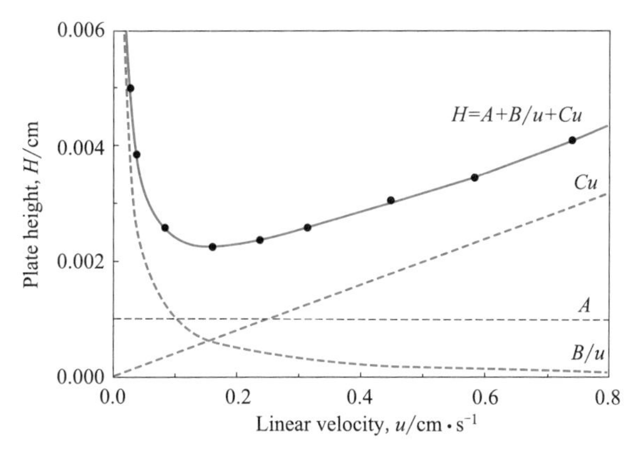

Fig.5-5 The relation curves of plate height and linear velocity

图 5-5 板高与线速度的关系曲线

(1) Multiple flow paths　The term A in the van Deemter equation (5.20) arises from multiple effects for which the theory is murky. Figure 5-6 is a pictorial explanation of one effect. Because some flow paths are longer than others, molecules entering the column at the same time on the left are eluted at different times on the right. For simplicity, we approximate many different effects by the constant A in equation (5.21). The term A was formerly called the eddy diffusion term.

（1）多径项（A）　Van Deemter 方程（5.20）中的 A 项，源于该理论很模糊的多重效应，如图 5-6。由于分子在色谱柱中走过的路径不同，在左边同时进入色谱柱的试样组分分子在右边的不同时间被洗脱。为简单起见，我们用式（5.21）中的常数 A 来近似该效应，A 项有时也被称为涡流扩散项。

$$A=2\lambda d_{p} \tag{5.21}$$

Where λ is packing non-uniformity factor, d_p is filler particle size. The smaller the stationary phase particles, the less serious this problem is. This process is absent in an open tubular column.

式中，λ 为填充不均匀因子，d_p 为固定相颗粒平均直径。固定相的颗粒越小，多径项越小。开管柱无多径项，即 A=0。

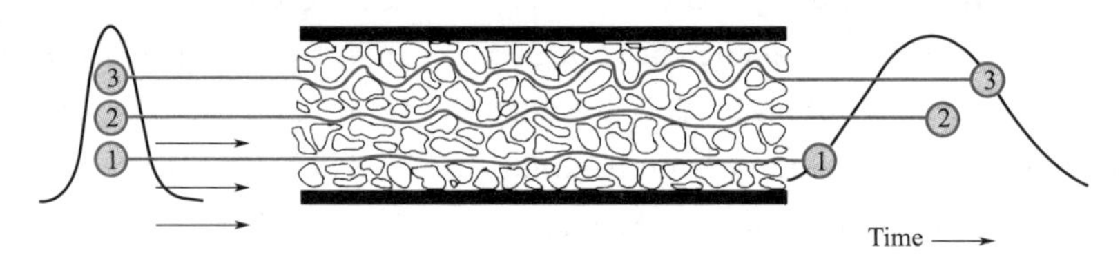

Fig.5-6　Band spreading from multiple flow paths

图 5-6　多径项示意图

(2) Longitudinal diffusion　The band would slowly broaden as molecules diffuse from the high concentration within the band to regions of lower concentration ahead of and behind the band. Diffusional broadening of a band, called longitudinal diffusion because it takes place along the axis of the column, occurs while the band is transported along the column by solvent flow.

（2）纵向扩散项（B/u）　当分子从浓度中心向流动相流动方向相同和相反的区域扩散时，形成了溶质分子超前和滞后，导致色谱带慢慢展宽，称为纵向扩散。由于其发生在色谱柱的轴向方向，故当溶剂沿色谱柱向前流动时，常产生纵向扩散。

$$\frac{B}{u}=2\gamma\frac{D_{m}}{u} \tag{5.22}$$

Where γ is a blocking factor, reflecting that the movement bends path of solute in the column and hinders molecular diffusion. D_m is coefficient of diffusion, $cm^2 \cdot s^{-1}$.

式（5.22）中 γ 为阻碍因子，反映了溶质在柱内运动路径弯曲阻碍分子扩散；D_m 为扩散系数。

The faster the linear flow, the less time is spent in the column and the less diffusional broadening occurs. Longitudinal diffusion in a gas is much faster than diffusion in a liquid, so the optimum linear velocity in gas chromatography is higher than in liquid

线性流动越快，分子在色谱柱中停留的时间就越短，纵向扩散展宽也就越小。气体中的纵向扩散比液体中的扩散要快得多，因此气相色谱法的最佳线速高于液相色谱法。开管柱不

chromatography.

存在路径弯曲，γ=1。

(3) Finite equilibration time between phases The term Cu in equation (5.20) comes from the finite time required for solute to equilibrate between mobile and stationary phases. Mass transfer is the movement of solute from one phase to another. Solute must diffuse from the mobile phase to the surface of the stationary phase for this equilibration to occur (figure 5-7). The time required depends on the distance that the solute must diffuse to get to the stationary phase, and inversely on how fast it diffuses. The faster the mobile phase velocity (u), the less time available for this transfer to occur.Plate height from finite equilibration time, also called the mass transfer term, is

（3）相间有限平衡时间（Cu） 式（5.20）中的 Cu 项源自溶质在流动相和固定相之间达到分配平衡所需的有限时间。传质是指溶质从一个相到另一个相的运动。溶质必须从流动相扩散到固定相的表面，才能实现这种平衡（图 5-7）。所需的时间取决于溶质扩散到固定相的距离，与扩散速度成反比。流动相速度（u）越快，传质的时间就越短。相间有限平衡时间又称传质阻力项，见式（5.23）。

$$H=Cu=(C_m+C_s)u \tag{5.23}$$

Where C_m describes mass transfer through the mobile phase and C_s describes mass transfer through the stationary phase. Specific equations for C_m and C_s depend on the type of chromatography and geometry of the column.

式中，C_m 为流动相传质阻力，C_s 为固定相传质阻力。C_m 和 C_s 的具体方程取决于色谱柱的类型和色谱柱的几何形状。

For gas chromatography in an open tubular column, the terms are as follows:

对于色相色谱，流动相和固定相上的传质阻力的计算式分别为式（5.24）和式（5.25）。

Mass transfer in mobile phase:

$$C_m=\frac{1+6k+11k^2}{24k(k+1)^2}\cdot\frac{r^2}{D_m} \tag{5.24}$$

Mass transfer in stationary phase:

$$C_s=\frac{2k}{3(k+1)^2}\frac{d^2}{D_s} \tag{5.25}$$

where k is the retention factor, r is the column radius, D_m is the diffusion coefficient of solute in the mobile phase, d is the thickness of stationary phase, and D_s is the diffusion coefficient of solute in the stationary phase. Figure 5-7 depicts C_m broadening. Decreasing column radius, r, reduces plate height by decreasing the distance through which solute must diffuse to reach the stationary phase. Slow diffusion of solute into and out of the stationary phase film causes C_s broadening

其中 k 为保留因子，r 为柱半径，D_m 为溶质在流动相中的扩散系数，d 为固定相的粒度，D_s 为溶质在固定相中的扩散系数。图 5-7 描绘了 C_m 峰展宽。通过减小溶质扩散到固定相的距离，减小色谱柱半径 r，可以减小塔板高度。溶质缓慢扩散进入和流出固定相液膜会导致 C_s 展宽［方程式（5.25)］。降低固定相液膜厚度 d，可以降低塔板

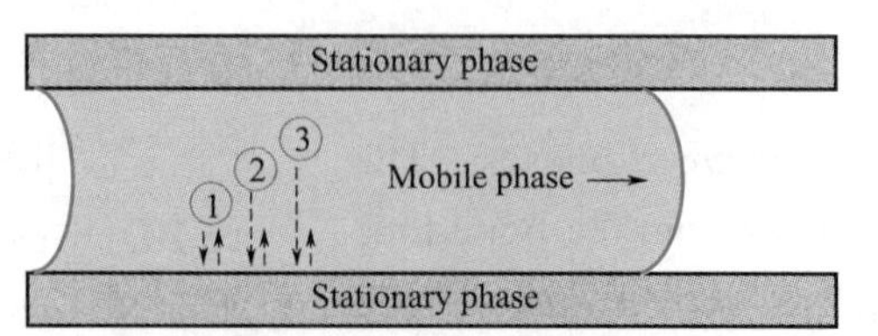

Fast mass transfer relative to mobile phase velocity

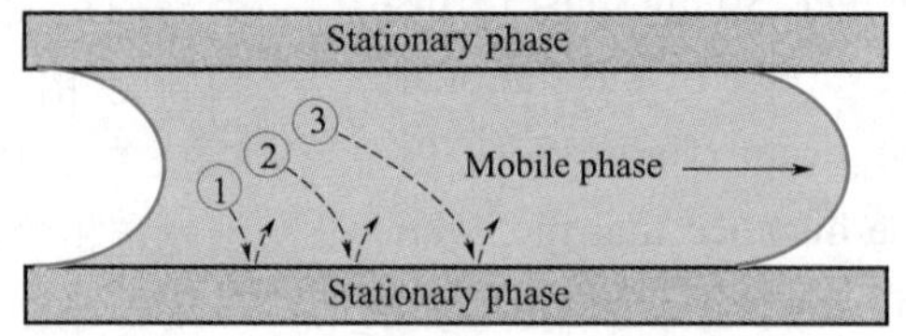

Slow mass transfer relative to mobile phase velocity

Fig.5-7 If mass transfer in the mobile phase (dashed arrows) occurs quickly, solute molecules do not experience additional broadening due to the C_m term. If mass transfer is slow relative to the mobile phase velocity, C_m broadening occurs.

图 5-7 如果流动相中的传质(虚线箭头)发生得很快，溶质分子不会因 C_m 项而展宽。如果传质速度相对于流动相速度较慢，则会发生 C_m 展宽。

[equation.(5.25)]. Decreasing stationary phase thickness, d, reduces plate height and increases efficiency because solute can diffuse faster from the depths of the stationary phase into the mobile phase.

高度并提高效率。

3. Chromatographic Resolution

3. 分离度

The goal of chromatography is to separate a sample into a series of chromatographic peaks, each of which representing a single component of the sample. Resolution is a quantitative measure of the degree of separation between two chromatographic peaks, and is defined as

色谱法的目的是将一个样品分离成一系列的色谱峰，每个峰代表样品的一个单一组分。分辨率，又称分离度，是两个色谱峰之间分离程度的定量表述，见式（5.26）。

$$R=\frac{t_{R2}-t_{R1}}{(W_2+W_1)/2}=\frac{2(t_{R2}-t_{R1})}{W_2+W_1}=\frac{2\Delta t_R}{W_2+W_1} \tag{5.26}$$

In this relationship, t_{R1}, and W_1, are the retention time and baseline width (both in the same units of time) for the first eluting peak, while t_{R2} and W_2 are the retention time and baseline width of the second peak. This produces a unitless value for R that represents the number of base-line widths that separate the centers of the two peaks. An important advantage of using the peak resolution instead of the separation factor is that

式中，t_{R1} 和 W_1 是第一个峰的保留时间和基线宽度（均以时间为单位），而 t_{R2} 和 W_2 是第二个峰的保留时间和基线宽度。R 是一个无单位的值，其既是反映柱效率又是反映选择性的综合性指标。用分辨率而不用分离因子的一个重要原因是，R 考虑了两种化合物之间的保留值差异（由 $t_{R2}-t_{R1}$ 表示）和带

R, considers both the difference in retention between two compounds (as represented by $t_{R2}-t_{R1}$), and the degree of band-broadening (as represented by W_2 and W_1). Figure 5-8 shows the overlap of two peaks with different degrees of resolution. When R=1, the degree of separation of the two peaks is 98%; When R=1.5, the degree of separation of the two peaks reaches 99.7%, which can be regarded as complete separation. For quantitative analysis, a resolution>1.5 (baseline resolution) is highly desirable. Baseline resolution means that the signal returns to zero between peaks.

展宽的程度（由 W_2 和 W_1 表示）。图 5-8 显示了具有不同分辨率的两个峰的重叠部分。当 R=1 时，两峰分离的程度为 98%；当 R=1.5 时，两峰分离的程度达 99.7%，可视为完全分离。对于定量分析，通常需要获得 1.5 以上的分辨率（基线分辨率）。

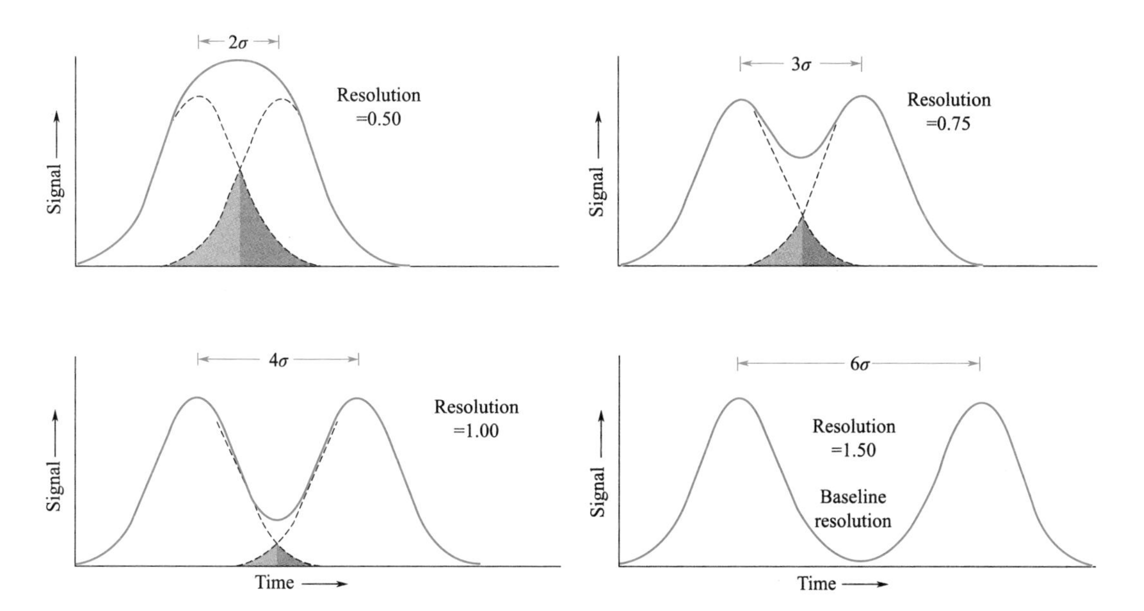

Fig.5-8 Resolution of Gaussian peaks of equal area and amplitude. Dashed lines show individual peaks, and solid lines are the sum of two peaks. Overlapping area is shaded.

图 5-8 等面积等振幅高斯峰分辨率，虚线表示单个的峰值，实线表示两个峰值的叠加。重叠区域被阴影覆盖。

Example 5-3

In a chromatographic analysis of lemon oil a peak for limonene has a retention time of 8.36 min with a baseline width of 0.96 min. γ-Terpinene elutes at 9.54 min, with a baseline width of 0.64 min. What is the resolution between the two peaks?

Solution

Using equation 10.25, we find that the resolution is

$$R=\frac{2(t_{R2}-t_{R1})}{W_2+W_1}=\frac{2\times(9.54-8.36)}{0.64+0.96}=1.48$$

4. Factors Affecting Resolution

Because the resolution between two peaks is a measure of both the difference in compound retention and band-broadening, any factors that affect retention or peak widths will also affect R. The effects these factors will have on the resolution is given by equation (5.27).

4. 分离度的影响因素

由于两个色谱峰的分离度是化合物保留值差异和谱带展宽差异的比值，因此影响保留值或谱带展宽的任何因素都会影响 R。这些因素对分离度的影响可用式（5.27）表示。

$$R=\frac{\sqrt{N_T}}{4}=\frac{\alpha-1}{\alpha}\left(\frac{k_2}{1+k_2}\right) \tag{5.27}$$

In this equation, k is the retention factor for the second peak, α is the separation factor between the first and second peaks, and N_T is the number of theoretical plates for the column being used in this separation. This relationship is called the **resolution equation** of chromatography and is simply a modified version of equation (5.26), where k, α, and N_T have been substituted in place of t_R and W. This equation is useful because it shows in a quantitative fashion that the degree of a separation in chromatography will be affected by three factors: ① the extent of band-broadening in the column (N_T), ② the overall degree of peak retention (k), and ③ the selectivity of the column's stationary phase in binding to one compound versus another (α). Figure 5-9 shows the change in resolution when these parameters are changed.

式中，k 是第二个峰的保留因子，α 是第一个峰和第二个峰之间的选择性因子，N_T 是该色谱柱的理论塔板数。该方程称为**色谱分离方程**。该方程表明，色谱的分离度受三个因素的影响：一是色谱柱中的谱带展宽程度 N_T，二是色谱峰的保留因子 k，三是色谱柱固定相与化合物结合的选择性 α。图 5-9 显示了改变这些参数时分辨率的变化。

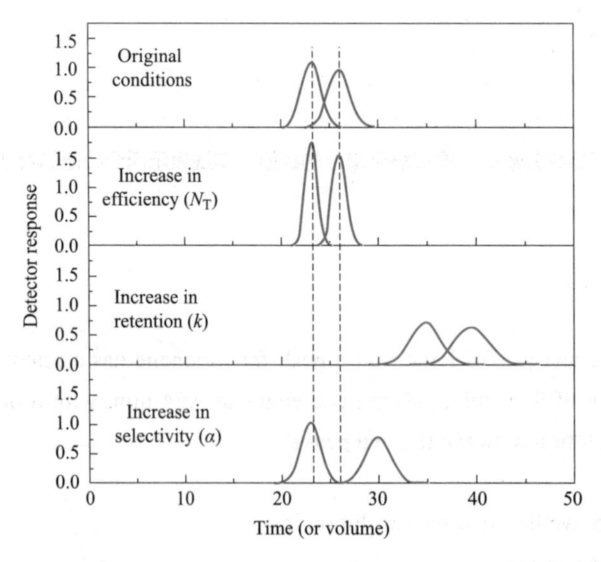

Fig.5-9 Effects of changes in the number of theoretical plates (N_T), retention factor(k), or separation factor (α) on the degree of separation that is obtained between neighboring peaks in a chromatogram

图 5-9 理论塔板数（N_T）、保留因子（k）或选择性因子（α）的变化对色谱图中相邻峰分离度的影响

Example 5-4

Two solutes have a relative retention of α=1.08 and retention factors k_1=5.0 and k_2=5.4. The number of theoretical plates is nearly the same for both compounds. How many plates are required to give a resolution of 1.5? Of 3.0? If the plate height is H=0.5 mm in gas chromatography, how long must the column be for a resolution of 1.5?

Solution

Using equation (5.27), we find that the resolution is

$$R=\frac{\sqrt{N_T}}{4}\frac{\alpha-1}{\alpha}(\frac{k_2}{1+k_2})\Rightarrow N_T=16R^2\times\left(\frac{\alpha}{\alpha-1}\right)^2\times\left(\frac{1+k_2}{k_2}\right)^2$$

If R=1.5 $\quad N_T=16\times1.5^2\times\left(\frac{1.08}{1.08-1}\right)^2\times\left(\frac{1+5.4}{5.4}\right)^2=9.2\times10^3$

If R=3.0 $\quad N_T=16\times3.0^2\times\left(\frac{1.08}{1.08-1}\right)^2\times\left(\frac{1+5.4}{5.4}\right)^2=3.68\times10^4$

To double the resolution to 3.0 requires four times as many plates =3.68×10^4 plates. For resolution =1.5, the required length is (0.5 mm/plate)(9.2×10^3 plates)= 4.6 m.

TEST YOURSELF

If α=1.08, k_2=5.4, and H= 6 μm in liquid chromatography, what length of column in cm gives a resolution of 1.5? (Answer: 5.5 cm)

5. Quantitative Analytical Method

(1) Area normalization method Suppose there are n components in the sample, the mass of each component is m_1, m_2,⋯, m_n respectively, the total mass m is 100%, if all the components in a simple can be eluted and the elution peaks are recorded, then the mass fraction of component i can be calculated according to the following equation:

5. 色谱的定量分析方法

（1）面积归一化法 假设试样中有n个组分，每个组分的质量分别为m_1，m_2，⋯，m_n，各组分质量的总和为m，且样品中所有组分都能在此色谱条件下出样，那么其中组分i的质量分数可按式（5.28）计算。

$$\begin{aligned}w_i&=\frac{m_i}{m}\times100\%\\&=\frac{m_i}{m_1+m_2+\cdots m_i+\cdots m_n}\times100\%\\&=\frac{A_if_i}{A_1f_1+A_2f_2+\cdots A_if_i+\cdots A_nf_n}\times100\%\end{aligned}\qquad(5.28)$$

Quality inspection of industrial tert butyl alcohol (normalization method)

工业叔丁醇质量检验（归一化法）

If the quality calibration factor (f) of each component is similar or same, for example the boiling points of homologs are close to each other, then above equation can be simplified into:

若各组分的质量校正因子f_i值近似或相同，例如同系物中沸点接近的各组分，则式（5.28）可简化为式（5.29）。

$$w_i=\frac{A_i}{A_1+A_2+\cdots A_i+\cdots A_n}\times 100\%=\frac{A_i}{\sum_{i=1}^{n}A_i}\times 100\% \quad (5.29)$$

The advantages of this method are simple and accurate. When the operation conditions change (such as sample amount or flow rate), the analytical result is slightly influenced.

该法优点是：简便、准确。当操作条件（如进样量、流速等）变化时，对结果影响小。

Example 5-5

The content of each component in the mixture of benzene, toluene, ethylene and xylene was analyzed by normalization method, and the peak height and the calibration factor of each component were measured in the table below. Calculate the content of each component in the sample.

component	benzene	toluene	ethylene	xylene
A/cm^2	103.8	119.0	66.8	44.0
f_i	1.00	1.99	4.16	5.21

Solution

The peak height replaced the peak area and peak height normalized:

$$w_i=\frac{A_i f_i}{\sum_{i=1}^{n}A_i f_i}\times 100\%$$

$$w_{benzene}=\frac{103.8\times 1.00}{103.8\times 1.00+119.0\times 1.99+66.8\times 4.16+44.0\times 5.21}\times 100\%=\frac{103.8}{848}\times 100\%=12.2\%$$

$$w_{toluene}=\frac{119.0\times 1.99}{848}\times 100\%=27.9\%$$

$$w_{thylene}=\frac{66.8\times 4.16}{848}\times 100\%=32.8\%$$

$$w_{xylene}=\frac{44.0\times 5.21}{848}\times 100\%=27.0\%$$

(2) Internal standard method The internal standard method uses a known amount of pure substance

（2）内标法 内标法是将一定量的纯物质作为内标物，加入准确称取的试

as internal standard substance and adds it into the accurately weighed sample. On the basis of the peak area ratio between the amount of analyte and the amount of internal standard substance present in the chromatogram, the concentration of the analyte in the original sample can be determined. For example, the mass fraction of component i (mass m_i) in the analyte sample is w_i, the mass of internal standard substance added into sample is m_s, the mass of sample is m, then:

样中，根据被测物和内标物的质量及其在色谱图上相应的峰面积比，求出某组分的含量。例如要测定试样中组分 i（质量为 m_i）的质量分数 w_i，可于试样中加入质量为 m_s 的内标物，试样质量为 m，则组分 i 的质量分数求解公式见式（5.30）。

Determination of toluene in industrail waste water (internal standard method)
工业废水中甲苯含量测定（内标法）

$$m_i=f_iA_i \qquad m_s=f_sA_s$$
$$\frac{m_i}{m_s}=\frac{f_iA_i}{f_sA_s} \longrightarrow m_i=\frac{f_iA_i}{f_sA_s}\times m_s$$
$$w_i=\frac{m_i}{m}\times 100\%=\frac{f_iA_i}{f_sA_s}\times\frac{m_s}{m}\times 100\% \tag{5.30}$$

Generally, the internal standard is used as the benchmark, f_s=1.

一般，内标物被作为基准，f_s=1。

Example 5-6

The contents of methane chloride, dichloromethane, and trichlormethane in the samples were determined by gas chromatography. Toluene was used as an internal standard, and the peak area and the calibration factor of each component were measured in the table below. The quality of the sample and toluene were 2.880 g and 0.2400 g, respectively. Calculate the content of each component in the sample.

component	toluene	methane chloride	dichloromethane	trichlormethane
A/cm^2	2.16	1.48	2.34	2.64
f_i	1.00	1.15	1.47	1.65

Solution

By $w_i=\frac{A_i}{A_s}\times\frac{m_s}{m}\times f_i\times 100\%$, then

$$w_{C_2H_5Cl}=\frac{1.48}{2.16}\times\frac{0.2400}{2.880}\times1.15\times100\%=6.57\%$$

$$w_{C_2H_4Cl_2}=\frac{2.34}{2.16}\times\frac{0.2400}{2.880}\times1.47\times100\%=13.27\%$$

$$w_{C_2H_3Cl_3}=\frac{2.64}{2.16}\times\frac{0.2400}{2.880}\times1.65\times100\%=16.80\%$$

(3) External standard method The external standard is a comparison method and is one of the most widely used methods in instrument analysis. In contrast with the internal standard, the standard substance is not added into the analyte sample in the external standard, rather, it is determined under the same chromatographic conditions as analyte sample. The concentration of the analyte can be obtained by comparing the peak areas of the standard substance and the analyte. Since the peak area (or peak height) is proportional to the concentration of the component, that is

（3）外标法　外标法是一种比较方法，是仪器分析中应用最广泛的方法之一。与内标法相比，外标法不把标准物质加入被测样品中，而是在与被测样品相同的色谱条件下单独测定，通过比较标准物质和被分析物的峰面积，可以得到被分析物质的浓度。

$$c_i=kA_i \qquad c_s=kA_s$$

Then,

$$c_i=c_s\frac{A_i}{A_s} \tag{5.31}$$

Where c_i, and c_s, are concentrations of the unknown and the external standard respectively;A_i, and A_s, are the corresponding areas.

式中，c_i、c_s 为被测物与外标物浓度，A_i、A_s 是对应面积。

The external standard substance is the same substance as analyte but with higher purity. In analysis, the concentration of the standard substance should be close to that of analyte to improve the accuracy of quantitative analysis. The error usually comes from the sample injection error, so the area repeatability (i.e. sample injection repeatability) experiment must be done before analysis.

外标物是与分析物是相同的物质，但纯度较高。分析时，标准物质的浓度应与被测物的浓度接近，以提高定量分析的准确性。误差通常来自于进样误差，因此在分析前必须进行面积重复性（即进样重复性）实验。

External standard method is usually used in routine test. Pure component I is selected to be external standard, from which a series of standard solutions are prepared. Based on the chromatographic data obtained, a

常规试验通常采用外标法。选择纯组分 I 作为外标物，并制备一系列标准溶液。根据所获得的色谱数据，绘制校准曲线，从校准曲线中也可以简单地求

calibration curve is plotted. Concentration of component I in the sample is also simply found out from the calibration curve.

出样品中组分 I 的浓度。

Section 4 Gas Chromatography

第 4 节 气相色谱法

1. Principle of GC

There are two phases in GC, one is stationary, called stationary phase; another is moving through the stationary phase, called mobile phase. Chromatography is a technique in which the components of a mixture are separated based upon the rates at which they are carried through a stationary phase by a gaseous or liquid mobile phase.

The mobile phase (gas, liquid) containing sample is driven by an external force to pass through the surface of the stationary phase which is fixed in a column or on a plate and does not dissolve with the mobile phase. When the mixture contained in the mobile phase is carried through the stationary phase, the various constituents of the mixture interact with the stationary phase. Since these constituents are different in property, structure, as well as the interaction force and strength with the stationary phase, along with the travel of mobile phase, the mixture is repeatedly distributed between two phases till balance and each constituent is held by the stationary phase for a different period of time, thus eluting from the stationary phase in a certain order. Combining with proper test methods downstream the column, the separation and detection of all the constituents are realized.

(1) Characteristics of GC

① High efficiency　In the packed column, the number of theoretical plates can be several thousands and capillary column one million.

1. 气相色谱法原理

在气相色谱法中存在两相，一相是固定不动的，叫固定相；另一相则不断流过固定相，叫流动相。色谱法的分离原理就是利用混合物中待分离的各组分在两相中的迁移速率不同来进行分离的。

使用外力使含有样品的流动相（气体、液体）通过一固定于柱中或平板上、与流动相互不相溶的固定相表面。当流动相中携带的混合物流经固定相时，混合物中的各组分与固定相发生相互作用。由于混合物中各组分在性质和结构上的差异，与固定相之间产生的作用力的大小、强弱不同，随着流动相的移动，混合物在两相间经过反复多次的分配平衡，使得各组分被固定相保留的时间不同，从而按一定次序由固定相中先后流出。与适当的柱后检测方法结合，实现混合物中各组分的分离与检测。

（1）气相色谱法特点

① 高效能　一般填充柱的理论塔板数可达数千，毛细管柱可达一百多万。

② High selectivity The substances with close partition coefficients or extremely complicated and hard to separated can be satisfactorily separated.

③ High sensitivity It can be used to identify substances of 10^{-11}-10^{-13} g, suitable for trace analysis.

④ Rapid analysis The analysis can be done within a few minutes or tens of minutes.

⑤ Wide application It can be applied for analysis of organic or inorganic substances with boiling points less than 400℃. The amount of analyte organics occupies around 20% of the total organics (about three million kinds).

⑥ Shortcomings Poor qualitative ability of the separated components.

(2) Applications of GC The gas chromatography is used for the analysis of most of the raw materials and products in petrochemical industry, for instance, the determination of the benzene and toluene content in motor gasoline and aviation gasoline; the measurement of the composition of hydrocarbon monomer in naphtha and the composition of cracked gas.

② 高选择性 可以使一些分配系数很接近的以及极为复杂、难以分离的物质，获得满意的分离。

③ 高灵敏度 可以检测 10^{-11} ～ 10^{-13}g 物质，适合于痕量分析。

④ 分析速度快 一个试样的分析可在几分钟到几十分钟内完成。

⑤ 应用广泛 可以用于沸点低于400℃的有机或无机物的分析。分析的有机物，约占全部有机物（约300万种）的 20%。

⑥ 不足之处 对被分离组分的定性能力较差。

（2）气相色谱法应用 在石油化学工业中大部分的原料和产品都可采用气相色谱法来分析。如车用汽油和航空汽油中苯及甲苯含量测定；石脑油中单体烃组成测定法；裂解气组成的测定等。

2. Components of A Gas Chromatograph (Fig.5-10)

The carrier gas flows out from the high pressure steel cylinder. After being de-pressurized to the wanted pressure by a release valve, it is purified by the purifying dryer, stabilized by a pressure maintaining valve and a rotameter, then the carrier gas flows at stable pressure and constant speed to vaporizer to mix with the vaporized sample. The mixture is separated in a chromatographic column. The separated constituents go into a detecter with the carrier gas, then the carrier gas is vented. The detector converts the change of the concentration or mass into electrical signals which are amplified and recorded on the recorder, an elution profile is plotted hereby. According to the retention time of each peak on the elution profile, a qualitative analysis can be made; according to the peak area and peak height,

2. 气相色谱仪组成（图 5-10）

载气由高压钢瓶中流出，经减压阀降压到所需压力后，通过净化干燥管使载气净化，再经稳压阀和转子流量计后，以稳定的压力、恒定的速度流经汽化室与汽化的样品混合，将样品气体带入色谱柱中进行分离。分离后的各组分随着载气先后流入检测器，然后将载气放空。检测器将物质的浓度或质量的变化转变为一定的电信号，经放大后在记录仪上记录下来，就得到色谱流出曲线。根据色谱流出曲线上得到的每个峰的保留时间，可以进行定性分析，根据峰面积或峰高的大小，可以进行定量分析。Agilent 8890 气相色谱仪见图 5-11。

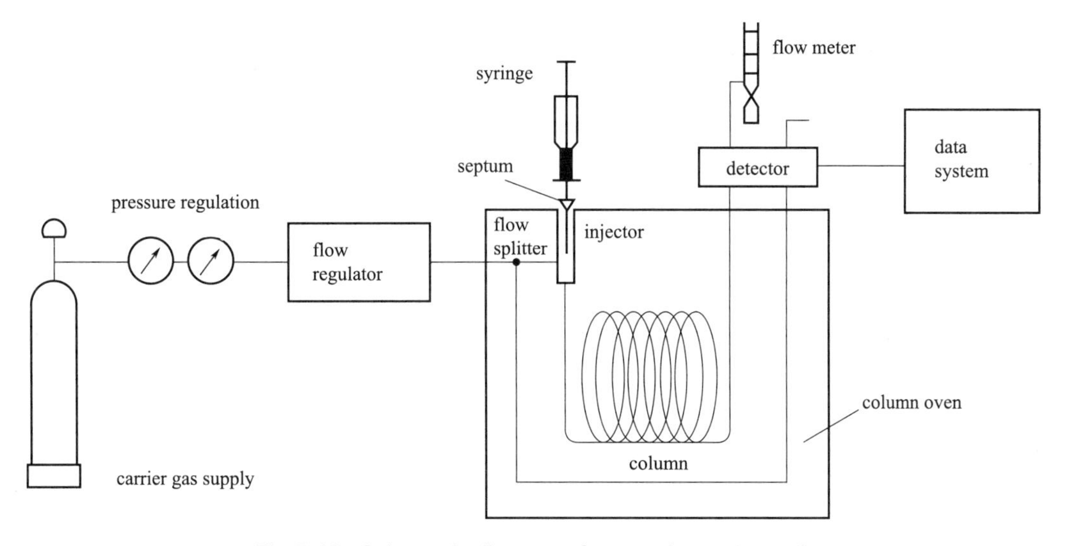

Fig.5-10 Schematic diagram of a gas chromatograph

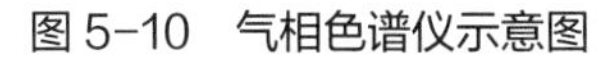

图 5-10 气相色谱仪示意图

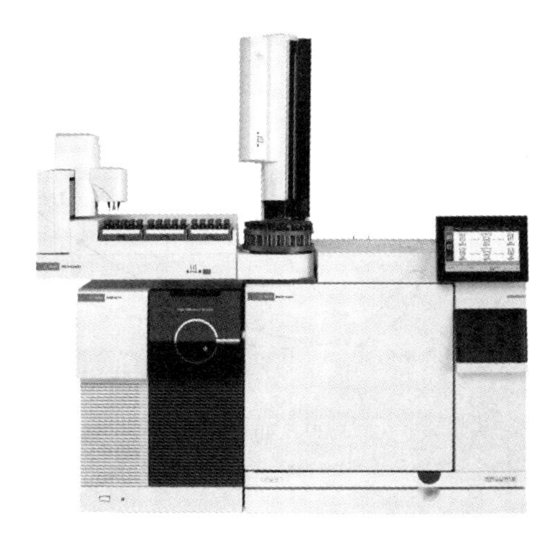

Fig.5-11 Agilent 8890 gas chromatograph

图 5-11 Agilent 8890 气相色谱仪

a quantitative analysis can be made. Fig.5-11 shows the Agilent 8890 gas chromatograph.

(1) Carrier gas system　It is used to obtain pure and stable flow carrier gas, including gas source, gas purifier, gas supply control valve and instruments.

① Requirements for carrier gas　The carrier gas should be chemically stable, highly pure, low in price, easily got and suitable for the detector.

② Commonly used carrier gas　Table 5-4 shows commonly used carrier gas. Helium is the most common carrier gas and is compatible with most

（1）载气系统　获得纯净、流速稳定的载气。包括气源、气体净化器、供气控制阀门和仪表。

① 载气要求　作为气相色谱载气的气体，要求要化学稳定性好；纯度高；价格便宜并易取得；能适合于所用的检测器。

② 常用载气　常用的载气有氮气（黑色钢瓶）、氦气（银灰色钢瓶）、氩

Operation of nitrogen high pressure cylinder and pressure reducing valve

氮气高压钢瓶与减压阀的操作

Operation of hydrogen high pressure cylinder and pressure reducing valve

氢气高压钢瓶与减压阀的操作

Table 5-4 Types of commonly used carrier gas

Carrier gas	Nitrogen	Helium	Argon	Hydrogen	Carbon dioxide
Color of cylinder	black	silver grey	white	dark green	black

表 5-4 常用载气种类

载气	氮气	氦气	氩气	氢气	二氧化碳
气瓶颜色	黑色	银灰色	白色	深绿色	黑色

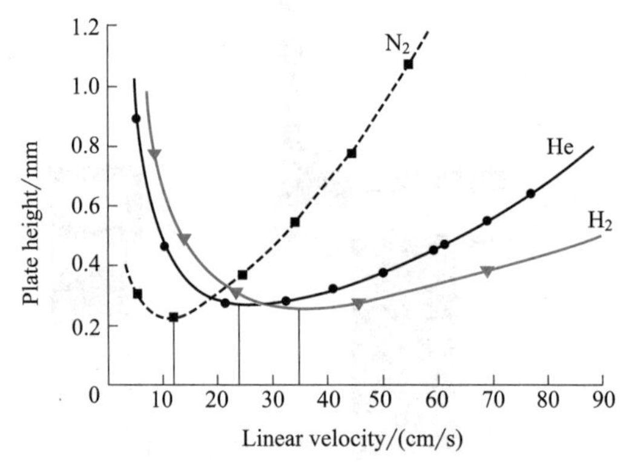

Fig.5-12 Van Deemter curves for gas chromatography of n-$C_{17}H_{36}$ at 175℃, using N_2, He, or H_2 in a 25m×0.25mm wall-coated column with nonpolar OV-101 stationary phase

图 5-12 175℃下，在装有非极性 OV-101 固定相的 25m × 0.25mm 涂壁空心柱中，用 N_2、He 或 H_2 作载气，n-$C_{17}H_{36}$ 的气相色谱法 Van Deemter 曲线

detectors. For a flame ionization detector, N_2 gives a lower detection limit than He. Figure 5-12 shows that H_2, He, and N_2 give essentially the same optimal plate height (0.3 mm) at significantly different velocities. Optimal velocity increases in the order N_2, He, H_2. Fastest separations can be achieved with H_2 as carrier gas, and H_2 can be run much faster than its optimal velocity with little penalty in resolution. Figure 5-13 shows the effect of carrier gas on the separation of two compounds on the same column with the same temperature program.

气（白色钢瓶）、氢气（深绿色钢瓶）、二氧化碳（黑色钢瓶）等，见表 5-4。其中，氮气是最常见的载气，其与大多数检测器兼容。对于火焰离子化检测器，N_2 的检测限低于 He。图 5-12 显示，H_2、He 和 N_2 在不同的速度下最佳塔板高度（0.3mm）基本相同。最佳流速的由小到大的顺序为 N_2、He、H_2。可见，使用 H_2 作为载气可以实现最快分离。图 5-13 显示了载气在相同温控程序下对同一色谱柱上分离的两种化合物的影响。

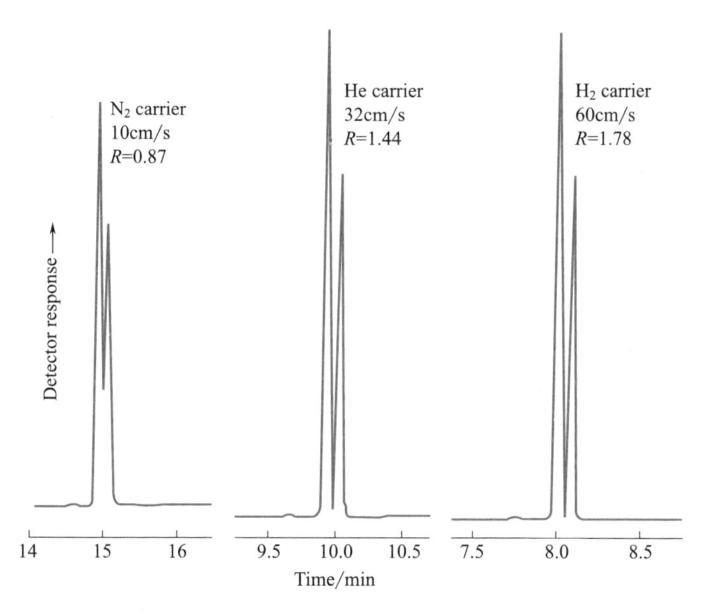

Fig.5-13 Separation of two polyaromatic hydrocarbons on a wall-coated open tubular column with different carrier gases(Resolution, *R*, increases and analysis time decreases as we change from N_2 to He to H_2 carrier gas.)

图 5-13 用不同载气在涂壁开管柱上分离两种多环芳烃（当从 N_2 变为 He 再变为 H_2 时，分离度 *R* 逐渐增大，分析时间逐渐减少。）

③ Purification of gas The purifier is usually metal tube with diameter 50 mm, length 200-250 mm. The type of packings inside the tube depends on the requirements for the purification of the carrier gas (for example, adsorbent silica gel and molecular sieve are used to remove moisture; active carbon is used to remove hydrocarbons.) The inlet and outlet of the purifier should be marked. The outlet should be plugged with some gauze or adsorbent cotton to prevent the depurative powder flowing from the purifier to the chromatograph.

③ 气体的净化 净化管通常为内径 50mm，长 200 ～ 250mm 的金属管。装填什么物质取决于载气纯度的要求（如除去水分可用吸附剂硅胶和分子筛，除去烃类化合物可用活性炭）。净化管的出口和入口应加上标志，出口应当用少量纱布或脱脂棉轻轻塞上，严防净化剂粉尘流出净化管进入色谱仪。

④ Flow controller Usually, there is a pressure maintaining valve downstream of the release valve to stabilize the pressure of the carrier gas (or fuel gas). When the temperature programming is used in chromatographic analysis, the rise of temperature column causes the increase of column resistance, leading to the change of flow rate of carrier gas. In order to maintain a stable carrier gas flow when the column resistance changes, a pressure maintaining valve is utilized to automatically control the flow rate of the carrier gas.

④ 气流调节阀 通常在减压阀输出气体的管线中还要串联稳压阀，用以稳定载气（或燃气）的压力。当用程序升温进行色谱分析时，由于色谱柱柱温不断升高引起色谱柱阻力不断增加，也会使载气流量发生变化。为了在气体阻力发生变化时，也能维持载气流速的稳定，需要使用稳流阀来自动控制载气的稳定流速。

(2) Sample injection system It includes an injector and a vaporizer. The latter is an apparatus used to instantly vaporize the liquid sample. Requirements for injecting: rapid injection, proper quantity of sample, simple and practicable method.

（2）进样系统 包括进样器、汽化室。汽化室是将液体试样瞬间汽化的装置。对进样的要求：进样速度快，进样量要适当，进样方法简单易行。

Column efficiency requires that the sample be of a suitable size and be introduced as a plug of vapor; slow injection or oversized samples cause band spreading and poor resolution. Calibrated microsyringes are used to inject liquid samples through a rubber or silicone septum into a head sample port located at the head of the column. The sample port is ordinarily about 50 ℃ above the boiling point of the least volatile component of the sample, shown in figure 5-14.

柱效要求样品的尺寸要合适，并作为蒸汽引入柱体；缓慢地注入或超大样品都会导致谱带展宽和分辨率变差。校准的微注射器通过橡胶或硅胶隔膜将液体样品注入位于柱头的头部样品端口。样品端口通常比样品中挥发性最小的组分的沸点高出约 50℃，如图 5-14 所示。

Gas circuit installation and leak detection
气路安装与检漏

Measurement of gas flow
气体流量的测定（皂膜流量计）

Operation of sample injection
气相色谱仪进样操作

Replacement of silicone pad
硅胶垫的更换

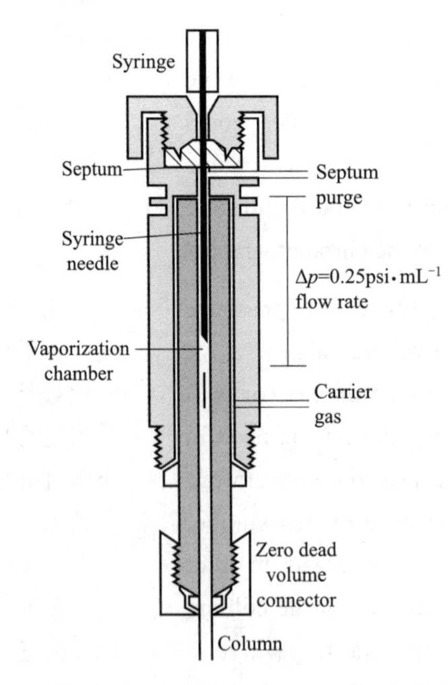

Fig.5-14 Typical gas injector system(1psi=6.895kPa)

图 5-14 典型的气体进样系统

① Gas injector (six-port switching valve, fig.5-15) The sample is firstly filled into a collection bottle. Upon switching, the carrier gas takes the sample along into the separation column.

① 气体进样器（六通阀，图5-15） 试样首先充满定量管，切换后，载气携带定量管中的试样气体进入分离柱。

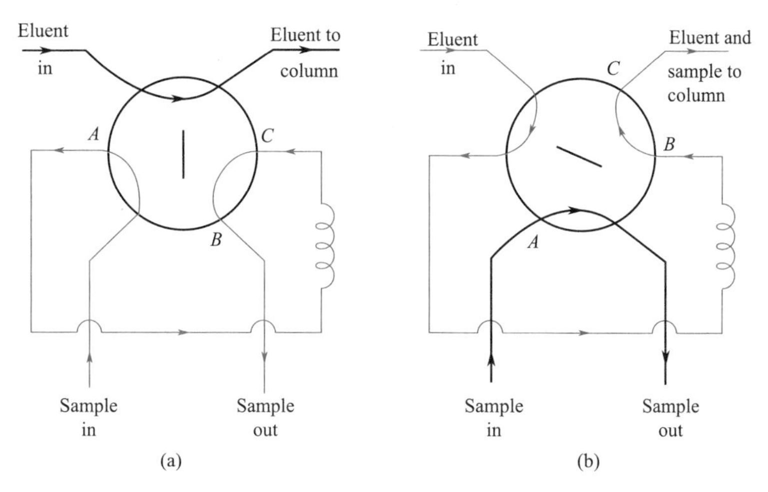

Fig.5-15 Six-port switching valve

图5-15 六通阀气体进样器

② Liquid injector (fig.5-16) The liquid samplers are special syringes of different sizes. The packed column uses 10μL syringe, while the capillary column uses 1μL syringe. The new model liquid sampler is fully automatic. The cleaning, rinsing, sampling, injecting and sample changing are done automatically. At most several tens of samples can be injected.

② 液体进样器（图5-16） 不同规格的专用注射器，填充柱色谱常用10μL；毛细管色谱常用1μL；新型仪器带有全自动液体进样器，清洗、润冲、取样、进样、换样等过程自动完成，一次可放置数十个试样。

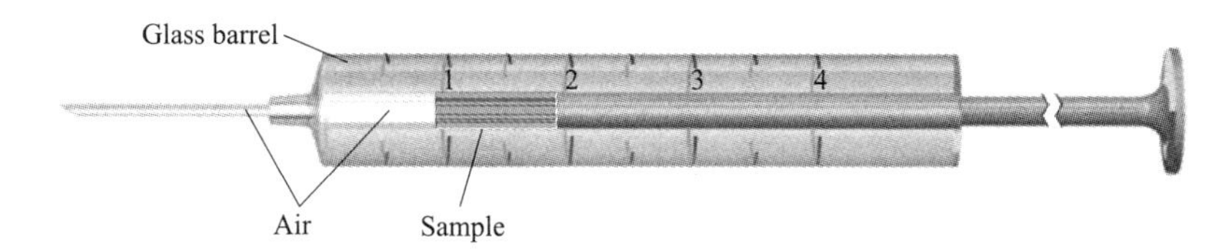

Fig.5-16 Microsyringe filled with 1μL of liquid sample.(Air is then drawn into the barrel to avoid premature evaporation of sample in the injection port.)

图5-16 装有1μL液体样品的微量注射器（吸入定量液体后再吸入一定量的空气，以免样品在进样口过早蒸发。）

(3) The separation system It includes the column which is the key component of the chromatograph with function of separating the sample.

（3）分离系统 色谱柱是色谱仪的核心部件，其作用是分离样品。

The chromatographic column is generalized into two types: one is filled with stationary phase, called packed column, usually an U-shaped or spiral metal

色谱柱一般分两种：一种是内装固定相的，称为填充柱，通常为用金属（铜或不锈钢）或玻璃制成的内径

(copper or stainless steel) or glass tube with diameter 2-6 mm, length 0.5-10 m; another is a hollow column coated on the inside surface with a liquid stationary phase called capillary column with diameter 0.1-0.5 μm, length 10-1000 m. The most common type of gas chromatography column is the wall-coated column in figure 5-17(c), with a 0.1 to 5 μm-thick film of high-molecular-weight stationary liquid phase coated on the inner wall. The less common porous-layer column has solid particles of stationary phase on the inner wall. Column inner diameters are typically 0.10-0.53 mm and lengths are 15-100 m, with 30 m being common. Columns are coiled [figure 5-17(b)]to fit within a compact temperaturecontrolled column oven. Narrow columns provide higher resolution than wider columns, shown in figure 5-18, but require higher operating pressure and have less sample capacity. The selection principle of a liquid stationary phase includes similar and dissolvable, similar function groups and main differences in component properties (e.g. difference in boiling point, difference in polarity).

2～6mm，长0.5～10m的U形或螺旋形的管子；另一种是将固定液均匀地涂敷在毛细管的内壁上形成中空的柱子，称为毛细管柱，长10～1000m，内径0.1～0.5μm。最常见的气相色谱柱类型是图5-17(c)中的涂壁开管柱（WCOT），其内壁涂有0.1至5μm厚的高分子量固定液膜。而不常见的多孔层开管柱（PLOT），其内壁上涂的是多孔性吸附剂固体微粒。这种色谱柱内径通常为0.10～0.53mm，长度为15～100m，一般为30m。色谱柱被卷绕[图5-17(b)]，以适应紧凑的温控箱。窄柱比宽柱可提供更高的分辨率（图5-18），但需要更高的工作压力，且样品容量较小。固定液的选择原则主要包括相似相溶原则、化学官能团相似原则及组分性质的主要差别（沸点差别；极性差别）。

Installation of capillary column

气相色谱仪毛细管色谱柱的安装

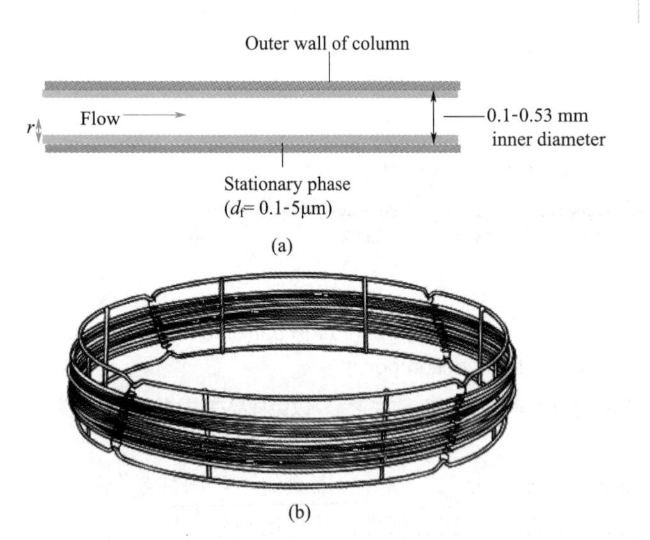

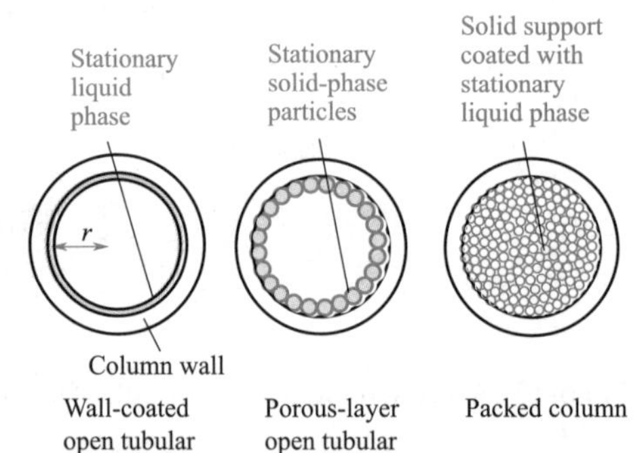

Fig.5-17 (a) Typical dimensions of open tubular gas chromatography column; (b) Fused-silica column with a cage diameter of 0.2m and column length of 15-100m; (c) Cross-sectional views of wall-coated, porous-layer, and packed columns

图5-17 （a）开管气相色谱柱的典型尺寸；（b）熔融石英柱直径0.2m，柱长15-100m；（c）涂壁空心柱、多孔空心柱和填充柱的横截面图

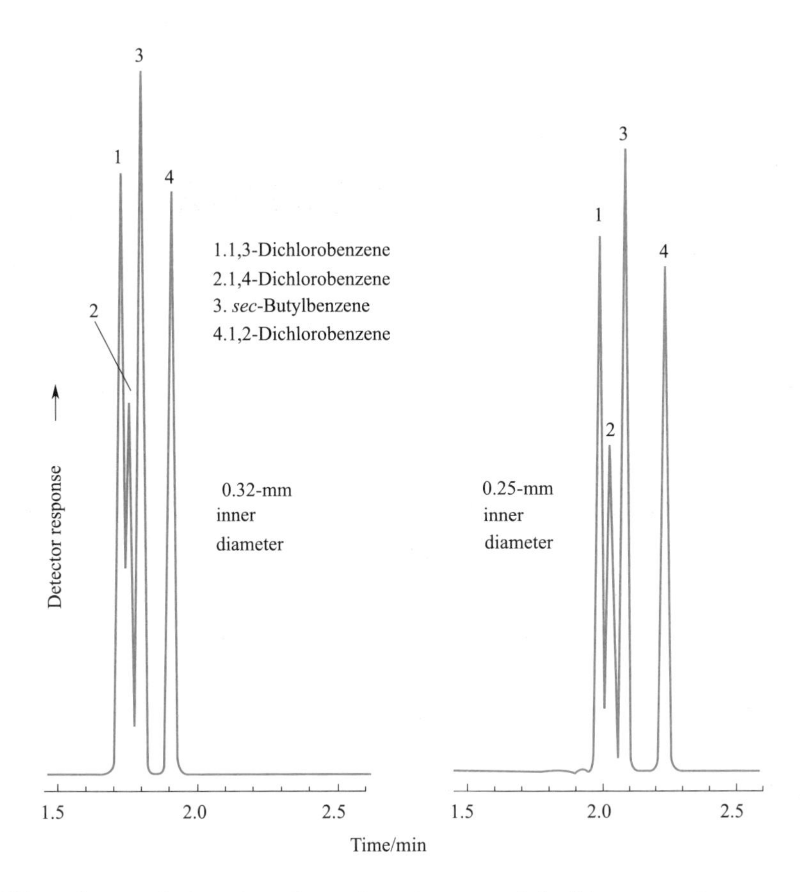

Fig.5-18 Effect of open tubular column inner diameter on resolution[Narrower columns provide higher resolution. Notice the increased resolution of peaks 1 and 2 in the narrow column. Conditions: DB-1 stationary phase (0.25 mm thick) in 15m-wall-coated column operated at 95℃ with He linear velocity of 34 cm/s.]

图 5-18 柱内径对分离度的影响[色谱柱内径越小，分离度越高（如图中峰 1 和峰 2 的分离度增加）。条件：DB-1 固定相（0.25mm），15m 长涂壁空心柱，在 95℃下操作，He 线速度为 34cm/s。]

(4) Detecting, amplifying and recording system It includes a detector, an amplifier and a recorder.

Detector is used to convert the component signals into electrical signals.

Requirements: rapid response, high sensitivity, good stability and wide linearity range.

(5) Temperature control system The temperature control system sets, controls and measures the temperatures of column oven, vaporizer and detector. The column temperature is an important operation parameter that directly affects separation efficiency and analysis speed, including constant temperature and temperature programming.

（4）检测和放大记录系统 包括检测元件、放大器、显示记录。

检测器：把组分信号转换成电信号。

要求：响应快，灵敏度高，稳定性好，线性范围广。

（5）温控系统 用于设定、控制、测量色谱柱箱、气化室和检测器三处的温度。柱温是一个重要的色谱操作参数，它直接影响分离效能和分析速度。柱温分为恒温和程序升温两种。

Section 5 Chromatographic Detectors

第5节 气相色谱检测器

1. Type of Detectors

According to the different response principle, the chromatographic detectors are classified into two categories: concentration detector and mass detector, shown in table 5-5. The concentration detector measures the instantaneous change of a component's concentration, that is, the response of the detector is directly proportional to the instantaneous concentration of the component, for instance, thermal conductivity detector (TCD) and electron capture detector (ECD). The mass detector measures the mass ratio of a component in the carrier gas, that is, the response of the detector is directly proportional to the mass of the component entering into the detector in unit time, for instance, flame ionization detector (FID) and flame photometric detector (FPD).

1. 检测器的类型

气相色谱检测器根据响应原理的不同可分为浓度型检测器和质量型检测器两类，见表5-5。浓度型检测器：测量的是载气中某组分瞬间浓度的变化，即检测器的响应值和组分的瞬间浓度成正比。如热导池检测器（TCD）和电子捕获检测器（ECD）。质量型检测器：测量的是载气中某组分质量比率的变化，即检测器的响应值和单位时间进入检测器的组分质量成正比。如氢火焰离子化检测器（FID）和火焰光度检测器（FPD）。

Table 5-5 Properties of common gas chromatography detectors

Detector Name	Compounds Detected	Detection Limits
General detectors		
Thermal conductivity detector (TCD)	Universal—all compounds	10^{-9} g
Flame ionization detector (FID)	All organic compounds	10^{-12} g carbon
Selective detectors		
Nitrogen-phosphorus detector (NPD)	Nitrogen and phosphorus-containing compounds	$10^{-14}\sim10^{-13}$ g nitrogen $10^{-14}\sim10^{-13}$ g phosphorus
Electron capture detector (ECD)	Compounds with electronegative groups	$10^{-15}\sim10^{-13}$ g
Structure-specific detectors		
Mass spectrometry	Universal—full-scan mode Selective—SIM mode	$10^{-10}\sim10^{-9}$ g (full-scan mode) $10^{-12}\sim10^{-11}$ g (SIM mode)

(1) TCD (fig.5-19)

Principle: TCD relies on the different thermal conductivity of different substances. As the sample

（1）热导检测器（图5-19）

原理：利用不同的物质具有不同的导热系数。气流中样品浓度发生变化，

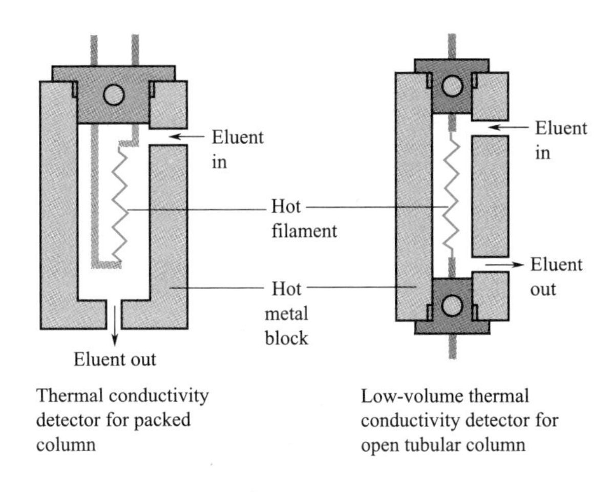

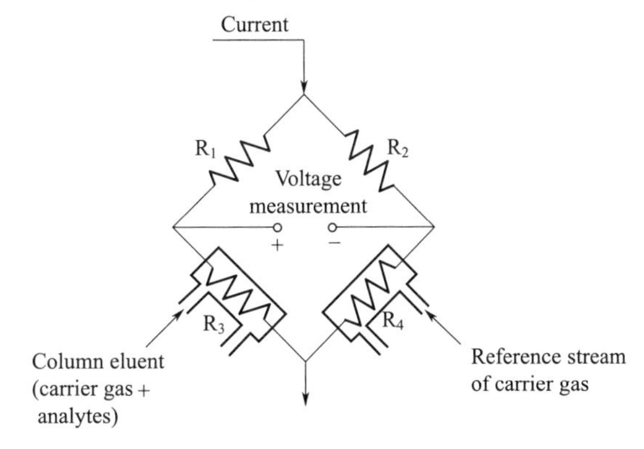

Fig.5-19 Thermal conductivity detectors and a simple Wheatstone bridge circuit used in TCD

图 5-19 TCD 热导检测器和惠斯顿电桥

concentration changes in the gas stream, the amount of heat carried off the sensitive element changes as well, resulting in the fluctuation of resistivity. Since the sensitive element composes the arm of Wheatstone bridge, as long as the resistance of an arm in the bridge circuit changes, signals will be sent out from the circuit immediately.

则从热敏元件上所带走的热量也就不同，从而改变热敏元件的电阻值，由于热敏元件为组成惠斯顿电桥之臂，只要桥路中任何一臂电阻发生变化，则整个线路就立即有信号输出。

Characteristics: this detector has response to almost all the volatile organic and inorganic substances, but its sensitivity is relatively low. The concentration of the analyte should not be lower than one ten-thousandths. It is non-destructive.

特点：此检测器几乎对所有可挥发的有机和无机物质均能响应。但灵敏度较低，被测样品的浓度不得低于万分之一，属非破坏性检测器。

(2) FID (fig.5-20) FID is mainly used for the detection of organic compounds (e.g. hydrocarbons) that are flammable in H_2-Air flame.

Principle: the carbon containing organic compounds produce fragment ions when pyrolyzed by the hydrogen/

（2）氢火焰离子化检测器（图 5-20）氢火焰离子化检测器主要用于可在 H_2-Air 火焰中燃烧的有机化合物（如烃类物质）的检测。其原理是，含碳有机物在 H_2-Air 火焰中燃烧产生碎片离子，

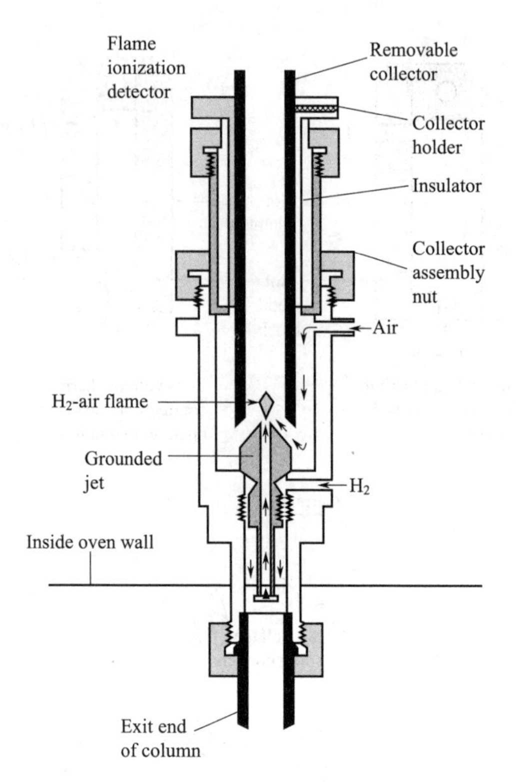

Fig.5-20 Principle structure of a flame-ionization detector

图 5-20 氢火焰离子化检测器的基本结构

air flame, which form ion current. The separated components by chromatographic column are detected according the signals generated by the ion current.

Characteristics: it has small size, high sensitivity, low dead volume and short response time, but it is only applicable to compounds which contain C-C or C-H bondings and has no response to the substances such as H_2, O_2, N_2, CO, CO_2, NO, NO_2, CS_2 and H_2O. It is destructive detector.

在电场作用下形成离子流，根据离子流产生的电信号强度，检测被色谱柱分离的组分。该检测器的特点是体积小、灵敏度高、死体积小、响应时间快，但只适用于测定含碳 - 碳或碳 - 氢键的有机化合物，对部分物质如 H_2、O_2、N_2、CO、CO_2、NO、NO_2、CS_2、H_2O 等无响应，其属于破坏性检测器。

$$C_nH_m \longrightarrow CH\cdot + CH_2\cdot + CH_3\cdot$$
$$2CH\cdot + O_2 \longrightarrow 2CHO^+ + e^-$$
$$CHO^+ + H_2O \longrightarrow H_3O^+ + CO$$

(3) FPD(fig.5-21) A flame photometric detector measures optical emission from phosphorus, sulfur, lead, tin, or other selected elements. When eluate passes through a H_2-air flame, as in the flame ionization detector, excited atoms emit characteristic light. Phosphorus emission at 526 nm or sulfur emission at 394 nm can be isolated by a narrow-band interference

（3）火焰光度检测器（图 5-21）火焰光度检测器测定的是磷、硫、铅、锡或其他选定元素的谱线发射。当洗脱液通过 H_2- 空气火焰时，就像在火焰电离检测器中一样，会激发出原子的特征谱线。526nm 处的磷发射谱线或 394nm 处的硫发射谱线均可通过窄带

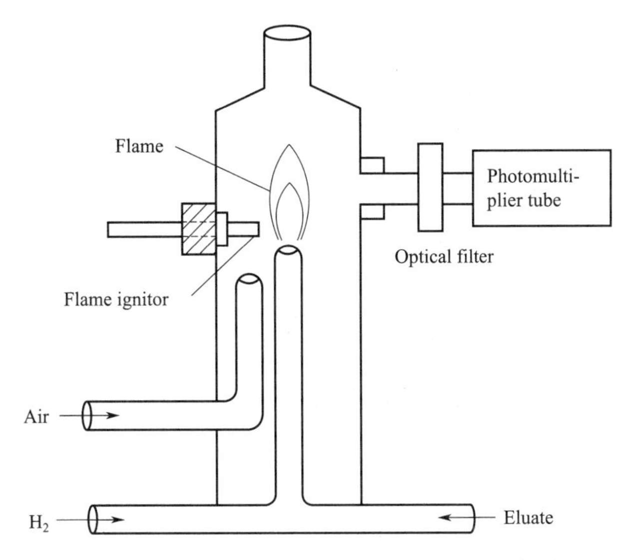

Fig.5-21 Principle structure of a flame photometric detector

图 5-21 火焰光度检测器的基本结构

filter and detected with a photomultiplier tube. The response to sulfur or phosphorus compounds is 10^5 times greater than to hydrocarbons.

Characteristics: It has high selectivity to phosphorus and sulfur compounds. Proper choice of the filter before photomultiplier tube may enhance selectivity and suppress interference.

干涉滤光片过滤并用光电倍增管检测出来。该检测器对硫或磷化合物的响应比对碳氢化合物的响应大 10^5 倍，具有很高的选择性。适当选择光电倍增管前的滤光片将有助于提高选择性，排除干扰。

$$RS+Air+O_2 \longrightarrow SO_2+CO_2$$
$$2SO_2+4H_2 \longrightarrow 2S+4H_2O$$
$$S+S \longrightarrow S_2^* \quad (390^{\circ}C)$$
$$S_2^* \longrightarrow S_2+h\nu \quad (\lambda=350\text{-}430nm)$$

(4) ECD (fig.5-22)

Principle: a radioactive foil such as ^{63}Ni emits a beta particle which collides with and ionizes the carrier gas to generate ions resulting in a current in the electric field. When analyte molecules with electronegative elements pass around, electrons are captured which results in a decrease in current generating a detector response.

Characteristics: it is a concentration detector with high selectivity and sensitivity. It has response to electronegative elements such as halogen, sulfur, phosphorus, nitrogen and oxygen. The stronger the electronegativity, the higher the sensitivity.

（4）电子捕获检测器（图 5-22）

原理：载气分子在 ^{63}Ni 辐射源中所产生的 β 粒子的作用下离子化，在电场中形成稳定的基流，当含电负性基团的组分通过时，俘获电子使基流减小而产生电信号。

特点：是一种具有选择性，高灵敏度的浓度型检测器。它对具有电负性的物质（卤素、硫、磷、氮、氧）有响应，电负性越强，灵敏度越高。

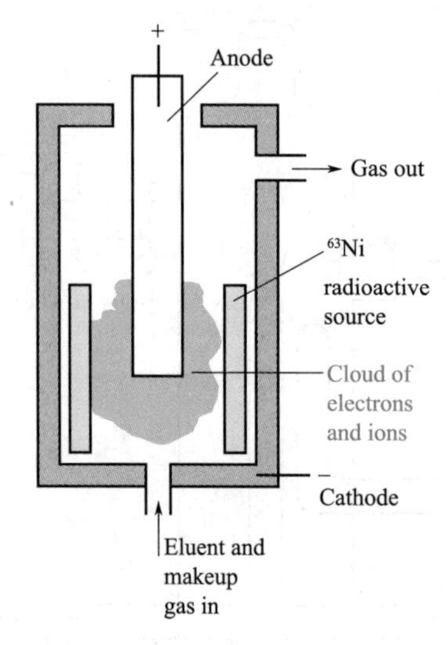

Fig.5-22　Principle structure of a electron capture detector

图 5-22　电子捕获检测器的基本结构

2. Performance Index of Detector

(1) Noise　The base line has fluctuating signals in a short time before sample entering into detector. These signals are called noise N, which is caused by the instrument itself or by other operation conditions, such as the fluctuation of carrier gas, temperature and voltage.

(2) Response time　The response time refers to the time taken by the output signal of a component in the detector when it reaches 63% of its true value. If a detector has small dead volume, there will be little hysteresis phenomenon in the circuit, then the speed of response is fast, usually less than 1s.

(3) Sensitivity　The sensitivity is change rate of the response signal to the quantity of sample (fig.5-23).

2. 检测器性能指标

（1）噪声　在没有样品进入检测器时，基线在短期内发生起伏的信号称为噪声 R_N。噪声是因仪器本身及其他操作条件所引起，如载气、温度、电压等的波动。

（2）响应时间　响应时间指进入检测器的某一组分的输出信号达到其真值的 63% 所需的时间。检测器的死体积小，电路系统的滞后现象小，响应速度就快，一般都小于 1s。

（3）灵敏度　灵敏度（S）即响应信号变化（ΔR）对通过检测器物质量变化（ΔQ）的比值，如图 5-23。

$$S=\frac{\Delta R}{\Delta Q} \tag{5.32}$$

(4) Linear range　The linear range is a range in which the detecting signal has linear relation with the quantity of the analyte, which decides the range of measurable concentration (or mass). Approximate injected analyte mass limits for detectors are shown in fig. 5-24.

（4）线性范围　检测信号的大小与被测组分的量呈线性关系的范围。其大小决定可测定的浓度（或质量）的范围。不同检测器进样量限值见图 5-24。

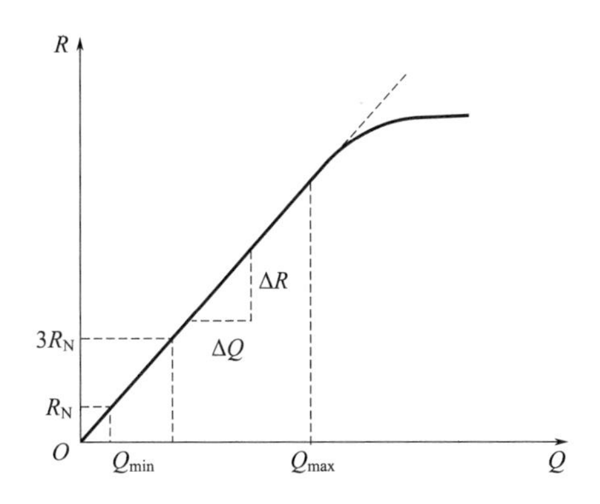

Fig.5-23 Response curve of detectors

图 5-23 检测器响应曲线

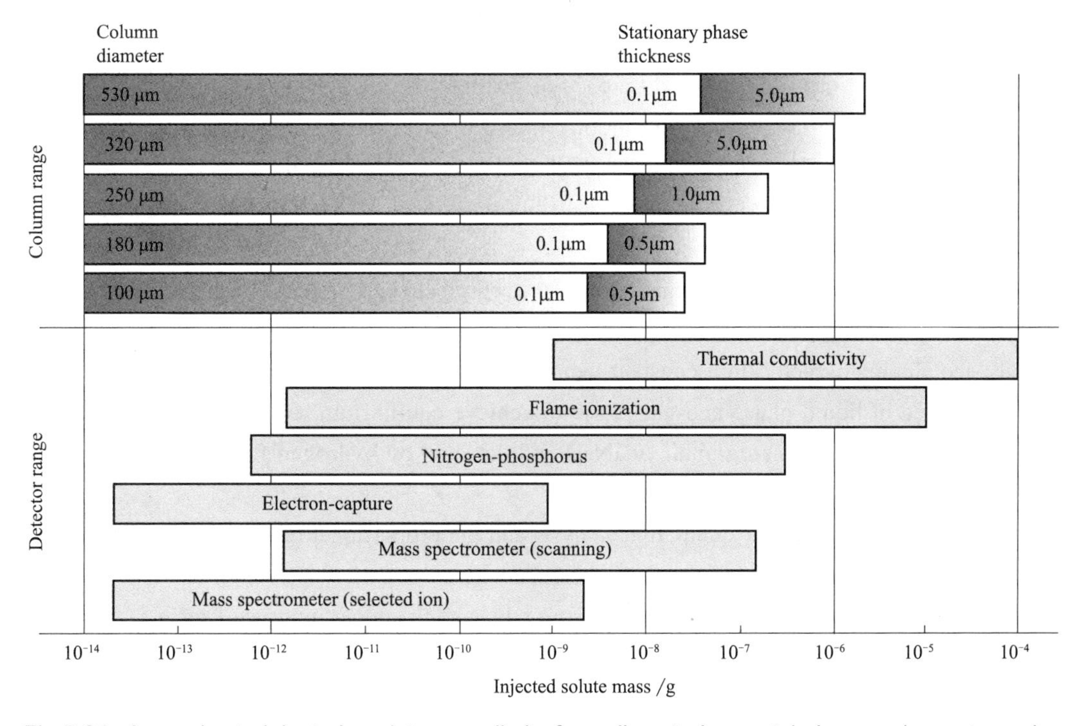

Fig.5-24 Approximate injected analyte mass limits for wall-coated open tubular gas chromatography columns and detectors

图 5-24 涂壁开管气相色谱柱及不同检测器的进样量限值

(5) Detection limit The detection limit means the quantity of sample introduced into detector in unit time (mass detector) or the content of sample in unit volume carrier gas (concentration detector) when the detector has three times of noise signal. Low noise indicates high sensitivity and good performance of the instrument.

（5）检出限 检测器产生 3 倍于噪声信号时，单位时间内进入检测器的样品量（质量型）或单位体积载气中样品的含量（浓度型）称为检出限。噪音越小，灵敏度越高，仪器性能越好。

$$D=\frac{3R_N}{S} \tag{5.33}$$

Experiment 4 Use of Gas Chromatograph and Determination of Benzene Series in Water

1. Purpose of the experiment

(1) Master the use of GC7820A gas chromatograph;

(2) Learn to use the separation degree to evaluate the separation performance of chromatograph;

(3) Master the determination method of benzene series in environmental water by headspace chromatography.

2. Experimental principle

Basic operation of GC7820

In gas chromatography analysis, the sample is vaporized and enters the column with carrier gas. Due to the difference of distribution coefficient between different components in the mobile phase (carrier gas) and the fixed phase, when the two phases move relative to each other, each component is separated through multiple distribution in the two phases and is detected in the detector successively.

Benzene series in water include benzene, toluene, ethylbenzene, xylene, isopropylbenzene and styrene.

Headspace chromatography: in a constant-temperature airtight container, benzene series in water are distributed in liquid phase and gas phase to achieve equilibrium. Gas phase samples on the liquid are taken for chromatographic analysis and detected by hydrogen flame detector. Typical chromatographic diagrams are as follows.

The retention time is used for qualitative analysis and the proportional relation between peak area and concentration of benzene series is used for quantitative analysis.

The minimum detected concentration is 0.005 mg • L^{-1} and the upper limit is 0.1 mg • L^{-1}.

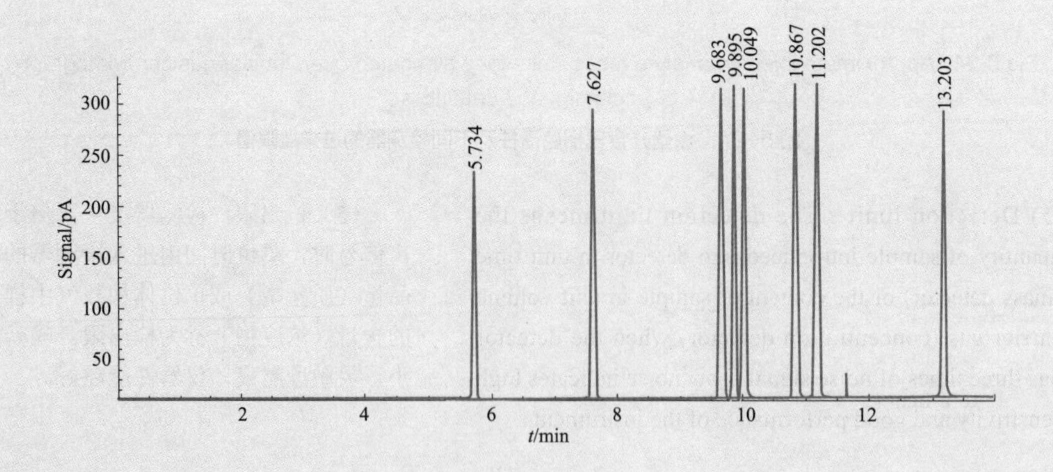

Number	Retention time/min	Name in Chinese	Name in English	CAS
1	5.734	苯	Benzene	71-43-2
2	7.627	甲苯	Toluene	108-88-3
3	9.683	乙苯	Ethylbenzene	100-41-4
4	9.885	对二甲苯	*p*-Xylene	106-42-3
5	10.049	间二甲苯	*m*-Xylene	108-38-3
6	10.867	异丙苯	Cumene	98-82-8
7	11.202	邻二甲苯	*o*-Xylene	95-47-6
8	13.203	苯乙烯	Styrene	100-42-5

Chromatogram of benzene series

3. Apparatus and reagents

(1) Apparatus　Gas chromatography (GC7820A), flame ion detector with hydrogen, headspace sampler (7694E), Agilent; Nitrogen cylinder;Hydrogen gas bottle;headspace bottle cap sealer and remover;graduated tube(10 mL); standard headempty bottles (22 mL, with caps and aluminium seals with inner PTFE film); capillary column (Dm-FFAP, 60 m×0.32 mm×1.0 nm) or capillary column (SE-30, 30 m×0.32 mm×1.0 nm).

(2) Reagents　Pure substance: benzene, toluene, ethylbenzene, xylene (containing au three xylene substances); Pure water, Chromatographic pure methanol.

4. Experimental steps

(1) Apparatus conditions

Injection port: temperature 150 ℃. Programmed temperature: 70 ℃ for 5 minutes, 10 ℃ · min^{-1} to 80 ℃ for 1 minute, 8 ℃ · min^{-1} to 110 ℃ for 3 minutes, 15 ℃ · min^{-1} to 150 ℃ for 3 minutes. Detector: FID detector. Temperature: 250 ℃. Hydrogen: 30 mL · min^{-1}. Air: 300 mL · min^{-1}. Tail blowing: 30 mL · min^{-1}.

Headspace temperature: 70 ℃. Balancing time: 15 minutes. Sampling time: 15 seconds.

(2) Determination of benzene series in water

① Sample collection and pretreatment

a. Sample: chemical wastewater

b. Sample collection

Container: glass bottle (250 mL)

Container washing method: wash with detergent once, wash with tap water three times and wash with distilled water once. Wash the container with the sample 2-3 times and fill it.

c. Sample storage: Store in the fridge for up to 7 days.

② Calibration curve drawing

a. Take a headspace bottle, add 10 mL pure water, then quickly add the mixed standard solution, concentration of each component is 5μg · L^{-1}, 20μg · L^{-1}, 40μg · L^{-1},60μg · L^{-1},80μg · L^{-1}, 100μg · L^{-1}, respectively.

b. Sample analysis: The above samples are placed in a headspace injector and balanced at 70 ℃ for 15 minutes. The headspace injector is automatically injected into the gas chromatography instrument for data acquisition and analysis, so as to qualitatively retain time and quantitatively quantify peak area. The headspace sampler automatically feeds the sample into the gas chromatography instrument for data acquisition and analysis, qualitative by retention time and quantitative by peak area.

c. The standard curve is drawn with the concentration of benzene series ($\mu g \cdot L^{-1}$) as abscissa and the peak area as ordinate. Check the correlation coefficient of standard curve, and the correlation coefficient $R \geqslant 0.999$; otherwise, the standard series should be reformulated.

d. Sample treatment: take 10 mL of water sample into a clean bottle, cover the sealing cap with the teflon film, and press the cap tightly with the gland holder. Analyze it according to standard sample conditions.

e. Result processing: The peak area of the sample is used to check the concentration of benzene series on the standard curve.

③ Determination of separation degree

Measure the retention time and half peak width of benzene, toluene and xylene, and calculate separation values of benzene and toluene and between toluene and xylene.

5. Notes

① When preparing the storage liquid of benzene series, it should be carried out in a well-ventilated environment to avoid health hazards.

② When preparing headspace samples, the temperature should be strictly controlled.

③ When preparing storage solution of benzene series, benzene series can be first dissolved into a small amount of methanol, and then prepare into an aqueous solution.

④ Before shutting down, the column temperature should be lowered to room temperature to prevent the loss of column fixing liquid.

6. Questions

① What is the principle of headspace analysis and what are the influencing factors?

② How is injector temperature determined for chromatographic analysis?

实验 4　气相色谱仪的使用及水中苯系物的测定

1. 实验目的

（1）掌握 GC7820A 气相色谱仪的使用方法；

（2）学会用分离度评价色谱仪分离性能的方法；

（3）掌握环境水中苯系物的顶空色谱测定方法。

GC7820 气相色谱仪的基本操作

2. 实验原理

在气相色谱分析中，试样被汽化后，随同载气进入色谱柱，由于不同组分在流动相（载气）和固定相间分配系数的差异，当两相作相对运动时，各组分在两相间经多次分配而被分离，依次进入检测器中被检测。水中苯系物包括苯、甲苯、乙苯、二甲苯、异丙苯和苯乙烯。苯系物的色谱图如下。

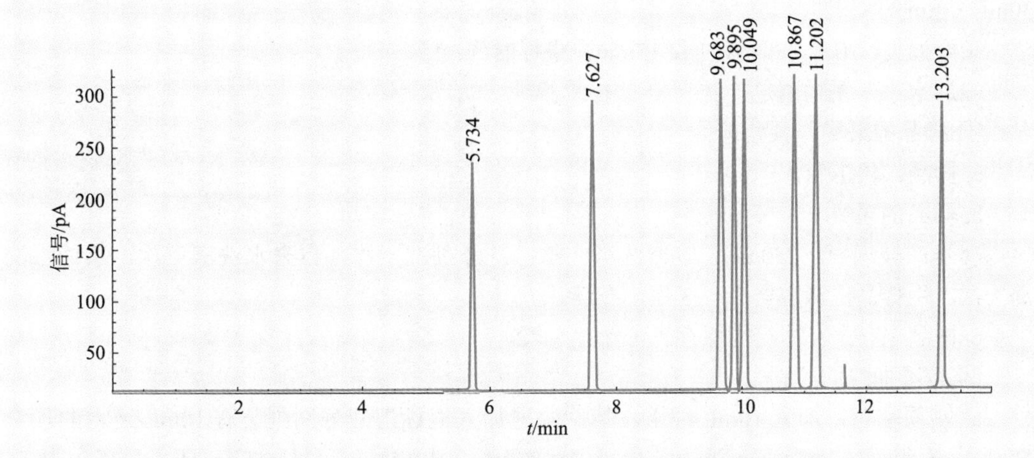

序列	保留时间/min	中文名称	英文名称	CAS
1	5.734	苯	Benzene	71-43-2
2	7.627	甲苯	Toluene	108-88-3
3	9.683	乙苯	Ethylbenzene	100-41-4
4	9.885	对二甲苯	*p*-Xylene	106-42-3
5	10.049	间二甲苯	*m*-Xylene	108-38-3
6	10.867	异丙苯	Cumene	98-82-8
7	11.202	邻二甲苯	*o*-Xylene	95-47-6
8	13.203	苯乙烯	Styrene	100-42-5

苯系物的色谱图

顶空色谱法：在恒温密闭容器中，水中的苯系物在液相、气相间分配，达到平衡，取液体上的气相样品用于色谱分析，用氢火焰检测器检测。用保留时间定性，用峰面积与苯系物浓度成正比的关系进行定量分析。该方法最低检出浓度为 $0.005mg \cdot L^{-1}$，检测上限为 $0.1mg \cdot L^{-1}$。

3. 实验仪器与试剂

（1）仪器　GC7820A 气相色谱仪，带氢火焰离子检测器，7694E 顶空进样器，Agilent；氮气瓶；氢气瓶；顶空瓶瓶盖密封器和拆卸器；10mL 刻度管；22mL 标准顶空瓶（带内涂四氟乙烯膜的瓶盖和铝密封盖）；DM-FFAP 毛细管柱（60m×0.32mm×1.0nm）或 SE-30 毛细管柱（30m×0.32mm×1.0nm）。

（2）试剂　纯物质：苯、甲苯、乙苯、二甲苯（基准物质，含有邻间对二甲苯三种物质）；纯水；色谱纯甲醇。

4. 实验内容与步骤

（1）仪器条件

进样口：温度150℃。程序升温：70℃保持5min，10℃·min^{-1}到80℃保持1min，8℃·min^{-1}到110℃保持3min，15℃·min^{-1}到150℃保持3min。

检测器：FID检测器，温度250℃，氢气：30mL·min^{-1}，空气：300mL·min^{-1}，尾吹：30mL·min^{-1}。

顶空温度：70℃，平衡时间：15min，进样时间：15s。

（2）水中苯系物测定

① 样品的采集与预处理

a. 样品：化工废水

b. 样品的采集

容器：玻璃瓶，250mL。容器洗涤方法：洗涤剂洗一次，自来水洗涤三次，蒸馏水洗涤一次。用样品洗涤容器2～3次，充满。

c. 样品保存：在冰箱内保存，可保存7天。

② 校准曲线的绘制

a. 取顶空瓶，加入10mL纯水，然后快速加入混标溶液分别到装有10mL纯水的顶空瓶中，各组分的浓度分别为5μg·L^{-1}、20μg·L^{-1}、40μg·L^{-1}、60μg·L^{-1}、80μg·L^{-1}、100μg·L^{-1}。

b. 进样分析：将上述样品置于顶空进样器中，70℃平衡15min，顶空进样器自动进样到气相色谱仪器中，进行数据采集分析，以保留时间定性，峰面积定量。

c. 以苯系物浓度（μg·L^{-1}）为横坐标，以峰面积为纵坐标，绘制标准曲线。进行标准曲线相关系数检查，相关系数$r \geqslant 0.999$，否则应重新配制标准系列。

d. 样品处理：移取10mL水样到干净瓶中，盖好带聚四氟乙烯薄膜的密封盖，用压盖器压紧盖好。按标准样品分析。

e. 结果处理：用样品的峰面积在标准曲线上查取各苯系物浓度。

③ 分离度的测定　测定苯、甲苯、二甲苯的保留时间和半峰宽，计算苯与甲苯，甲苯与二甲苯的分离度。

5. 注意事项

① 配制苯系物贮备液时，要在通风良好的环境下进行，以免危害健康。

② 顶空样品制备时，要严格控制温度。

③ 配制苯系物贮备液时，可先将苯系物溶入少量甲醇中，再配制成水溶液。

④ 关机前，应使柱温下降至室温，防止色谱柱固定液损失。

6. 思考题

① 顶空分析的原理是什么，影响的因素有哪些？

② 色谱分析的进样器温度是怎么确定的？

Exercises

5-1 Fill in the blanks.

(1) The selectivity of the chromatographic column can be expressed by the total separation efficiency index. When $R \geqslant$________, the two components can be completely separated.

(2) For gas chromatography FID, in general, H_2 is________, air is________, and N_2 is________.

(3) The principle of selecting a stationary liquid for chromatographic analysis is______. If non-polar substances are to be separated, a________stationary liquid is used, and the sequence of components flowing out of the chromatographic column generally conforms to the boiling point law, that is, the________components flow out first, and the________ components flow out last.

(4) According to different measurement principles, gas chromatography detectors are often divided into________and________detectors, and FPD belongs to________.

(5) The performance indicators of gas chromatographic detectors mainly refer to ________, ________,________,________.

(6) Thermal conductivity detectors are based on components that have different thermal conductivity between the________.

(7) The carrier gas is the mobile phase of gas chromatography, and the commonly-used carrier gases are________and________. (at least two)

5-2 The main components of gas chromatography include (   ).

A. carrier gas system, spectroscopic system, chromatographic column, detector

B. carrier gas system, sampling system, chromatographic column, detector

C. carrier gas system, atomization device, chromatographic column, detector

D. carrier gas system, light source, chromatographic column, detector

5-3 The main factor affecting the sensitivity of the hydrogen flame detector is (   ).

A. detector temperature   B. carrier gas flow rate

C. the ratio of the three gases   D. polarization voltage

5-4 The position of the chromatographic peak in the chromatogram is expressed by (   ).

A. retention value   B. peak height value   C. peak width value   D. sensitivity

5-5 True or false

(1) FID responds to inorganic and organic compounds, and the sample is destroyed during detection.

(2) The absolute correction factor is equal to the peak area divided by the amount of the component.

(3) The greater difference in the *K* value of the distribution coefficients of each component in the chromatographic analysis, the more difficult it is to separate them.

5-6 Briefly describe the basic principles of gas chromatographic separation?

5-7 Determine the content of *X* in the sample with GC, take 1.800 g sample, add 0.4000 g internal standard S in it, inject the sample after mixing well, obtain A_x=27.00 cm², A_s=25.00 cm² from the GC spectrum. It is known that f'_{wx}=1.11, f'_{ws}=1.00, calculate the percentage of the composition X in the sample.

(26.64%)

5-8 A mixture of benzene, toluene and air was injected into a gas chromatograph. Air gave a sharp peak in 30 s, whereas benzene requires 223 s and toluene was eluted in 262 s. Find the adjusted retention time and capacity factor for each solute. Also, find the relative retention value (selectivity factor) of benzene and toluene.

(1.20)

5-9 A sample is analyzed by packed gas chromatography. When the column length is 1 m, the retention times for components A and B are 11.6 min and 12.2 min respectively. Meanwhile, the theoretical plate number for the column is 3600. Calculate:

(1) The corresponding baseline widths for components A and B;

(2) The resolution;

(3) Assume that the the peak widths of A and B are equal, the column length for the baseline separation of the two components.

[(1) A 0.77 min; B 0.81 min (2) 0.76 (3) 1.73 m]

5-10 With GC to analyze a sample and get the adjusted retention times of two components are 13 min and 16 min respectively, the peak widths at the baseline of the two are 1 min, calculate: (1) The number of the effective plates; (2) The selectivity factor of the two components, α; (3) If the resolution R_s is demanded to be 1.5, how many effective plates are needed? How long is the column at least? (H_e=0.1 cm)

[(1) 4096 (2) 1.23 (3) 1.03 m]

5-11 A sample is analyzed by packed gas chromatography. When the column length is 1 m, the retention times for components A and B are 5.80 min and 6.60 min respectively, meanwhile, the corresponding baseline widths are 0.78 min and 0.78 min, the dead time is 1.10 min. Calculate: (1) The average linear velocity for the carrier-gas; (2) The capacity factor of component B; (3) The resolution R_s; (4) The effective plate number; (5) The column length for the baseline separation of the two components (R_s=1.5).

[(1) 0.91 m · min^{-1} (2) 5 (3) 1.03 (4) 803 (5) 2.12 m]

5-12 To determine the content of H_2O in a sample with GC, take 4.586 g sample, add 0.0213 g internal standard S in it, inject the sample after mixing well, obtain A (H_2O)=150.00 mm², A_s=174.00 mm² from the GC spectrum. It is known that f (H_2O)=0.550, f'_s=0.580, calculate the percentage of H_2O in the sample.

(0.38%)

5-13 A mixed sample is analyzed by gas chromatography. The peak areas and the mass correction factors for all components are given as follows:

Component i	A_i/cm^2	f_i'	Component i	A_i/cm^2	f_i'
ethanol	8.4	0. 64	cyclohexane	5.6	0.68
n-heptane	9.2	0. 73	benzene	10. 7	0. 78

To calculate the contents of the four compounds respectively.
(ethanol 22.17%; n-heptane 27.70%; cyclohexane 15.71%; benzene 34.42%)

Chapter 6 High Performance Liquid Chromatography

第6章 高效液相色谱分析法

Study Guide 学习指南

High performance liquid chromatography (HPLC) is a new type of separation and analysis technique developed on the base of the traditional liquid chromatography (LC). Traditional LC uses unevenly distributed coarse particles as stationary phase and relies on the force of gravity to pass the mobile phase through the column. Therefore, LC is low in analysis speed and separation efficiency. The inventions of highly efficient stationary phase, high pressure pump, gradient elution technology and various highly sensitive detectors have promoted the rapid development of HPLC. This chapter mainly focuses on the principle, basic concepts, commonly used detectors and quantitative analytical methods of HPLC. The learning objective of this chapter is to understand the analytical procedure of HPLC.

高效液相色谱法（HPLC）是在经典液相色谱法基础上发展起来的一种新型分离、分析技术。经典液相色谱法由于使用粗颗粒的固定相，填充不均匀，依靠重力使流动相流动，因此分析速度慢，分离效率低。新型高效的固定相、高压输液泵、梯度洗脱技术以及各种高灵敏度的检测器相继发明，高效液相色谱法迅速发展起来。本章重点介绍高效液相色谱法原理、基本概念、常用检测器及定量分析方法等内容。通过本章的学习，了解高效液相色谱法的分析过程。

Section 1 Introduction on HPLC

第1节 高效液相色谱简介

1. Principle and Type of HPLC

(1) Principle　In HPLC the mobile phase (a single solvent of different polarity or the mixture of solvents or buffering agent with different ratio) is forced by

1. 高效液相色谱法原理及分类

（1）原理　高效液相色谱法是用高压输液泵将具有不同极性的单一溶剂或不同比例的混合溶剂、缓冲液等流动相

a high pressure pump into a column that is packed with a stationary phase. The sample to be analyzed is introduced into the stream of mobile phase via an injection valve and separated into various components which in turn enter a detector. Finally the chromatographic signals are recorded by a recorder or an integrator.

压入装有固定相的色谱柱，经进样阀注入分析样品，由流动相带入柱内，在柱内各成分被分离后，依次进入检测器，色谱信号由记录仪或积分仪记录。

(2) Type HPLC is divided into the following different types based on the separation mechanisms: liquid-solid adsorption chromatography, liquid-liquid partition chromatography, chemically bonded phase chromatography, ion-exchange chromatography and molecule-exclusion chromatography, shown in figure 6-1.

（2）分类 根据分离机制不同，可将高效液相色谱分为：液固吸附色谱、液液分配色谱、化合键合相色谱、离子交换色谱以及分子排阻色谱等类型，见图 6-1。

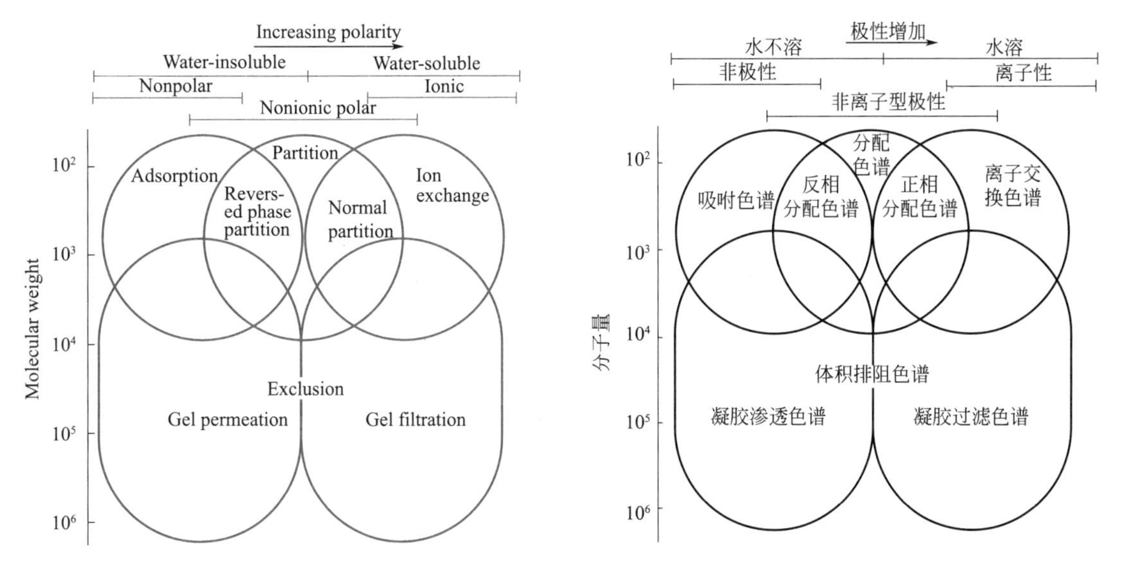

Fig.6-1 The approximate range of samples applicable to the various HPLC methods

图 6-1 各种 HPLC 方法适用的大致试样范围

2. Characteristics of HPLC

2. 高效液相色谱法特点

(1) High efficiency Thanks to the fine packing particles, highly efficient stationary phase and even packing technology, HPLC enjoys very high separation efficiency of which the column efficiency can reach 10^4 theoretical plates per meter. The micro-packed column (ID 1 mm) and the capillary liquid phase column (ID 0.05 μm) appeared in recent years can even reach than 10^5 theoretical plates per meter and have realized highly efficient separation.

（1）高效 由于使用了细颗粒、高效率的固定相和均匀填充技术，高效液相色谱法分离效率极高，柱效一般可达每米 10^4 理论塔板。近几年来出现的微型填充柱（内径 1mm）和毛细管液相色谱柱（内径 0.05μm），理论塔板数甚至超过每米 10^5，能实现高效的分离。

(2) High speed　Since employing high pressure pump to transfer mobile phase, using gradient elution apparatus and detecting the components eluting from the column directly after column, HPLC can complete a separation and analysis operation in a few minutes or several tens of minutes, much faster than the traditional LC.

（2）高速　由于使用高压泵输送流动相，采用梯度洗脱装置，用检测器在柱后直接检测洗脱组分等，HPLC 完成一次分离分析操作一般只需几分钟到几十分钟，比经典液相色谱快得多。

(3) High sensitivity　The use of UV, fluorescent, electrochemical and mass spectrometric detectors enables HPLC to determine the minimum inspection quality to 10^{-9}-10^{-12} g.

（3）高灵敏度　紫外、荧光、电化学、质谱等检测器的使用，使 HPLC 的最小检测量可达 10^{-9} ～ 10^{-12}g。

(4) High automation　The use of computer makes it possible for HPLC to not only process data, plot graphs and print analytical results automatically, but also automatically control chromatographic conditions to let the chromatographic system work under optimal conditions from beginning to end. Thus the chromatographic system becomes a fully automated instrument.

（4）高度自动化　计算机的应用，使 HPLC 不仅能自动处理数据、绘图和打印分析结果，而且还可以自动控制色谱条件，使色谱系统自始至终都在最佳状态下工作，成为全自动化的仪器。

(5) Wide applications　Comparing with GC, HPLC can be used to separate and analyze the organic compounds of high boiling points, high relative molecular mass and poor thermal stability and various ions.

（5）应用范围广　与气相色谱法相比，HPLC 可用于高沸点、分子量大、热稳定性差的有机化合物及各种离子的分离分析。

(6) Wide range of mobile phase　Various kinds of solvents can be used as mobile phase. The separation efficiency can be improved by adjusting the composition of mobile phase. Therefore, the substances with similar properties and structures can be better separated by HPLC than by GC.

（6）流动相可选择范围广　它可用多种溶剂作流动相，通过改变流动相组成来改善分离效果，因此对于性质和结构类似的物质分离的可能性比气相色谱法更大。

3. Comparison of HPLC and GC (table 6-1)

3. 高效液相色谱法与气相色谱法的比较（表 6-1）

Table 6-1　Comparison of High-Performance Liquid Chromatography and Gas-Liquid Chromatography

表 6-1　高效液相色谱法和气相色谱法的比较

Characteristics of both methods 两种方法的共同特点	Advantages 优点	
	HPLC 高效液相色谱法	GC 气相色谱法
Efficient, highly selective, widely applicable 高效、高选择性、应用广泛	Can accommodate nonvolatile and thermally unstable compounds 适用于非挥发性和热不稳定的化合物	Simple and inexpensive equipment 设备简单廉价

续表

Characteristics of both methods 两种方法的共同特点	Advantages 优点	
	HPLC 高效液相色谱法	GC 气相色谱法
Only small sample required 样品用量少	Generally applicable to inorganic ions 适用于无机离子测定	Rapid 快速
May be nondestructive of sample 对样品破坏性小		Unparalleled resolution (with capillary columns) 高分辨率（使用毛细管柱）
Readily adapted to quantitative analysis 适用于定量分析		Easily interfaced with mass spectrometry 与质谱仪连接方便

4. Applications of HPLC

The HPLC is a chromatographic technique applicable for the qualitative and quantitative analysis of polymer compounds and ionic compounds of low volatility, poor thermal stability and large molecular weight. Typical Applications of HPLC was shown in table 6-2.

4. 高效液相色谱法应用

高效液相色谱法是一种分离分析方法，适用于挥发性低、热稳定性差、分子量大的高分子化合物以及离子型化合物等的定性、定量分析。其典型应用见表 6-2。

Table 6-2 Typical Applications of High-Performance Partition Chromatography

表 6-2　高效液相色谱法的典型应用

Field 应用领域	Typical Mixtures Separated 分离的典型混合物
Pharmaceuticals 药品	Antibiotics, sedatives, steroids, analgesics 抗生素、镇静剂、类固醇、镇痛剂
Biochemicals 生化品	Amino acids, proteins, carbohydrates, lipids 氨基酸、蛋白质、碳水化合物、脂质
Food products 食物产品	Artificial sweeteners, antioxidants, aflatoxins, additives 人工甜味剂、抗氧化剂、黄曲霉毒素、添加剂
Industrial chemicals 工业化学品	Condensed aromatics, surfactants, propellants, dyes 缩合芳烃、表面活性剂、推进剂、染料
Pollutants 污染物	Pesticides, herbicides, phenols, polychlorinated biphenyls (PCBs) 农药、除草剂、酚类、多氯联苯（PCBs）
Forensic chemistry 法医化学	Drugs, poisons, blood alcohol, narcotics 毒品、毒药、血液酒精、麻醉品
Clinical medicine 临床医学	Bile acids, drug metabolites, urine extracts, estrogens 胆汁酸、药物代谢物、尿提取物、雌激素

Section 2 High Performance Liquid Chromatograph

第 2 节 高效液相色谱仪

1. The working procedure of the HPLC

There are many types of HPLC instruments. But no matter which type of HPLC chromatograph it is, a instrument typically includes four parts: high pressure pumping system, sample injection system, separation system and detection system. In addition to them, some auxiliary apparatus may be equipped such as gradient elution, auto-injection, auto-collection and data processing upon special requirements. The working procedure of the HPLC chromatograph (fig.6-2) is as follows: high pressure pump sends the solvent in collecting bottle to enter into column via sampler and emerges from the outlet of detector. The sample to

1. 高效液相色谱仪的工作过程

高效液相色谱仪种类繁多，但不论何种类型的高效液相色谱仪，基本上都分为四个部分：高压输液系统、进样系统、分离系统和检测系统。此外，还可以根据一些特殊的要求，配备一些附属装置，如梯度洗脱、自动进样、自动收集及数据处理装置等。图 6-2 是高效液相色谱仪的结构示意图，其工作过程如下：高压泵将贮液罐的溶剂经进样器

Operation of Shimadzu LC-20A HPLC
岛津 LC-20A 高效液相色谱仪的使用

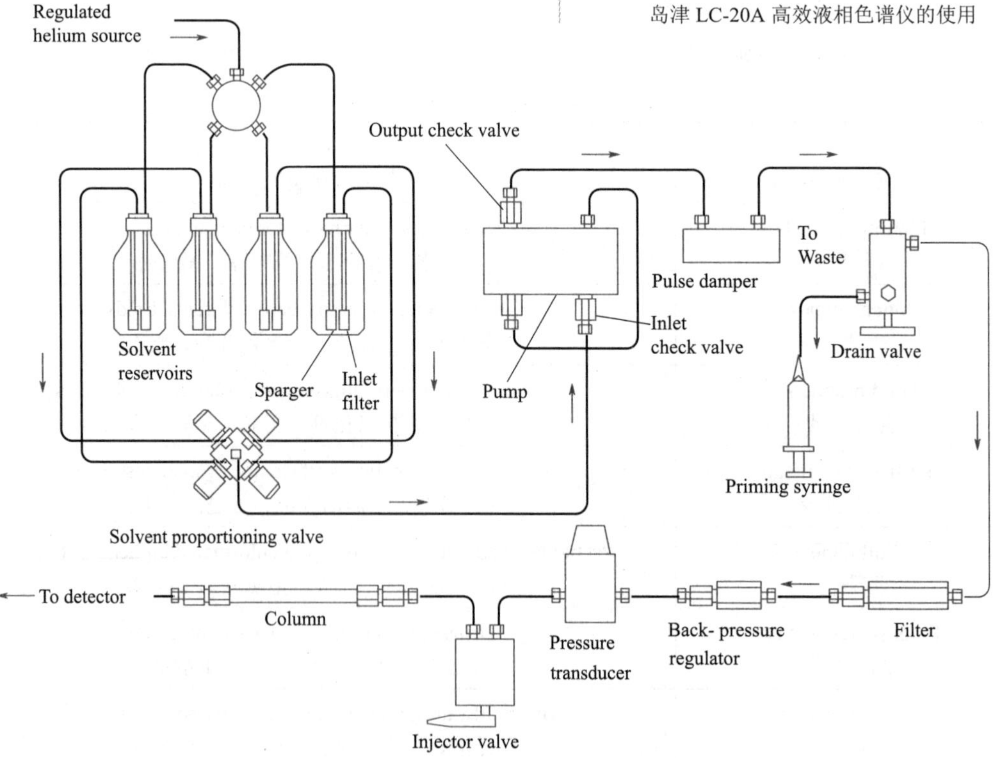

Fig.6-2 The working procedure of the HPLC

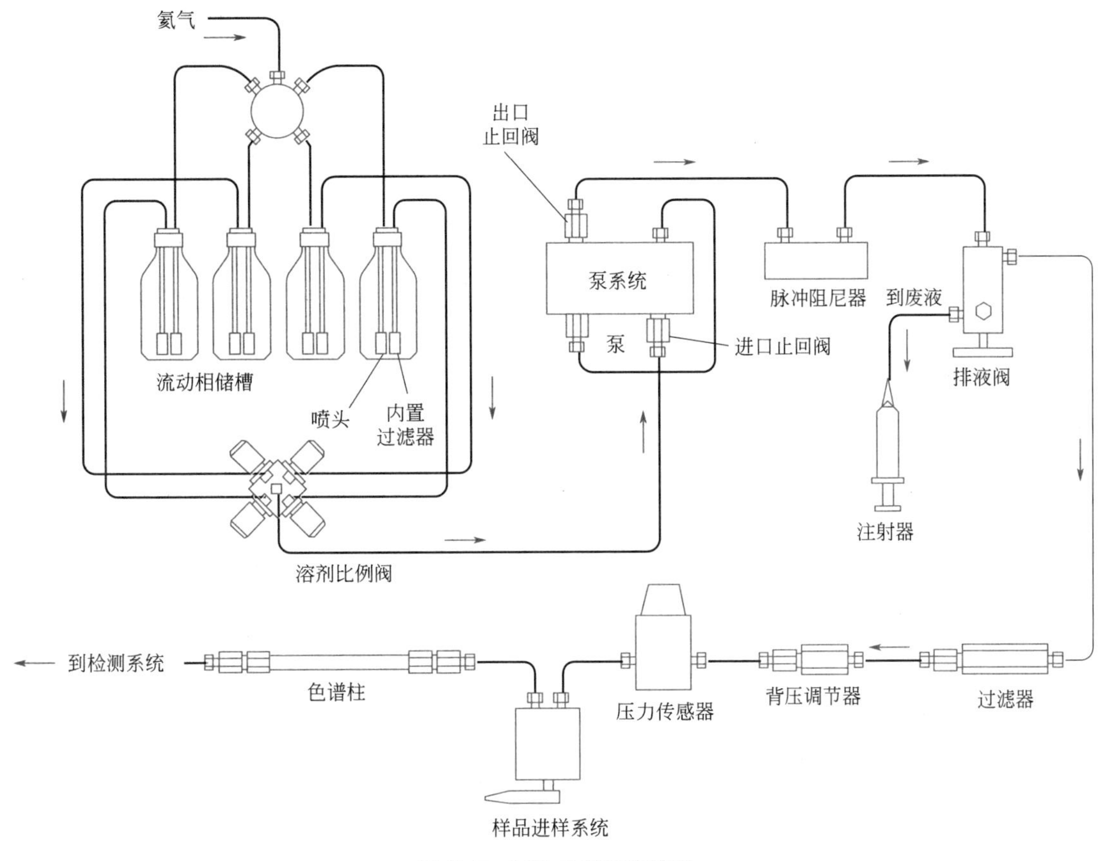

图 6-2　HPLC 的工作流程

be analyzed is introduced into the stream of mobile phase via sampler and separated in column into various components which in turn enter a detector. Finally the chromatographic signals are recorded by a recorder and plotted into liquid phase chromatogram.

送入色谱柱中，然后从检测器的出口流出。当待分离样品从进样器进入时，流经进样器的流动相将其带入色谱柱中进行分离，然后以先后顺序进入检测器，记录仪将进入检测器的信号记录下来，得到液相色谱图。

2. Structure of Instrument

2. 仪器结构

(1) High pressure pumping system　The high pressure pumping system is composed of liquid storage container, degasser, high pressure pump, filter and gradient elution unit.

（1）高压输液系统　由贮液罐、脱气装置、高压输液泵、过滤器、梯度洗脱装置等组成。

① Liquid storage container　The liquid storage container is made of corrosion resistant materials such as glass,

① 贮液罐　由玻璃、不锈钢或氟塑料等耐腐蚀材料制成。贮液罐的放置位置要高于泵体，以保持输液静压差，

stainless steel or fluoroplastic. The position of the container should be higher than the pump to maintain the static pressure difference. In use, the container should be air-tight to avoid the composition change of mobile phase caused by volatilization and the entrance of air.

使用过程应密闭，以防止因蒸发引起流动相组成改变，还可防止气体进入。

② Mobile phase The mobile phase usually uses solvent such as methanol-water or acetonitrile-water as base agent. The mobile phase must be degassed before use, otherwise gas bubbles may easily escape at the low pressure section and influence the separation efficiency of column and the sensitivity or even normal function of detector. The degassing methods include heating reflux, vacuum degassing, ultrasonic degassing and on-line vacuum degassing.

② 流动相 流动相常用甲醇 - 水或乙腈 - 水为底剂的溶剂系统。流动相在使用前必须脱气，否则很易在系统的低压部分逸出气泡，气泡的出现不仅影响柱分离效率，还会影响检测器的灵敏度甚至不能正常工作。脱气的方法有加热回流法、抽真空脱气法、超声脱气法和在线真空脱气法等。

③ High pressure pump The high pressure pump is one of the key components for high performance liquid chromatograph, used to transfer mobile phase. The requirements for the pump are: resistant to corrosion and high pressure, free of pulsation, wide range of discharge flow rate, constant velocity of flow and easy to clean and maintain. The high pressure pump can be classified into two types: constant pressure pump and constant flowrate pump, in which the latter is commonly used whose flowrate remains constant as its pressure changes with the system resistance.

③ 高压输液泵 高压输液泵是高效液相色谱仪的关键部件之一，用以完成流动相的输送任务。对泵的要求是耐腐蚀、耐高压、无脉冲、输出流量范围宽、流速恒定，且泵体易于清洗和维修。高压输液泵可分为恒压泵和恒流泵两类，常使用恒流泵（其压力随系统阻力改变而流量不变）。

Figure 6-3 shows the basic design of a pump that produces a programmable, constant flow rate up to 10 mL • min^{-1} at pressures up to 40 MPa (400 bar) for HPLC and up to 100 MPa (1000 bar) for UHPLC. A sapphire piston draws solvent from a reservoir into the piston chamber and then delivers the solvent to the column. Two check valves govern the direction of solvent flow through the pump. The quality of a pump for HPLC is measured by how steady and reproducible a flow it can produce.

图 6-3 显示了泵的基本设计，该泵在高效液相色谱中压力高达 40MPa（400bar）和超高压液相色谱中压力高达 100MPa（1000bar）的情况下产生可编程的恒定流速，最高可达 10mL • min^{-1}。活塞将溶剂从储液罐吸入活塞室，然后将溶剂输送至色谱柱。两个止回阀控制溶剂流经泵的方向。高效液相色谱泵的质量通过其产生的流量的稳定性和可重复性来衡量。

Preparation of mobile phase
流动相配制

Treatment of mobile phase
流动相处理

Structure and application of six-port switching valve injector
六通进样阀结构及使用

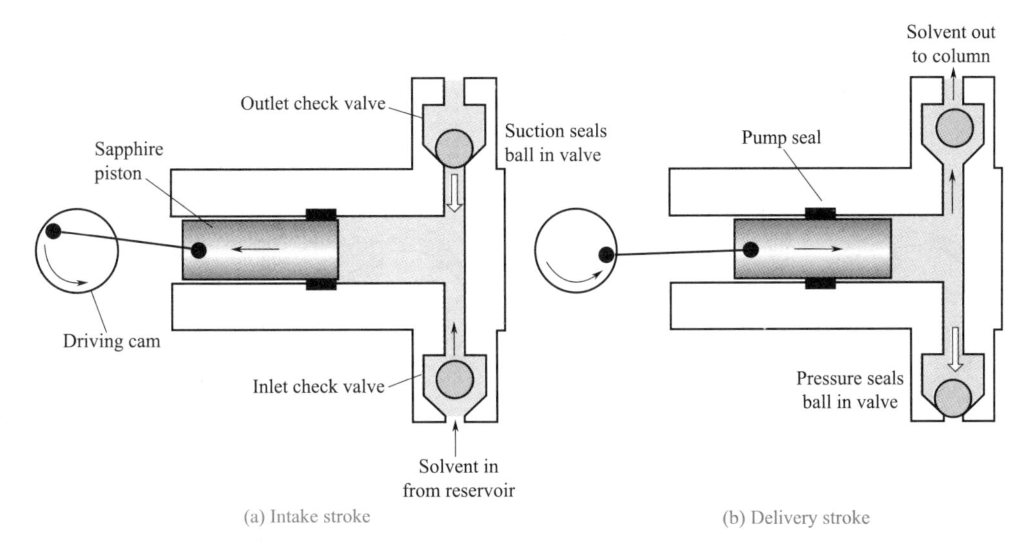

Fig.6-3 High-pressure reciprocating piston pump for HPLC (The sapphire piston is moved back and forth by a rotating cam. Check valves open or close depending on the direction of flow. When the piston moves to the left, solvent is drawn into the piston chamber through the inlet port. Suction causes the ball in the outlet valve to block the outlet port. When the piston moves to the right, solvent is forced out of the piston chamber to the column. Flow causes the ball in the inlet valve to block the inlet port. Flow lifts the ball in the outlet check valve, allowing liquid to be delivered to the column. Delivery rate is controlled by stroke frequency or volume.)

图 6-3 用于 HPLC 的高压往复活塞泵（蓝宝石活塞通过旋转凸轮前后移动，止回阀根据流动方向打开或关闭。当活塞向左移动时，溶剂通过入口吸入活塞腔，吸力导致出口被阀球堵塞。当活塞向右移动时，溶剂被挤出活塞腔进入色谱柱，同时入口被阀球阻塞。输送速率由冲程频率或容量控制。）

(2) Sample injection system The sample is injected by a six-port switching valve injector, and the sample size is determined by an injection loop. In operation, place the switch of the injector to the position of LOAD, thus the inlet of the injector only connects with the loop and is under atmospheric pressure. Using a microsyringe (with volumn larger than that of loop) to inject sample solution which stays in the loop. Then switching to the position of INJECT so that the injection loop connects with the liquid circuit and the sample is brought into the mobile phase and carried into the column.

The injection valve in figure 6-4 has interchangeable sample loops, each of which holds a fixed volume. Loops of different sizes hold volumes that range from 2μL to 1000μL. In the load position, a blunt-tipped syringe is used to rinse and load the loop with fresh sample at atmospheric pressure. High-pressure flow from the pump to the column passes through the lower left segment of the valve. When the valve is rotated

（2）进样系统 常用六通阀进样器进样，进样量由定量环确定。操作时先将进样器手柄置于采样位置（LOAD），此时进样口只与定量环接通，处于常压状态，用微量注射器（体积应大于定量环体积）注入样品溶液，样品停留在定量环中。然后转动手柄至进样位置（INJECT），使定量环接入输液管路，样品由高压流动相带入色谱柱中。

图 6-4 中六通阀具有可替换的样品环，每个样品环具有固定的体积。不同尺寸的样品环容纳的体积范围为 2～1000μL。在加载位置，使用平头注射器在大气压力下用鲜样冲洗和加载

Selection and installation of chromatographic column of HPLC
高效液相色谱仪色谱柱的选择与安装

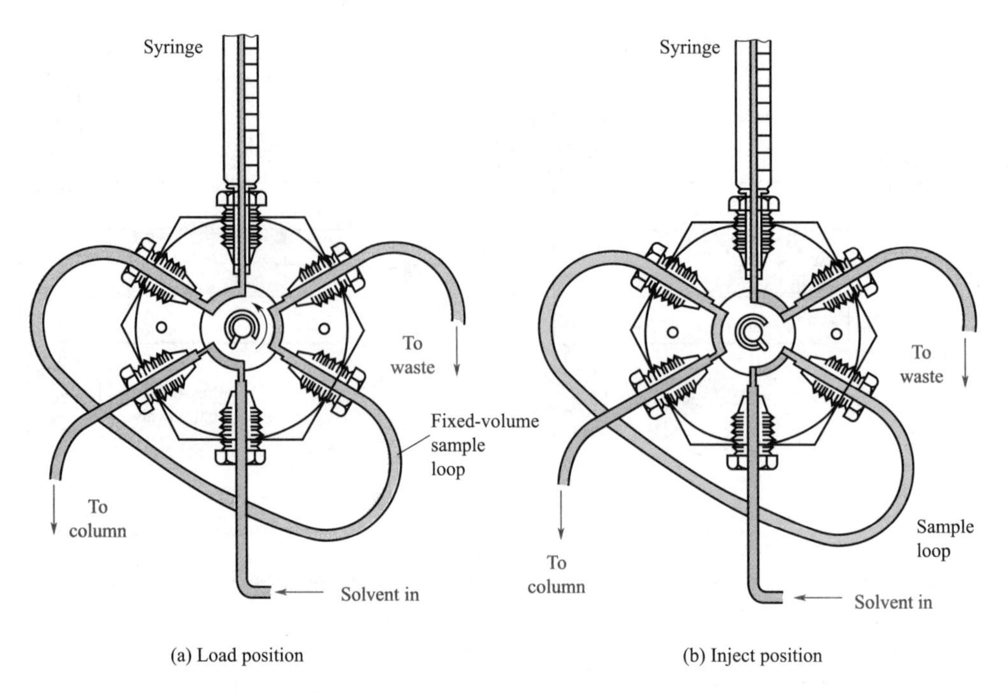

Fig.6-4 Injection valve for HPLC in the filled-loop injection mode

图 6-4 配有样品环的 HPLC 六通阀进样器

60° counterclockwise in figure 6-4(b), the pump is connected to the right side of the sample loop and flow from the pump transfers the content of the sample loop onto the column.

样品环。在图 6-4（b）中，将阀逆时针旋转 60°，此时泵连接至样品环的右侧，并通过流动相将样品环内的物质带入色谱柱。

(3) Chromatographic column The chromatographic column is the most important component in HPLC, including the column tube and the stationary phase. The column tube can be made of glass, stainless steel, aluminium, copper and other metals inner-lined with smooth polymers. Since the glass column has limited pressure resistance, the metal columns are used more commonly. In general, the column has length 5-30 cm and ID 4-5 mm, in which gel column has ID 3-15 mm and preparative column has ID more than 25 mm. The column is filled with silica gel and bonded stationary phase. In the bonded stationary phase, there are octadecyl silicane (so called ODS column or C_{18} column), octyl silicane (C_8 column), amino or cyano bonded silica gels.

（3）色谱柱 色谱柱是液相色谱的心脏部件，它包括柱管与固定相两部分。柱管材料有玻璃、不锈钢、铝、铜及内衬光滑的聚合材料的其他金属。玻璃管耐压有限，故金属管用得较多。一般色谱柱长 5 ～ 30cm，内径为 4 ～ 5mm，凝胶色谱柱内径 3 ～ 12mm，制备柱内径较大，可达 25mm 以上。柱内填充剂有硅胶和化学键合固定相。在化学键合固定相中有十八烷基硅烷键合硅胶（又称 ODS 柱或 C_{18} 柱）、辛烷基硅烷键合硅胶（C_8 柱）、氨基或氰基键合硅胶等。

① Guard column The entrance to the main column is protected by a short guard column containing the same

① 保护柱 主柱的入口处常安装一个很短的保护柱，其包含与主柱相同

stationary phase as the main column. Fine particles and strongly adsorbed solutes are retained in the guard column, which is periodically replaced when column pressure increases or after a set number of injections or time in service. HPLC column with replaceable guard column shows in figure 6-5.

的固定相。保护柱中截留着细颗粒和强吸附的溶质，当柱压力增加时，或经过一定量的注射次数或使用时间后，应定期更换。具有可更换保护柱的 HPLC 柱见图 6-5。

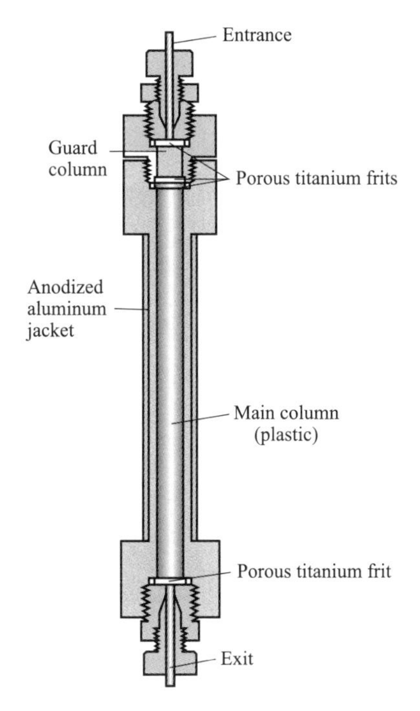

Fig.6-5 HPLC column with replaceable guard column to collect

图 6-5 具有可更换保护柱的 HPLC 柱

② The stationary phase The most common support is highly pure, spherical, microporous particles of silica (figure 6-6) that have a surface area of several hundred square meters per gram. Greater than 99% of this surface area is inside the pores. Pores must be wide enough for the solute and solvent to enter freely. Most silica cannot be used above pH 8, because it dissolves in base.

② 固定相 最常见的载体是高纯度的球形微孔二氧化硅颗粒（图 6-6），其表面积为每克几百平方米，超过 99% 的表面积位于孔隙内。孔隙必须足够宽，以便溶质和溶剂自由进入。大多数二氧化硅不能在 pH 值高于 8 时使用，因为它可溶于碱。

Type B silica which has fewer exposed silanol groups and fewer metallic impurities, is the most common form used today. Bare silica can be used as the stationary phase for adsorption chromatography. Most commonly, liquid-liquid partition chromatography is conducted with a bonded stationary phase covalently attached to the silica surface by reactions such as:

B 型二氧化硅暴露的硅醇基团和金属杂质较少，是目前最常用的形式。裸露的二氧化硅可用作吸附色谱的固定相。最常见的是，液 - 液分配色谱法是通过以下反应将键合固定相共价连接到二氧化硅表面上：

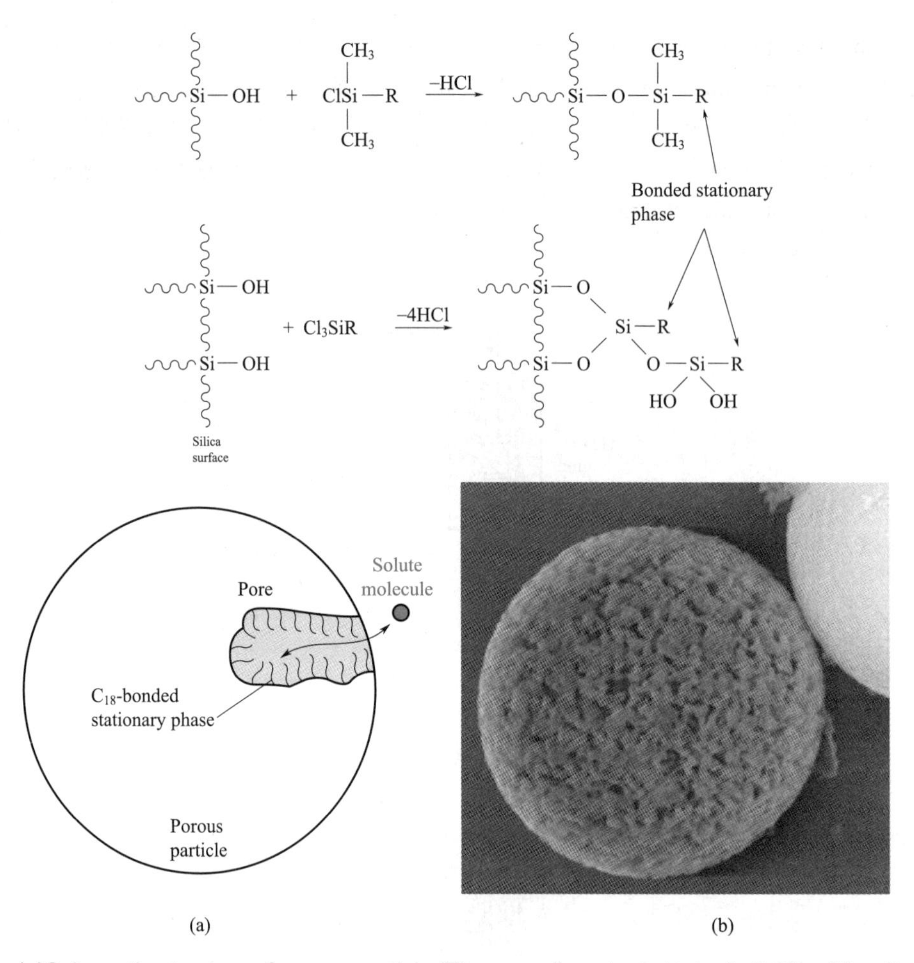

Fig.6-6 (a)Schematic structure of porous particle (The pore diameter is typically 0.2% of the diameter of the particle); (b) Scanning electron micrograph of 4.4-mm-diameter microporous silica chromatography particle from an experimental batch made by K. Wyndham at Waters Corporation

图 6-6 （a）多孔颗粒示意图（孔径通常为颗粒直径的 0.2%）；（b）直径为 4.4 mm 的微孔二氧化硅色谱颗粒的扫描电镜照片

There are about 4 μmol of R groups per square meter of support surface area, with little bleeding of the stationary phase from the column during chromatography. By far, the most commonly used stationary phase is octadecyl (C_{18}, often abbreviated to ODS). Retention factors for a given solute on different nonpolar bonded phases (such as C_4, C_8, and C_{18}) are different. Retention factors for a given solute on C_{18} columns from different manufacturers can vary, in part due to differences in surface area.

每平方米的载体表面约有 4μmol 的 R 基团，分离过程中色谱柱上的固定相几乎没有渗出。十八烷基（C_{18}）固定相（通常缩写为 ODS）是目前最常见的。特定溶质在不同非极性键合相（如 C_4、C_8 和 C_{18}）上的保留因子不同。不同制造商的 C_{18} 色谱柱上给定溶质的保留因子可能也不同，部分原因是表面积不同。

(4) Detector The detector is one of three critical components (pump, column and detector) in HPLC. The commonly used detectors are UV/Vis absorbance, refractive index and electrical

（4）检测器 在高效液相色谱仪中，检测器是三大关键部件（高压输液泵、色谱柱、检测器）之一。常用的检测器有紫外吸收检测器、示差折光检测

conductivity detectors.

(5) Date recording and processing system.

3. Precautions for Instrument

(1) High pressure pumping system

① Although pump can work under maximum pressure 6000 psi (1 psi=6895 Pa), it is better to keep it under pressure below 4000 psi. If the pump stands idle for more than two days, fill the liquid phase with methanol to avoid the possible multiplying of microorganism in the hydrosolvent system.

② Do not let the pump turn idly. Do not operate the injection valve if there is no mobile phase passing through. Even in operation, shift the switch as less as possible to prevent the wear of sealing gasket.

③ Use solvents and reagents with high purity and high quality. The various solvents must be mutually soluble, which is very important to buffer the mobile phase because the deposition of salt will soon damage the components to be maintained.

④ The first choice of mobile phase is methano-water system. If it proves to be unsuitable, choose other solvents. For the purpose of protecting the instrument, use the mobile phase without the buffering agent. If the mobile phase contains buffering agent, flush the instrument circuits, pump, injector, column and detection cell with the mobile phase liquid without buffering agent or fresh purified water after use each day.

⑤ If there exist in the liquid phase the unfiltered eluent, unfiltered sample or buffering eluent, the system may be blocked or the pump plunger may be scratched. Therefore, the mobile phase and the sample must be filter with a 0.45 μm micropore filter membrane and the mobile phase must be degassed. When the pump stops, do not allow the buffering eluent stay in the system. Flush with filtered fresh purified water until the buffering agent is completely cleaned out.

器、电导检测器等。

（5）数据记录处理系统。

3. 仪器使用注意事项

（1）高压输送泵系统

① 虽然泵能耐压 6000psi（1psi=6895Pa）的压力，但不要使泵处于 4000psi 以上的压力为宜。如果泵闲置 2 天以上，应将甲醇注满液相系统，以避免水溶剂体系中可能发生的微生物繁殖。

② 严禁开空泵，在无流动相通过时不要扳动进样阀的操作杆，使用时要注意尽可能少扳动，以免磨损内部的密封垫圈。

③ 应使用高纯度、高质量的溶剂和试剂。使用的所有溶剂必须是互溶的。这对于缓冲流动相来说是非常重要的，盐的沉积将很快损坏要维护的部件。

④ 流动相首选甲醇 - 水系统，如经试用不适合时，再选用其他溶剂。为保护仪器，应尽可能少用含有缓冲液的流动相。如果流动相中含有缓冲剂，每日使用后应用不含缓冲剂的流动相或新鲜纯化水将仪器管路、泵、进样阀、色谱柱及检测池等充分冲洗干净。

Sample filtration
样品过滤

⑤ 如果液相系统使用未过滤的洗脱液、注入未过滤的样品、系统中滞留缓冲洗脱液都能堵塞系统或划伤泵柱塞。所以流动相、样品使用前必须用 0.45μm 的微孔滤膜过滤；流动相并先经脱气处理后使用。停泵后决不允许缓冲洗脱液滞留在系

统中，须用经过滤后的新鲜纯化水进行清洗，并保证将缓冲剂冲洗干净。

(2) Chromatographic column

（2）色谱柱

① Pay attention to the orientation of column when assembling or disassembling it. The reversely mounted column may cause decrease of efficiency or even damage of the column.

① 安装及拆卸色谱柱时应注意柱的连接方向，千万不能接反，否则会导致柱效降低，甚至损坏色谱柱。

② Avoid the overruning of pH value and control it in the range of 2.2-7.5. Low or high pH value will corrode the stainless steel material, damage the packing structure and inactivate the packing of column.

② 控制 pH 值在 2.2 ～ 7.5 之间。pH 值偏低或偏高都会腐蚀液相系统的不锈钢材料，破坏色谱柱填料的结构，使填料失活。

③ The column temperature cannot surpass the stipulated figure. High column temperature will expedite the ageing of packing and damage its structure.

③ 色谱柱温不能超过规定要求，柱温过高会加速色谱柱填料老化，破坏其结构。

④ The column should be preserved with absorbent in humid condition and sealed up at both ends. The reversed-phase column (e.g. C18 column) can be kept in methanol, while normal-phase column (e.g. silica gel column) can be kept in *n*-hexane.

④ 色谱柱保存时应使填充剂处在润湿状态，两端密封。反相柱（如十八烷基硅烷键合硅胶柱）可在甲醇中保存；正相柱（如硅胶柱）可在正己烷中保存等。

⑤ Never use hydrochloric acid. Generally, any halide of any concentration will corrode the non-passivated stainless steel material.

⑤ 决不能使用盐酸溶液。一般来说，任何浓度的卤化物都会腐蚀未经钝化的不锈钢材料。

⑥ To avoid the use of metallic ions that will cause corrosion during electrochemical process, such as manganese, chromium, zinc, copper, nickel, molybdenum and iron typically.

⑥ 尽量避免使用在电化学过程中引起腐蚀的金属离子，如锰、铬、锌、铜、镍、钼、铁。

(3) Detection system　To prolong the life time of detector light, switch on the detector when the pump runs steadily and switch it off immediately after the analysis.

（3）检测系统　为了延长检测器光源的使用寿命，要在色谱泵稳定后，再打开检测器电源开关，分析结束后立即关闭检测器。

Section 3　Detectors for HPLC

第 3 节　高效液相色谱检测器

There are two basic types of detectors for HPLC, one is solute detector, responding only to the physical or

在液相色谱中，有两种基本的检测器类型。一种是溶质型检测器，它仅

chemical properties of the separated compositions, such as UV/Vis, fluorescence and electrochemical detectors; another is general detector responding to all the physical or chemical properties of the reagent and eluent, such as refractive index and conductivity detectors.

对被分离组分的物理或化学特性有响应，这类检测器有紫外、荧光、电化学检测器等。另一类是通用型检测器，它对试样和洗脱液总的物理或化学性质有响应，如示差折光检测器、电导检测器等。

Table 6-3 Properties of common liquid chromatography detectors

Detector Name	Compounds Detected	Gradient Compatible or Not	Detection Limits
General Detectors			
Refractive-index detector	Universal (all compounds)	No	0.1-1μg
UV/Vis absorbance detector	Compounds with chromophores	Yes	0.1-1 ng
Evaporative light-scattering detector	Nonvolatile compounds	Yes	10μg
Selective Detectors			
Fluorescence detector	Fluorescent compounds	Yes	1-10 pg
Conductivity detector	Ionic compounds	No	0.5-1 ng
Electrochemical detector	Electrochemically active compounds	No	0.01-1 ng
Structure-Specific Detectors			
Mass Spectrometry	Universal (Full-scan mode) Selective (SIM mode)	Yes	0.1-1 ng

表6-3 常用液相色谱检测器的特性

检测器名称	检测的化合物	同梯度是否兼容	检出限
通用型检测器			
折射率检测器	所有化合物	否	0.1 ～ 1μg
紫外可见吸收检测器	带有发色团的化合物	是	0.1 ～ 1 ng
蒸发光散射检测器	非挥发性化合物	是	10μg
选择性检测器			
荧光检测器	荧光化合物	是	1 ～ 10 pg
电导检测器	离子化合物	否	0.5 ～ 1 ng
电化学检测器	电化学活性化合物	否	0.01 ～ 1 ng
结构特异性检测器			
质谱检测器	通用（全扫描模式） 选择性（SIM 模式）	是	0.1 ～ 1 ng

1. General Detectors

(1) UV/Vis absorbance detector An ultraviolet detector using a flow cell such as that in figure 6-7 is the most common HPLC detector because many solutes absorb ultraviolet light. Simple systems employing the intense 254 nm emission of a mercury vapor lamp were the backbone of early HPLC systems, but are

1. 通用型检测器

（1）紫外吸收检测器（UV） 许多溶质都有紫外吸收，配有流通池（如图 6-7 所示）的紫外检测器是最常见的 HPLC 检测器。早期的 HPLC 系统，常用 254nm 汞蒸气灯作光源，但目前很少使用。具备更多功能的可变波长检测

little used today. More versatile, variable-wavelength detectors have broadband deuterium, xenon, or tungsten lamps and a monochromator, so you can choose an optimum wavelength for your analytes. At wavelengths above 210 nm, detection is selective for compounds with an absorbing chromophore. Many compounds absorb wavelengths below 210 nm and so ultraviolet detection<210 nm is nearly universal. The system in figure 6-8 uses a photodiode array (PDAD) to record the spectrum of each solute as it is eluted. The spectrum can be matched with library spectra to help identify the peak. Spectra collected across a peak can be compared to evaluate peak purity and detect coeluting components. Characteristics of this detector are high sensitivity, wide linear range, insensitive to

器装有宽带氘灯、氙灯或钨灯和一个单色器，故可以根据分析需要选择一个最佳吸收波长。当波长超过 210nm 时，检测器对含生发色团的化合物具有选择性。许多化合物吸收波长低于 210nm 的光，因此 <210nm 的紫外检测器应用普遍。图 6-8 中的系统使用光电二极管阵列（PDAD）记录被洗脱的每个组分的吸收光谱。将该光谱与光谱文献相比对，可协助识别色谱峰。通过一个色谱峰收集的光谱，可以评估峰纯度并检测共洗脱组分。该检测器具有灵敏度高、线性范围宽、对流动相的流速和温度变化不敏感，波长可选，易于操作，可用于梯度洗脱等特点。

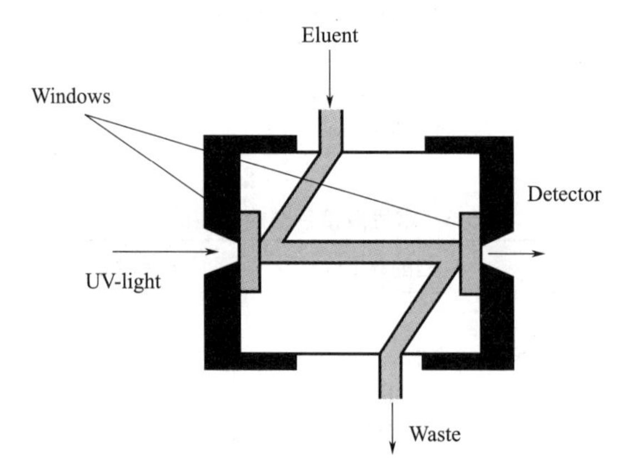

Fig.6-7 Liquid flow cell for UV detection in HPLC

图 6-7 HPLC 紫外检测器的 Z 型液体流通池

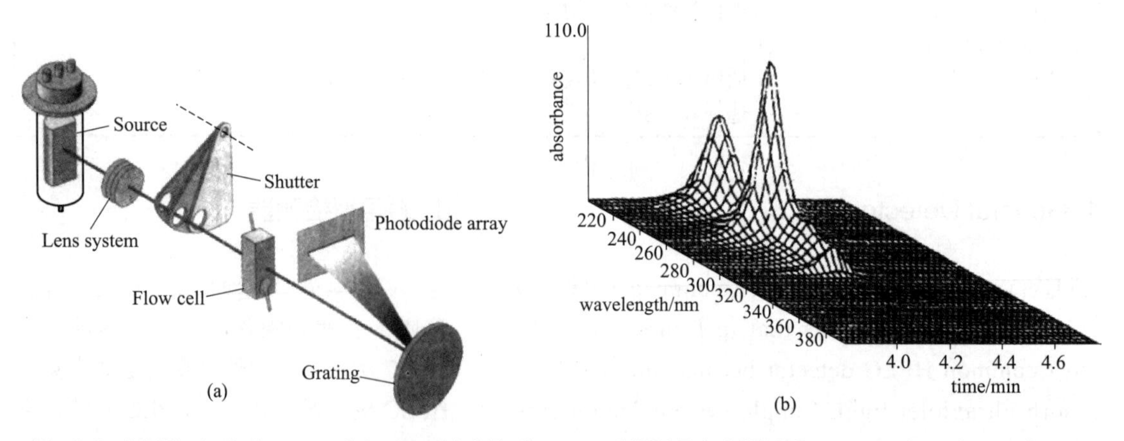

Fig.6-8 (a) Photodiode array detector; (b) 3-D diagram of PDAD in HPLC for measuring phenanthrene

图 6-8 （a）二极管阵列检测器的光路图；（b）PDAD 检测菲的 3-D 色谱光谱图

the flow rate and temperature of the mobile phase, selectable wavelength, easy to operate and applicable to gradient elution.

(2) Evaporative light-scattering detector An evaporative light-scattering detector responds to any analyte that is significantly less volatile than the mobile phase. In figure 6-9, eluate enters the detector at the top. In the nebulizer, eluate is mixed with nitrogen gas and forced through a small-bore needle to form a uniform dispersion of droplets. Solvent evaporates from the droplets in the heated drift tube, leaving a fine mist of solid particles to enter the detection zone at the bottom. Particles are detected by the light that they scatter from a diode laser to a photodiode. Detector response is related to the mass of analyte, not to the structure or molecular mass of the analyte.

（2）蒸发光散射检测器（ELSD） 蒸发光散射检测器对挥发性明显低于流动相的任何分析物都有响应。在图 6-9 中，洗脱液首先进入检测器顶部的雾化器中，与氮气混合，并强制通过小孔针头形成均匀分散的液滴。溶剂从加热漂移管中的液滴中蒸发掉，留下一层由固体颗粒组成的细雾进入底部检测区。固体颗粒通过一个激光束，发生光散射，并被与气流形成 90° 的方向的光电二极管检测出来。检测器响应信号与分析物的质量有关，而与分析物的结构或分子量无关。

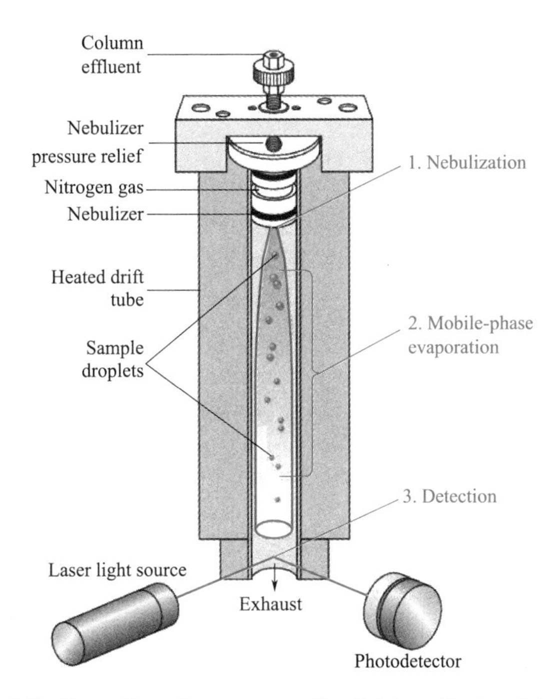

Fig.6-9 Operation of an evaporative light-scattering detector

图 6-9 蒸发光散射检测器的原理

(3) Refractive-index detector RID is the second widely used detector (fig.6-10). RID works by continuously monitoring the differential refractive index of solutions in reference cell and measuring cell to determine the concentration of analyte. The sensitivity of the detector

（3）示差折光检测器（RID） 示差折光检测器是除紫外检测器之外应用最多的检测器（图 6-10）。其通过连续检测参比池和测量池中溶液的折射率之差，以确定被分析物的浓度。检测器的

is related to the natures of solvent and solute. The refractive index difference between the mobile phase and the mixture of mobile phase and sample solution indicates the concentration of sample solution in the mobile phase. It shows that the response signal of refractive index detector is directly proportional to the concentration of solute, indicating that RID belongs to concentration detector. Each substance has a specific reflective index. Therefore, RID is a general type detector, and all the samples different from solvent can be detected by RID in principle. The higher the concentration of sample is, the differential refractive index between solute and solvent is, the greater the response signal of detector is.

灵敏度与溶剂和溶质的性质有关系，溶有样品的流动相和流动相本身之间折射率之差反映了样品在流动相中的浓度，即示差折光检测器的响应信号与溶质的浓度成正比，说明它属于浓度型检测器。每种物质都有一定的折射率，原则上只要与溶剂有差别的样品都可以用该检测器检测。因此，示差折光检测器是一种通用型检测器。样品的浓度越高，溶质与溶剂的折射率差别越大，检测器的响应信号越大。

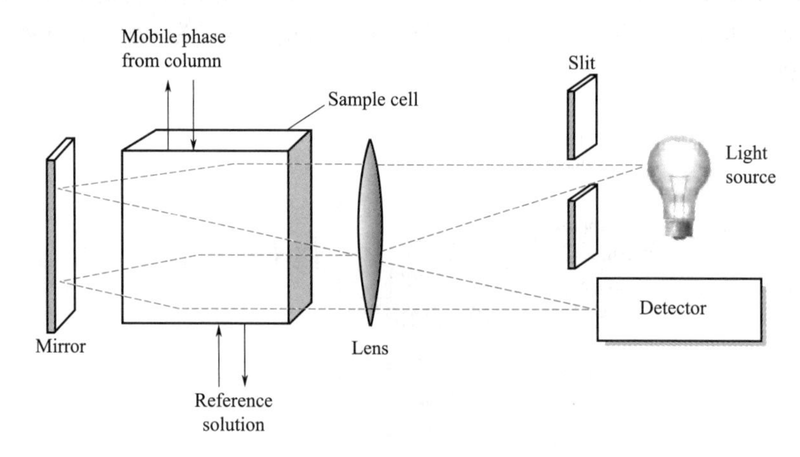

Fig.6-10 General design of a refractive-index detector

图 6-10 示差折光检测器的原理

2. Selective Detectors

(1) Electrolytic Conductivity detector A electrolytic conductivity detector is a device that can monitor ionic compounds in HPLC by measuring the ability of the mobile phase and its contents to conduct a current when placed in an electrical field. The design of such a detector is shown in figure 6-11. This design consists of a flow cell and two electrodes. The electrodes apply an electric field to the solution in the flow cell and measure the resulting current. Electrolytic conductivity detectors can be used to detect any compound that is ionic. These detectors are widely used in ion chromatography and

2. 选择性检测器

（1）电导检测器（ECD） 电导检测器是一种通过测量流动相及其含量在电场中传导电流的能力来检测 HPLC 中的离子化合物的装置，如图 6-11 所示。该检测器由一个流通池和两个电极组成。电极在流通池溶液中施加一个电场，并测量其所产生的电流。电导检测器可用于检测任何含离子的化合物，被广泛应用于离子色谱及食品、工业样品和环境样品的离子成分的分析。只要流动相的离子强度（可能是 pH）保持不

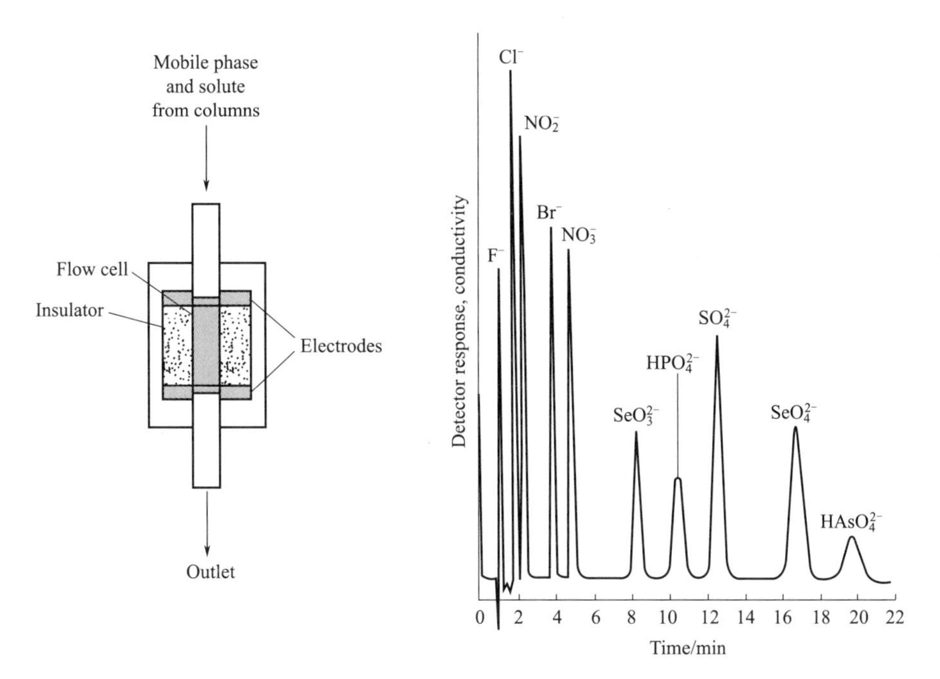

Fig.6-11 General design of a conductivity detector and an example of the use of this detector in ion chromatography(The chromatogram is reproduced with permission of C.F.)

图 6-11 电导检测器的原理及其在离子色谱中的应用示例

in the analysis of ionic components of foods, industrial samples, and environmental samples. This type of device can be used with gradient elution as long as the ionic strength (and possibly pH) of the mobile phase is kept constant. It is also necessary for the background conductance of the mobile phase to be sufficiently low so that sample ions can be detected.

变，该装置就可以进行梯度洗脱。流动相的背景电导也必须足够低，以便能够检测到样品离子。

(2) Electrochemical detector An electrochemical detector is another device used to monitor specific compounds in HPLC (see figure 6-12). This combination is known as liquid chromatography/electrochemical detection (or LC/EC). An electrochemical detector can be used to measure the ability of an analyte to undergo either oxidation or reduction. One way such a reaction can be monitored is by measuring the change in current (i.e., electron flow) that a reaction produces when present at a constant potential. Another way is to measure the change in the potential when a constant current is applied. Examples of compounds that may be detected by their reduction include aldehydes, ketones, oximes, conjugated acids, esters, nitriles,

（2）电化学检测器（ED） 电化学检测器是用于检测 HPLC 中特定化合物的另一种装置（见图 6-12），被称为液相色谱 / 电化学检测（或 LC/EC）。电化学检测器可用于测定被分析物的氧化或还原能力。检测方法有两种：一种是通过测定氧化还原反应在恒定电位存在时产生的电流变化；另一种方法是测定施加恒定电流时的电位变化。通过还原检测到的化合物包括醛、酮、肟、共轭酸、酯、腈、不饱和化合物、芳香烃和活性卤化物。通过氧化检测到的化合物包括酚类、硫醇、过氧化物、芳香胺、二胺、嘌呤和一些碳水化合物。电化学

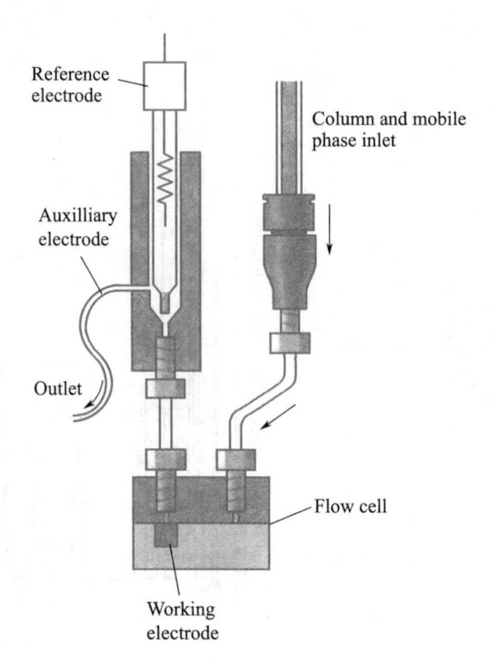

Fig.6-12 Design of an electrochemical detector for liquid chromatography

图 6-12 液相色谱电化学检测器的结构

unsaturated compounds, aromatics, and activated halides. Compounds that may be detected by their oxidation include phenols, mercaptans, peroxides, aromatic amines, diamines, purines, and some carbohydrates. The response of an electrochemical detector depends on the extent of oxidation or reduction that occurs at the given potential of the electrode.

检测器的响应取决于电极在给定电位下发生的氧化或还原的程度。

(3) Fluorescence detector There isa specific detector, fluorescence detector, that can also be used in HPLC. This device measures the ability of chemicals to absorb and emit light at a particular set of wavelengths. Because these wave-lengths are characteristic of a given chemical, this method can provide a signal that has a low background and is reasonably specific for the analyte of interest. Fluorescence can be used to selectively detect any analyte absorbing and emitting light at the given excitation and emission wavelengths. Although relatively few chemicals are fluorescent, those that do fluoresce are frequently of great importance. Examples include many drugs and their metabolites, food additives, and environmental pollutants. Fluorescence can also be used to detect analytes that are

（3）荧光检测器（FD） 有一种特殊的检测器——荧光检测器，也可用于 HPLC。该装置测量化学物质在特定波长下吸收和发射光的能力。由于这些波长是给定化学物质的特征波长，故其可以提供具有低背景的信号，并且对感兴趣的分析物具有合理的特异性。荧光可用于选择性地检测任何分析物在给定的激发和发射波长下的吸收和发射。虽然具有荧光的化学物质相对较少，但这类物质往往非常重要。例如，许多药物及其代谢物、食品添加剂和环境污染物等。荧光也可以用来检测可以转化为荧光衍生物的分析物，包括醇、胺、氨基酸和蛋白质。

first converted to a fluorescent derivative; compounds that can be detected this way include alcohols, amines, amino acids, and proteins.

(4) Charged aerosol detector   The charged aerosol detector is a sensitive, almost universal detector with nearly equal response to equal masses of nonvolatile analytes. At the top left of the charged aerosol nebulizer in figure 6-13, eluate and N_2 gas enter a nebulizer similar to the premix burner in AAS. Fine mist from the nebulizer reaches the drying tube, while larger droplets fall to the drain. In the drying tube, solvent evaporates at ambient temperature, leaving an aerosol containing about 1% of the original analyte. Meanwhile, part of the N_2 stream passes over a Pt needle held at about +10 kV with respect to the outer case of the corona charging chamber to form N_2^+. A chain of events, perhaps as written for atmospheric pressure chemical ionization, transfers positive charge to aerosol particles that flow from the charging chamber through a small-ion trap.

（4）电雾式检测器（CAD） 电雾式检测器是一种高灵敏度的通用型检测器，能为难挥发性化合物提供一致响应性。在图 6-13 电雾式检测器的左上角，洗脱液和 N_2 气体进入类似于 AAS 的雾化室。在 N_2 的辅助下，雾化室中的气溶胶到达干燥管，大液滴直接落入废液管；在干燥管中，溶剂在环境温度下蒸发，留下含有约 1% 原始分析物的气溶胶。同时，第二路 N_2 经过 +10kV 高压电晕放电针（高压铂金丝电极），变成带正电的 N_2^+，在碰撞室与干燥的目标化合物进行正交碰撞后，带电的 N_2 包裹在颗粒物表面，把电荷转移给目标化合物颗粒并使其带上电，然后离开碰撞室。迁移速度较快的多余带电 N_2^+ 被一个低压离子

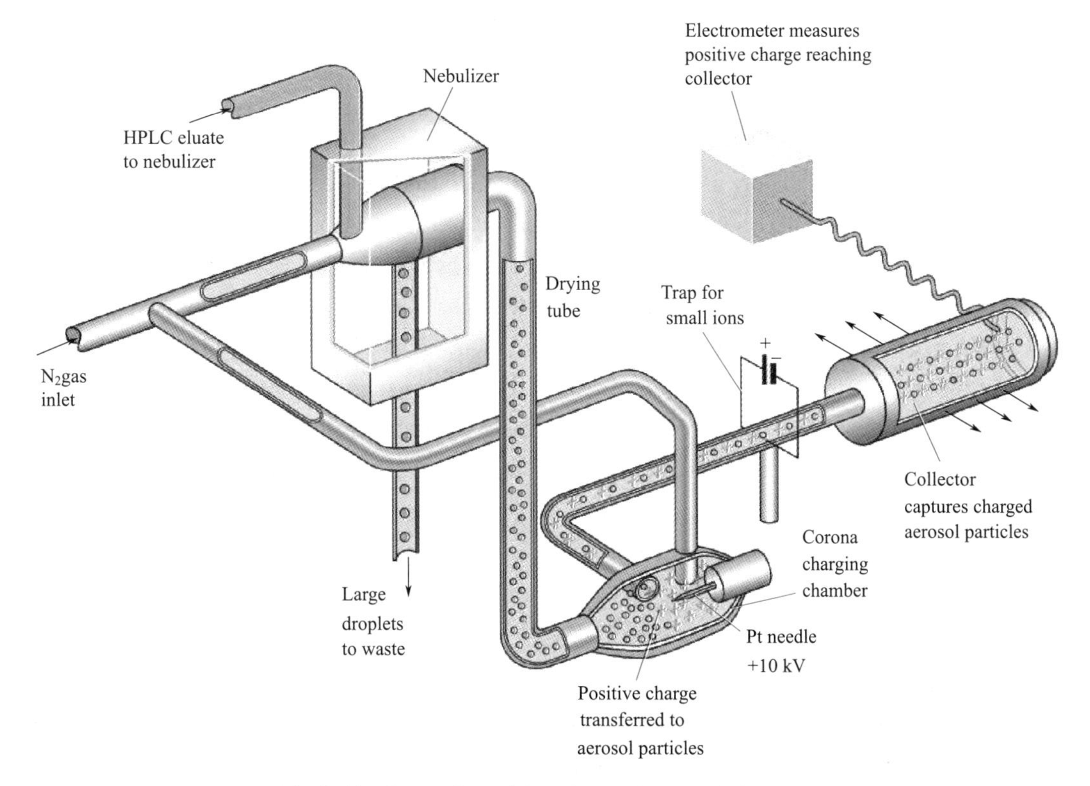

Fig.6-13 Operation of the charged aerosol detector

图 6-13 电雾式检测器的基本原理

Charged plates of the trap attract and remove small mobile ions. Aerosol particles are too massive to be deflected and pass through the trap to the collector. Total charge reaching the collector is measured by an electrometer, which produces the detector signal for the chromatogram. The dynamic range of the detector spans 4-5 orders of magnitude in concentration. Response depends on solvent composition, with higher response for higher percentage of volatile organic solvent and lower response for water. As the organic composition of a gradient increases, so does the response. However, the detector can still be used with gradient elution.

阱捕集移除；迁移率较慢的多电荷目标化合物颗粒通过离子捕集阱，撞向颗粒物收集器，并把电荷转移到过滤网上，由此产生的电流被高灵敏度的静电计测得，静电计产生色谱图的检测信号。电雾式检测器的检出限较低，灵敏度较高；动态检测范围宽，可达4～5个数量级。响应取决于溶剂组成，对挥发性有机溶剂百分比越高，响应越高，对水的响应越低。随着梯度中有机成分的增加，响应随之增大。然而，该探测器仍然可以用于梯度洗脱法。

Experiment 5 Analysis of *p*-hydroxybenzoate mixtures by reversed-phase HPLC

1. Purpose of the experiment

(1) Learn the qualitative by retention value and quantification by normalization method of High performance liquid chromatography;

(2) Be familiar with Agilent 1260 HPLC operation.

Operation of Agilent 1260 HPLC

2. Experimental principle

Paraben is one of the commonly used preservatives in food, widely existing in soy sauce, vinegar and cosmetics. *P*-hydroxybenzoic acid ester includes: *p*-hydroxybenzoic acid methyl ester, *p*-hydroxybenzoic acid ethyl ester, *p*-hydroxybenzoic acid propylene ester, *p*-hydroxybenzoic acid butyl ester. They are strongly polar compounds and can be analyzed by reverse-phase liquid chromatography with non-polar C_{18} alkyl bonded phase as stationary phase and methanol aqueous solution as mobile phase.

Retention values of each component of esters remain constant under certain experimental conditions. Therefore, under the same conditions, the existence of each component in the unknown can be determined by comparing the retention time of each component of the unknown with that of the known pure ester.

Quantitative normalization method is adopted in this experiment.

$$c=\frac{f_i A_i}{\sum_{i=1}^{n} f_i A_i}\times 100\%$$

The mixture of *p*-hydroxybenzoic acid esters is a homologue with the same chromophore and chromophore, so they have the same correction factor on the ultraviolet detector, the above equation can be simplified as:

$$c=\frac{A_i}{\sum_{i=1}^{n} A_i}\times 100\%$$

3. Apparatus and reagents

(1) Apparatus Agilent 1260 Infinity LC (liquid chromatograph) [reversed-phase bonded phase ODS column (5 μm, 4.6 mm×150 mm)]; ultrasonic cleaning instrument;ultrapure water preparation instrument; analytical balance.

(2) Reagents Methyl paraben; ethyl paraben; propyl paraben; butyl paraben; methanol; ultrapure water.

(3) Preparation of standard solution

① The standard storage solution: Methanol solutions of the four ester compounds are prepared in four 100 mL volumetric bottles at a concentration of 1000μg · mL^{-1}.

② Standard solution: Methanol solutions of four ester compounds with the concentration of 100μg · mL^{-1} are prepared in four 10 mL volumetric bottles using the four standard storage liquids.

③ Standard mixed solution: Methanol solutions containing 100μg · mL^{-1} ester mixtures are prepared in a 10 mL volumetric flask using the four standard storage liquids.

4. Experimental steps

(1) Experimental conditions

① Column: length 15 cm, the inner diameter 3 mm, filled with C_{18} alkyl bonded phase and the stationary phase with particle size of 10 μm.

② Mobile phase: methanol: water (40 ∶ 60), flow rate 1 mL · min^{-1}.

③ Detector: UV detector (254 nm), sensitivity 0.04.

④ The injection volume: 1μL.

(2) Preparation before experiment

① Pretreatment for mobile phase: 2000 mL of HPLC-grade anhydrous methanol and 2000 mL of secondary distilled water are respectively filtered by 0.45 μm organic filter membrane and then pour them into mobile phase liquid storage tank, degas them with an ultrasonic cleaner for 10-20 min.

② Turn onthe chromatographic workstation, set the experimental conditions, and adjust to the injection state according to apparatus' operation steps. When the liquid circuit and circuit system of the apparatus reach balance, and the baseline of recorder is horizontal, the sample can be injected.

③ 1μL of four standard solutions, standard mixed solutions and unknown test solutions are respectively injected, and the chromatograms are recorded and repeat twice.

5. Data and processing

① Record experimental conditions.

② Measure the retention time (t_R) of chromatographic peaks of four *p*-hydroxybenzoate compounds and fill them in the table below.

Component	t_R /min			
	1	2	3	The average
Methyl parabolic benzoate				
Ethyl *p*-hydroxybenzoate				
Propyl *p*-hydroxybenzoate				
Butyl *p*-hydroxybenzoate				

③ Measure the retention time (t_R) of each chromatographic peak on the chromatogram of standard mixed solution in turn and fill in the following table. Then compare with the values in the above table to determine what compounds each chromatographic peak represents and fill in the following table.

Chromatographic peak	t_R/ min				Name of the corresponding compound
	1	2	3	The average	
Peak 1					
Peak 2					
Peak 3					
Peak 4					

④ The percentage of c_i for each component is calculated according to the data of peak area for the chromatographic peak.

6. Questions

① What are the advantages and disadvantages of normalization for HPLC analysis?Why the relative quality correction factor is not necessary in this experiment?

② Why is the mobile phase degassed during HPLC analysis?How does the non-degassing of the mobile phase interfere with the experiment?

实验 5 对羟基苯甲酸酯类混合物的反相高效液相色谱分析

1. 实验目的

（1）学习高效液相色谱保留值定性方法和归一化法定量；

（2）熟悉安捷伦 1260 高效液相色谱仪操作。

安捷伦 1260 高效液相色谱仪的操作

2. 实验原理

对羟基苯甲酸酯类是食品中常用的防腐剂之一，广泛存在于酱油、醋、化妆品中。对羟基苯甲酸酯类包括对羟基苯甲酸甲酯、对羟基苯甲酸乙酯、对羟基苯甲酸丙酯、对羟基苯甲酸丁酯等。它们都是强极性化合物，可采用反相液相色谱进行分析，选用非极性的 C_{18} 烷基键合相作固定相，甲醇的水溶液作流动相。

在一定的实验条件下，由于酯类各组分的保留值保持恒定，因此在同样条件下，将测得的未知物的各组分保留时间，与已知纯酯类各组分的保留时间进行对照，即可确定未知物中各组分存在与否。

本实验采用归一化法定量：

$$c=\frac{f_iA_i}{\sum_{i=1}^{n} f_iA_i}\times 100\%$$

对羟基苯甲酸酯类混合物属同系物，具有相同的生色团和助色团，因此它们在紫外光度检测器上具有相同的校正因子，故上式可简化为：

$$c=\frac{A_i}{\sum_{i=1}^{n} A_i}\times 100\%$$

3. 仪器和试剂

（1）仪器　安捷伦高效液相色谱仪 LC1260，反相键合相 ODS 色谱柱（5μm，4.6mm×150 mm）、超声波清洗仪、超纯水制备仪、万分之一天平。

（2）试剂　对羟基苯甲酸甲酯、对羟基苯甲酸乙酯、对羟基苯甲酸丙酯、对羟基苯甲酸丁酯、甲醇（均为分析纯）、超纯水。

（3）标准溶液的配制

① 标准贮备液　分别于四只 100mL 容量瓶中，配制浓度均为 $1000\mu g\cdot mL^{-1}$ 的上述四种酯类化合物的甲醇溶液。

② 标准使用液　用上述四种标准贮备液分别于四只 10mL 容量瓶中，配制浓度均为 $100\mu g\cdot mL^{-1}$ 的四种酯类化合物的甲醇溶液。

③ 标准混合使用液　于一只 10mL 容量瓶中，用上述四种标准贮备液，配制浓度均含 $100\mu g\cdot mL^{-1}$ 的酯类混合物的甲醇溶液。

4. 操作步骤

（1）实验条件

① 色谱柱　长 15cm、内径 3mm，装填 C_{18} 烷基键合相，颗粒度为 10μm 的固定相。

② 流动相　甲醇∶水 =40 ∶ 60，流量 $1mL\cdot min^{-1}$。

③ 检测器　紫外光度检测器，254nm，灵敏度 0.04。

④ 进样量　1μL。

（2）准备工作

① 流动相的预处理　取 HPLC 级无水甲醇 2000mL，二次蒸馏水 2000mL，分别用

0.45μm 的有机滤膜过滤后，装入流动相贮液器内，用超声波清洗器脱气 10 ～ 20min。

② 开启色谱工作站，设置实验条件，按照仪器的操作步骤调节至进样状态，待仪器液路和电路系统达到平衡时，记录仪基线呈平直，即可进样。

③ 依次分别吸取 1μL 的四种标准使用液、标准混合使用液和未知试液进样，记录各色谱图，并重复两次。

5. 数据及处理

① 记录实验条件。

② 测量四种对羟基苯甲酸酯化合物色谱峰的保留时间 t_R 值，并填入下表中。

组分	t_R /min			
	1	2	3	平均值
对羟基苯甲酸甲酯				
对羟基苯甲酸乙酯				
对羟基苯甲酸丙酯				
对羟基苯甲酸丁酯				

③ 请依次测量标准混合溶液色谱图上各色谱峰的保留时间 t_R 值，并填入下表中，然后与上表中的值对照，确定各色谱峰代表何种化合物，填入下表中。

色谱峰	t_R/min				相应化合物的名称
	1	2	3	平均值	
峰 1					
峰 2					
峰 3					
峰 4					

④ 根据色谱峰峰面积数据计算各组分含量 c_i。

6. 思考题

① 高效液相色谱分析采用归一化法定量有何优缺点？为什么本实验中不需要用相对质量校正因子？

② 在高效液相色谱分析中流动相为何要脱气？流动相不脱气对实验有何妨碍？

Exercises

6-1 In gas-liquid chromatography, (　　) of flame ionization detector is superior to thermal conductivity detector.

A. device simplification　　B. sensitivity

C. scope of application D. separation effect

6-2 The position of the chromatographic peak in the chromatogram is indicated by ( ).

A. retention value B. peak height value C. peak width value D. sensitivity

6-3 In a gas-liquid chromatography column, the factor independent of resolution is to ( ).

A. increase the column length B. change to a more sensitive detector

C. adjust the flow rate D. change the chemical properties of the fixative

6-4 In chromatographic analysis, the advantage of normalization is ( ).

A. that precise injections are not required B. no correction factor

C. that qualitative are not required D. no standard sample

6-5 The purpose of stationary phase aging is to ( ).

A. remove the moisture adsorbed on the surface

B. remove the powdery substances in the stationary phase

C. remove the residual solvent and other volatile substances in the stationary phase

D. improve the separation efficiency

6-6 To determine the molecular weight and molecular weight distribution of polyethylene, the chromatography that should be selected is ( ).

A. Liquid-Liquid Partition Chromatography B. Liquid-Solid Adsorption Chromatography

C. Bonded Phase Chromatography D. Gel Chromatography

6-7 In liquid chromatography, the most effective way to improve column efficiency is to ( ).

A. increase column temperature B. decrease plate height

C. decrease mobile phase flow rate D. decrease particle size

6-8 The general-purpose detector in liquid chromatography is ( ).

A. Ultraviolet absorption detector B. Differential refractive index detector

C. Thermal conductivity cell detector D. Hydrogen flame detector

6-9 True or false

(1) Relative retention values are only related to column temperature and stationary phase properties, not to operating conditions.

(2) The selectivity of a chromatographic column can be expressed by total separation efficiency index, which can be defined as the ratio of the difference between the retention times of two adjacent chromatographic peaks and the sum of the widths of the two chromatographic peaks.

(3) Three peaks appear on the chromatogram of a sample with a maximum of three components.

(4) Set the column each time when a new column is installed.

(5) When determining the liquid-to-charge ratio, the type of the carrier, the boiling point of the sample, and the injection volume should be considered.

(6) The choice of flow is more important than that of column temperature in LC analysis.

(7) The commonly-used mobile phase in reversed-phase bonded liquid chromatography is water-methanol.

(8) In high performance liquid chromatography, the pre-column in front of the column will reduce efficiency of the column.

6-10 What is chemically bonded phase chromatography? What are its characteristics?

6-11 (a) Why does mobile phase strength increase as solvent becomes less polar in reversed-phase chromatography, whereas mobile phase strength increases as solvent becomes more polar in normal-phase chromatography? (b) What kind of gradient is used in supercritical fluid chromatography?

6-12 Why are the relative eluent strengths of solvents in adsorption chromatography fairly independent of solute?

6-13 In hydrophilic interaction chromatography (HILIC), why is eluent strength increased by increasing the fraction of water in the mobile phase?

6-14 (a) Why is high pressure needed in HPLC? (b) For a given column length, why do smaller particles give a higher plate number? (c) What is a bonded phase in liquid chromatography?

6-15 A known mixture of compounds A and B gave the following HPLC results:

Compound	Concentration ($mg \cdot mL^{-1}$ in mixture)	Peak area (arbitrary units)
A	1.03	10.86
B	1.16	4.37

A solution was prepared by mixing 12.49 mg of B plus 10.00mL of unknown containing just A and diluting to 25.00mL. Peak areas of 5.97 and 6.38 were observed for A and B, respectively. Find the concentration of A ($mg \cdot mL^{-1}$) in the unknown.

6-16 The figure shows the separation of two enantiomers on a chiral stationary phase.

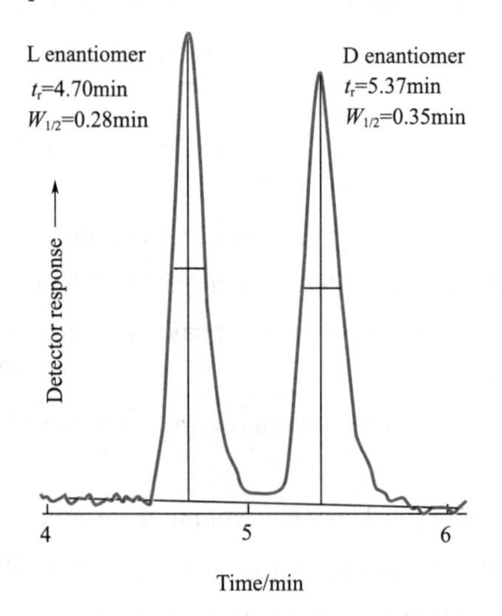

(a) From t_r and $W_{1/2}$, find N for each peak. (b) From t_r and $W_{1/2}$, find the resolution.

Chapter 7 Analysis of Petroleum Products

第 7 章　石油产品分析

Study Guide　学习指南

This chapter covers the basic concepts and the significance of correlated indexes in the determination of petroleum products. The target of this chapter is to familiarize with the meanings, expressions and significance of the basic properties of petroleum products such as physical and chemical properties, vaporizability, cryogenic flowability, flammability, corrosiveness and stability.

本章主要介绍石油产品性能测定中所涉及的基本概念及相关指标的测定意义。通过本章的学习主要掌握油品基本理化性质、蒸发性能、低温流动性能、燃烧性能、腐蚀性能、安定性能等指标的表示方法及含义，了解各项指标的测定意义。

Section 1　Introduction

第 1 节　概述

1. Oil Composition

(1) Elemental composition of oil　**Crude oil is mainly composed of carbon, hydrogen, sulfur, nitrogen and oxygen (with some trace elements)**. In crude oil, carbon occupies 83%-87% in mass fraction, hydrogen 11%-14%, sulfur 0.05%-8%, nitrogen 0.02%-2%, oxygen 0.05%-2%.

(2) Compound composition of oil　In terms of chemical composition, The compound compositions in oil belong to two categories: hydrocarbon and non-hydrocarbon (derivatives of hydrocarbon and inorganics).

1. 石油的组成

（1）石油的组成元素　原油主要由碳、氢、硫、氮、氧五种元素（还有一些微量元素）所组成。原油中：碳的质量分数一般为 83% ～ 87%；氢的质量分数为 11% ～ 14%；硫的质量分数为 0.05% ～ 8%；氮的质量分数为 0.02% ～ 2%；氧的质量分数为 0.05% ～ 2%

（2）石油的化合物组成　从化学组成来看，石油中主要含有烃类和非烃类（烃的衍生物和无机物）这两大类。

Hydrocarbon can be broken down to paraffin hydrocarbons, naphthenic hydrocarbons and aromatic hydrocarbons, and the hydrocarbon mixture of the above three hydrocarbons.

石油中烃类主要是由烷烃、环烷烃和芳香烃以及在分子中兼有这三类烃结构的混合烃构成。

2. Purpose of Analysis of Petroleum Products

The analysis of petroleum products is a scientific experiment by using stipulated or generally accepted methods to analyze the physical and chemical properties and applications of oil products. The main tasks of oil analysis are as follows:

(1) Familiarize with the composition and property of raw oil and raw materials in petroleum process, furnish with the basic data for processing plan.

(2) Monitor the production process, supply the data for process condition control.

(3) Check the quality of oil products when they leave the factory.

(4) Evaluate the performances of petroleum products.

(5) Provide arbitration basis for oil quality.

2. 石油产品分析的任务

油品分析是用统一规定或公认的方法，分析检验石油和石油产品理化性质、使用性能的科学试验。其主要任务如下：

（1）了解用于石油加工的原料油及原材料的组成和性能，为确定加工方案提供基础数据。

（2）对生产过程进行监控，为控制工艺条件提供数据。

（3）检验出厂油品质量。

（4）评价油品使用性能。

（5）对油品质量仲裁。

Section 2 Determination of Basic Physiochemical Properties of Petroleum Products

第 2 节 油品基本理化性质的测定

1. Density and Relative Density

(1) Concepts

① Density is the mass of a material per unit volume, the symbol is ρ, with unit $g \cdot mL^{-1}$ or $kg \cdot m^{-3}$.

② Relative density is the ratio of the viscosity of a solution under certain temperature to the viscosity of a standard material under specific temperature.

1. 密度和相对密度

（1）相关概念

① 单位体积物质的质量称为密度，符号 ρ，单位 $g \cdot mL^{-1}$ 或 $kg \cdot m^{-3}$。

② 物质的相对密度是指物质在给定温度下的密度与规定温度下标准物质的密度之比。

(2) Correlation of oil density and temperature

（2）油品密度与温度的关系

$$\rho_{20}=\rho_t+\gamma(t-20) \tag{7.1}$$

ρ_{20}—standard density, g · cm^{-3};

ρ_t—observant density, g · cm^{-3};

t—measuring temperature, ℃ ;

γ—average density, g · cm^{-3} · ℃$^{-1}$.

ρ_{20}—标准密度，g · cm^{-3}；

ρ_t—视密度，g · cm^{-3}；

t—测量温度，℃；

γ—平均密度温度系数，g·cm^{-3}·℃$^{-1}$。

(3) Significance of determination of oil density

① Calculate oil properties.
② Evaluate oil quality.
③ Affect performance of fuel oil.

（3）测定油品密度的意义

① 计算油品性质。
② 判断油品质量。
③ 影响燃料的使用性能。

(4) Method for determination of oil density (density-hydrometer method) *Crude Petroleum and Liquid Petroleum Products—Laboratory Determination of Density(Hydrometer Method)* (GB/T 1884—2000). The schematic diagram and the appearance of density hydrometer are in fig.7-1 and fig.7-2.

（4）测定油品密度的方法 《原油和液体石油产品密度实验室测定法（密度计法）》（GB/T 1884—2000）。密度计示意图和外形图见图 7-1、图 7-2。

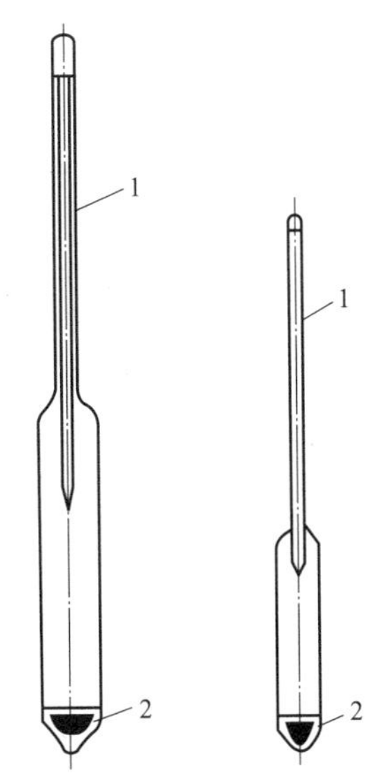

Fig.7-1 Schematic diagram of the density-hydrometer
1—scaleplate; 2—pressure load room

图 7-1 密度计的示意图
1—刻度；2—压载室

Fig.7-2 SYD-1884 petroleum density tester

图 7-2 SYD-1884 型石油产品密度计

2. Viscosity

Viscosity is one of the physiochemical properties of fluids, reflecting the friction between the molecule. It is used to evaluate the flowability of crude oil and its products. Viscosity decreases with increasing temperature. It is a very useful physical constant in the processing, transportation, management, sales and use of the crude oil and the petrochemical products. For instance, viscosity has great influence on flowrate and pressure drop in transportation process. It is an important quality index in the inspection of petrochemical products.

(1) Expressions of viscosity

① Dynamic viscosity　The dynamic viscosity is the frictions produced by the relative motion of two liquids with area 1cm^2, perpendicular distance 1cm and speed 1 cm · s^{-1}. It is denoted by η, with unit Pa · s.

② Kinematic viscosity　The kinematic viscosity is the ratio of the dynamic viscosity to its density of the fluid at the same temperature. It is usually denoted by v_t, with unit m^2 · s^{-1}.

2. 黏度

黏度是流体的理化性质之一，它反映了液体内部分子间的摩擦力，是评价原油及其产品流动性能的指标。黏度值随温度的升高而降低。在原油和石油化工产品加工、运输、管理、销售及使用过程中，黏度是很有用的物理常数。如输送过程中，黏度对流量、压力降影响很大。在石油加工中，黏度是检验许多石油产品的重要质量指标。

（1）黏度的表示方法

① 动力黏度　它是两液体层面积为 1cm^2，垂直距离为 1cm，以 1cm · s^{-1} 的速度做相对运动时所产生的内摩擦力。用符号 η 表示，单位为 Pa · s。

② 运动黏度　在同一温度下液体的动力黏度与该液体的密度之比，称为该流体的运动黏度，用符号 v_t 表示，单位为 m^2 · s^{-1}。

$$v_t=\frac{\eta}{\rho_t} \qquad (7.2)$$

③ Specific viscosity　The specific viscosity is measured in conditional units by different specific viscometers. The specific viscosity can relatively measure the flowability of oil products, but it does not have any physical meaning. It is just a nominal value. The specific viscosity includes Engler viscosity, Saybolt viscosity and Redwood viscosity.

Engler viscosity is the ratio of the time required to flow 200 mL test oil from Engler viscometer at the specified temperature to the time required to flow the same volume of distilled water at 20 ℃ . Engler viscosity is expressed by the symbol °E. Engler viscosity is used to evaluate the viscosity of some dark lubricating oils and residual oils in China.

③ 条件黏度　采样不同的特定黏度计所测得的以条件单位表示的黏度。条件黏度可以相对地衡量油品的流动性，但它不具有任何物理意义，只是一个公称值。有恩氏黏度、赛氏黏度、雷氏黏度等。

恩氏黏度（恩格勒黏度）是在规定温度下，从恩氏黏度计中流出 200mL 试油所需要的时间与在 20℃下流出相同体积的蒸馏水所需时间（s）的比值。恩氏黏度用符号°E 表示。我国用恩氏黏度评价一些深色润滑油及残渣油的黏度。

(2) Significance of the determination of viscosity

① important parameters in calculation.
② main quality index of lubrication oil.
③ important quality index of jet fuel.
④ important quality index of diesel oil.

(3) Method for determination of viscosity *Petroleum Products-Determination of Kinematic Viscosity* (GB/T 265—88). The kinematic viscosity tester for petroleum products is shown in fig.7-3.

（2）测定油品黏度的意义

① 黏度是工艺计算的重要参数。
② 黏度是润滑油的主要质量指标。
③ 黏度是喷气燃料的重要质量指标。
④ 黏度是柴油的重要质量指标。

（3）油品黏度的测定方法 《石油产品运动粘度测定法和动力粘度计算法》（GB/T 265—88）。石油产品动力黏度测定仪见图 7-3。

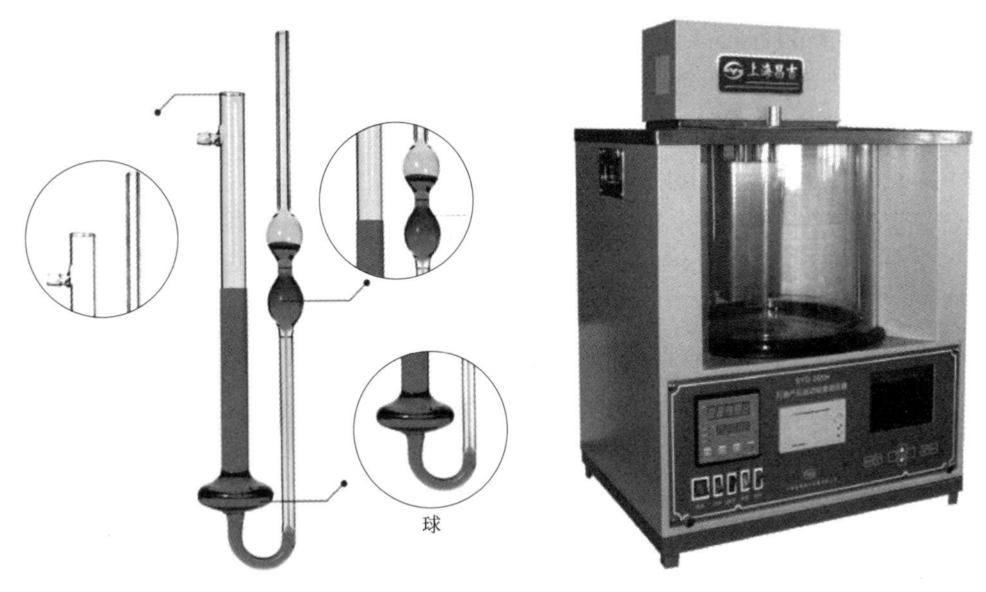

Fig.7-3 Kinematic viscosity tester for petroleum products

图 7-3 石油产品动力黏度测定仪

3. Flash Point, Ignition Point and Spontaneous Ignition Point

Petroleum products are mostly used as fuels. Furthermore, petroleum and its products are highly flammable and explosive materials. Therefore, flash point, ignition point and spontaneous ignition point are critical quality indexes with relation to the explosion, inflammation and combustion of the fuels. These indicators are of great significance to the storage, transportation and use of oil products. These indexes can also estimate the weight of oil fraction composition and guide oil production.

3. 闪点、燃点与自燃点

石油产品大多数是作为燃料，而且石油及其产品又是极易着火爆炸的物质。与燃料的爆炸、着火、燃烧有关的性质如闪点、燃点、自燃点等都是极其重要的质量指标。这些指标对油品的储存、运输和使用有着重要的意义。这些指标还可以判断油品馏分组成的轻重，指导油品生产。

(1) Concepts

① Flash point is the lowest temperature at which an open flame causes the mixed gas of oil vapor and air to ignite under specified conditions. Methods for determination of flash point:

a. closed cup method The evaporation of oil is carried out in the closed container, suitable for light oil products;

b. open cup method The evaporation of oil is carried out in the open container, from where oil vapor can disperse to air, suitable for heavy oil products.

② Ignition point is the minimum temperature at which oil products can be ignited by an open flame and burn for more than 5 s after being preheated under specified conditions.

③ Spontaneous ignition point is the minimum temperature at which oil products ignite by themselves instead of by an open flame after they are preheated to a certain temperature in a closed container, ℃ .

It is known from above definitions that flash point and ignition point need external ignition, but spontaneous ignition point does not.

(2) Correlations of flash point, ignition point and spontaneous ignition point with compositions of petroleum products

① Correlation with hydrocarbons.

② Correlation with boiling range of petroleum products.

(3) Significance of determination

① Evaluate the weight of oil fraction in the production.

② Appraise the hazards of breaking of fire.

③ Evaluate the quality of lubrication oil.

(4) Methods for determination of flash points and fire points

① Closed cup method *Determination of Flash Point—Pensky—Martens Closed Cup Method* (GB/T 261—2021). The closed up flash point tester is shown in fig.7-4、fig.7-5.

② Open cup method *Petroleum Products—Determination of Flash and Fire Points—Cleveland Open Cup Method*

（1）相关概念

① 闪点指石油产品在规定条件下加热到它的蒸气与空气形成的混合气与明火接触，发生瞬间闪火的最低温度。测定油品闪点的方法有：

a. 闭口杯闪点测定法　油品的蒸发是在密闭的容器中进行的，适用于轻质油品；

b. 开口杯闪点测定法　油品的蒸发是在敞开的容器中进行的，油品蒸气可自由扩散到空气中去，这只能测定重质油品的闪点。

② 燃点指石油产品在规定条件下加热到能被接触的火焰点着并燃烧 5s 以上的最低温度。

③ 自燃点指将油品在密闭容器中预先加热至某一温度，然后与空气接触，则不需要点火能发生自动着火燃烧时的最低温度，℃。

由定义知：闪点、燃点都需要外部引火，而自燃点无须外部引火。

（2）油品闪点、燃点与自燃点与组成的关系

① 与烃类组成有关。

② 与油品馏程有关。

（3）测定意义

① 判断油品馏分组成的轻重，指导油品生产。

② 鉴定油品发生火灾的危险性。

③ 评定润滑油质量。

（4）闪点和燃点的测定方法

① 闭口杯闪点测定法 《闪点的测定　宾斯基 - 马丁闭口杯法》（GB/T 261—2021）。闭口杯法测定仪器见图 7-4、图 7-5。

② 开口杯闪点测定法 《石油产品

(GB/T 3536—2008). The Cleveland open cup is shown in fig.7-6.

闪点和燃点的测定 克利夫兰开口杯法》(GB/T 3536—2008)。克利夫兰开口杯见图 7-6。

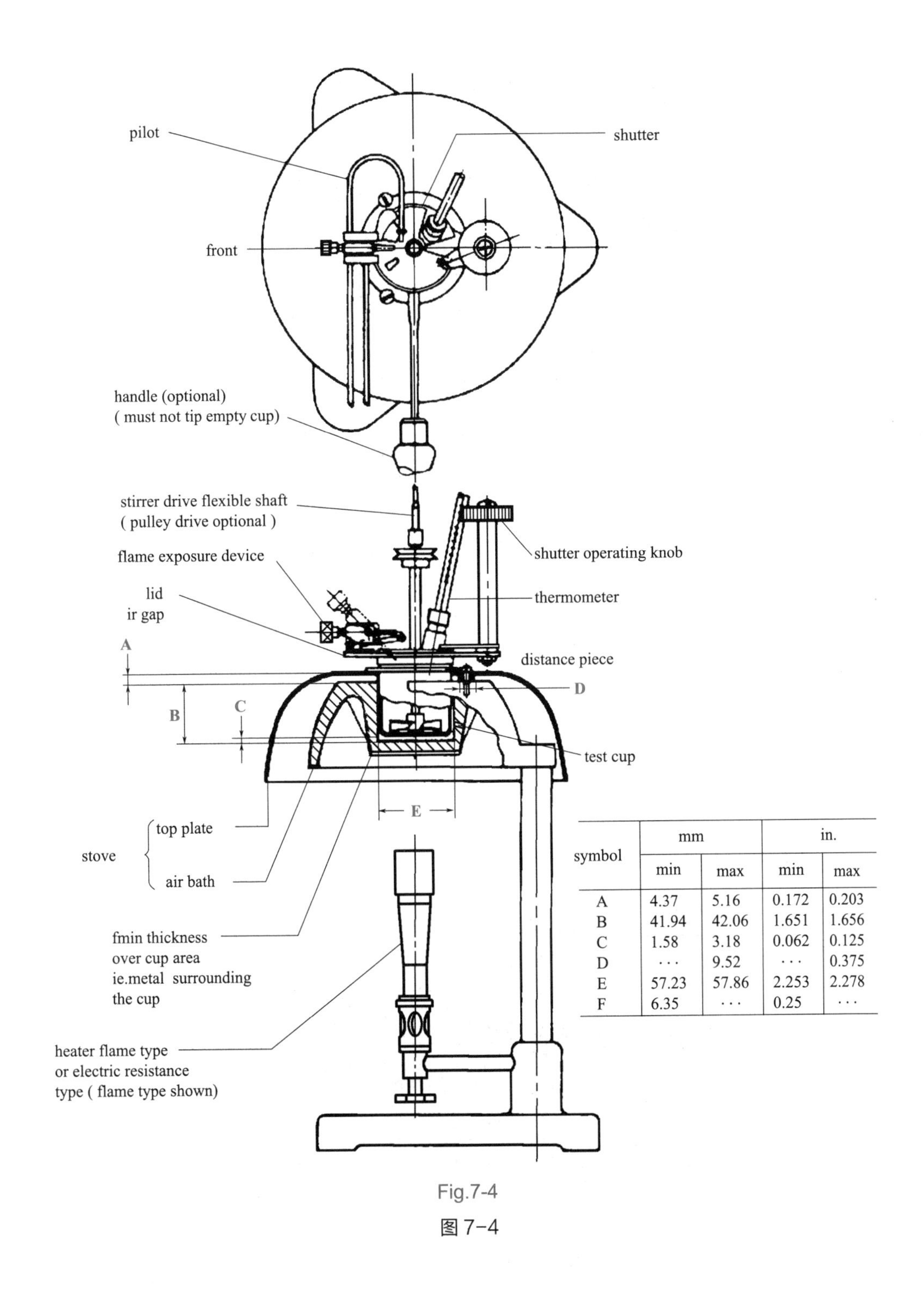

symbol	mm		in.	
	min	max	min	max
A	4.37	5.16	0.172	0.203
B	41.94	42.06	1.651	1.656
C	1.58	3.18	0.062	0.125
D	···	9.52	···	0.375
E	57.23	57.86	2.253	2.278
F	6.35	···	0.25	···

Fig.7-4

图 7-4

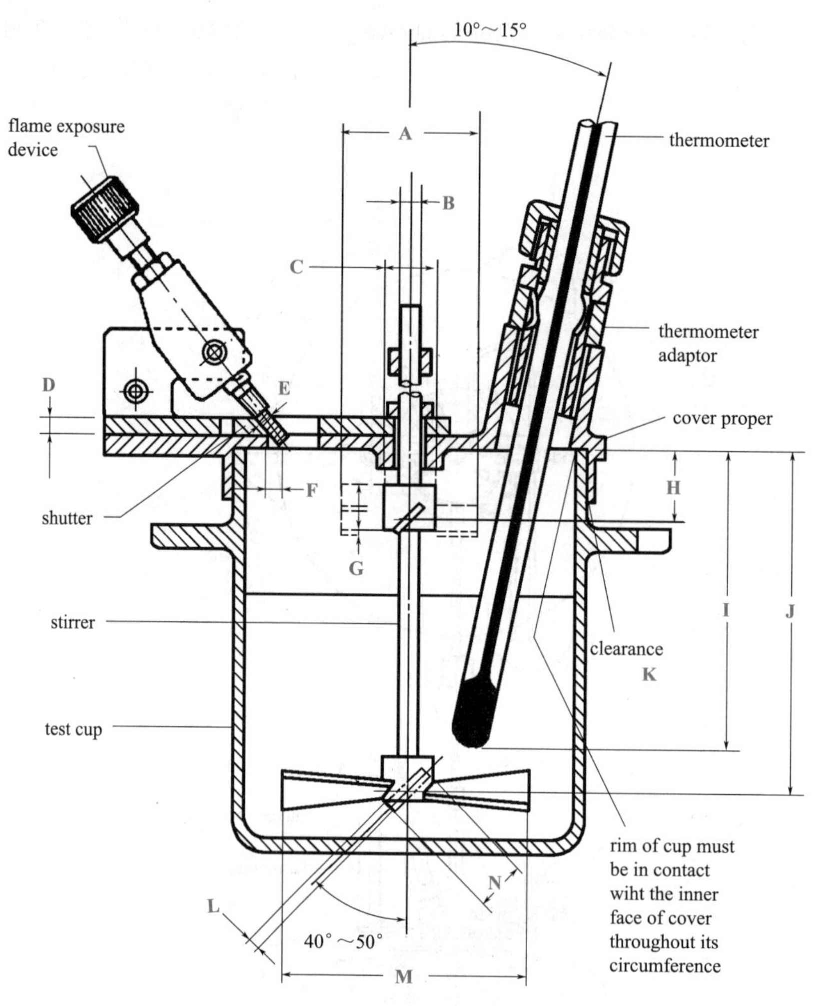

Fig.7-4 Pensky-Martens closed flash point tester

图 7-4 宾斯基－马丁闭式闪点测定仪

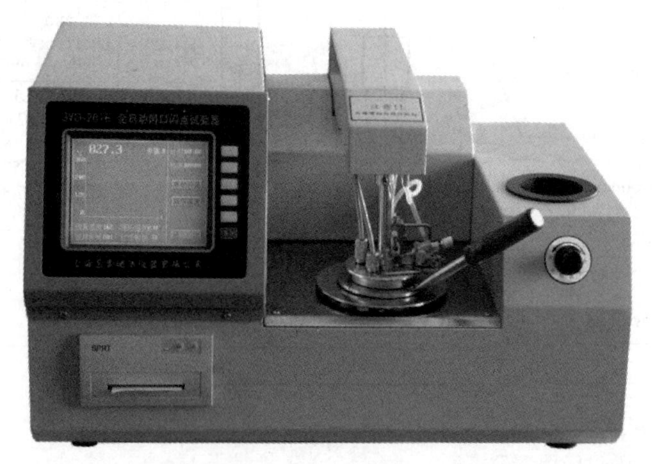

Fig.7-5 Automatic Martin closed cup flash point tester SYD-261E

图 7-5 SYD-261E 型马丁闭口杯闪点测定仪

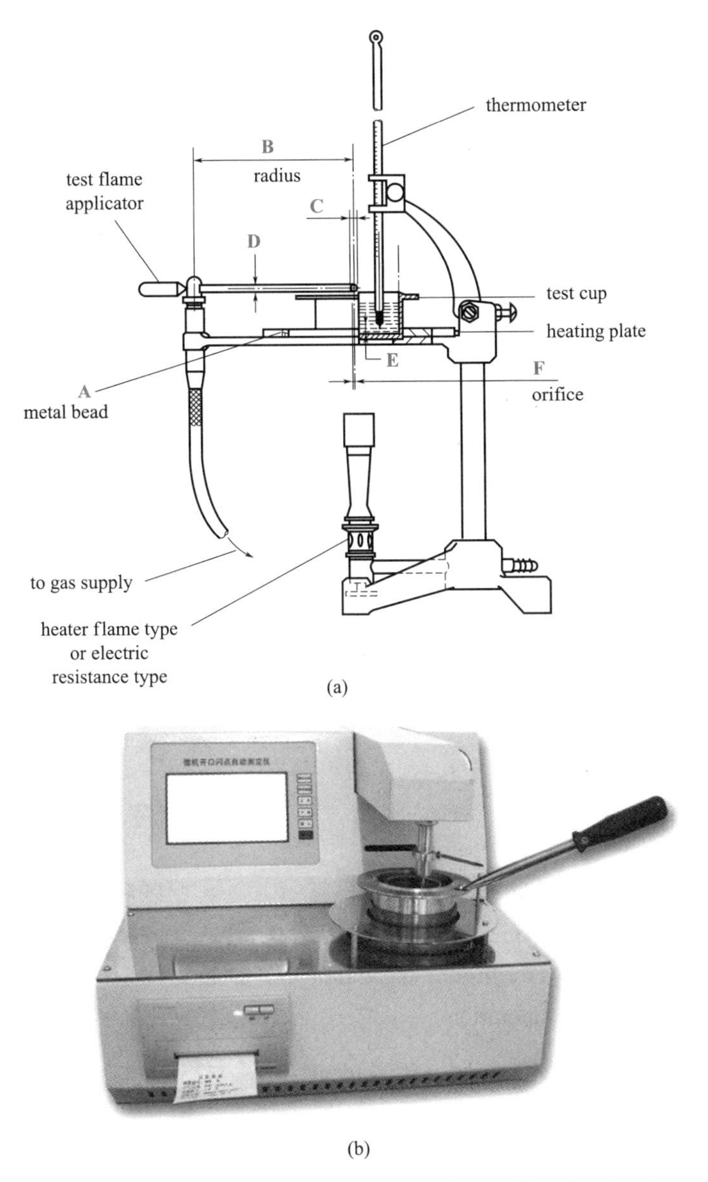

Fig.7-6 Test equipment for flash and fire points—Cleveland open cup

图 7-6 克利夫兰开口杯闪点 - 燃点测定仪

4. Carbon Residue

(1) Concept Carbon residue is the black residue formed by evaporation and thermal degradation of the sample oil in the carbon residue tester under the condition of non-air. According to different test methods, carbon residue is classified into Conradson carbon residue, electric furnace carbon residue and Ramsbottom carbon residue.

4. 残炭

（1）概念 在残炭测定器中加热试油在不通空气的条件下经蒸发分解，缩合后剩余的焦黑色残留物称为残炭。根据测定方法的不同，有康氏残炭、电炉残炭、兰氏残炭。

(2) Significance of determination The carbon residue can be used as a rough approximation of the fuel to deposits for mation in vaporizing pot-type and sleeve-type burners.

（2）测定意义 可用来粗略地估计燃料在蒸发式的釜型和套管型燃烧器中形成沉积物的倾向。

(3) Methods for determination *Petroleum Products Determination of Carbon Residue—Micro Method* (GB/T 17144—2021).

（3）测定方法 《石油产品 残炭测定法 微量法》（GB/T 17144—2021）。

Section 3 Determination of Evaporation of Petroleum Products

第3节 油品蒸发性能的测定

The evaporation property of petroleum and petroleum products is an important property to reflect difficulty of vaporization and evaporation, described by vapor pressure and boiling range.

石油和石油产品的蒸发性能是反映其汽化、蒸发难易的重要性质，用蒸气压、沸程来描述。

1. Vapor Pressure

1. 蒸气压

Vapor pressure is the saturated pressure exerted by a vapor in thermodynamic equilibrium with its liquid phase at a given temperature. The higher the vapor pressure, the easier the liquid's evaporation.

For hydrocarbons of homology, those with larger relative molecular mass have lower vapor pressure at the same temperature. For a pure hydrocarbon, the vapor pressure increases with rising temperature.

在某一温度下一种物质液相与其上方的气相呈平衡状态时，该蒸气所产生的压力称为饱和蒸气压，简称蒸气压。蒸气压愈高的液体愈易于汽化。对同族烃类，在同一温度下，分子量较大的烃类的蒸气压较小。对某一纯烃而言，其蒸气压是随温度的升高而增大。

Methods for determination: *Standard test method for vapor pressure of petroleum products Reid Method* (GB/T 8017—2012). The test equipment for vapor pressure (Reid method is shown in fig.7-7).

测定方法：《石油产品蒸气压的测定 雷德法》（GB/T 8017—2012）。雷德法测定石油产品蒸气压的仪器见图 7-7。

2. Boiling Range

2. 馏程（沸程）

(1) Concept

（1）概念

① The boiling point is the temperature at which the saturated vapor pressure of a pure liquid equals the ambient pressure that is specified. When the ambient pressure is fixed,

① 对于纯物质，它的沸点等于在一定外压下，其饱和蒸气压与外界压力相等时的温度。外压一定时，纯化合物

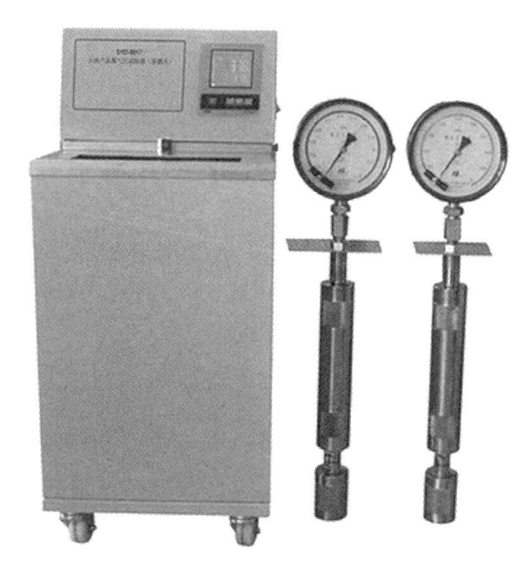

Fig.7-7　Test equipment for vapor pressure (Reid method)

图 7-7　蒸气压测定仪（雷德法）

the boiling point of a pure compound is a constant. However, for the petroleum products that are mixtures of organic compounds with varying boiling points, the temperature range (boiling range) is used to describe the evaporation.

的沸点是一个恒定值，但对于由沸点不等的有机化合物的混合物所组成的石油产品来说，需用某一温度范围——沸程来表示其蒸发性。

② Boiling range is a temperature range covering the boiling points from the initial boiling point to the final boiling point of the sample oil distilled under the specified conditions to show its evaporation property. Characteristic: when the ambient temperature is fixed, the boiling point of a mixed hydrocarbon is not a constant, but rather a temperature range in the order from low to high.

② 沸程指在规定条件下蒸馏所测得的从初馏点到终馏点表示试油蒸发特性的温度范围。特点：外压一定时，混合烃的沸点不是一个恒定值，而是一个温度由低到高的温度范围。

③ Distillation range is the relation established between V and t when a 100 mL specimen of the sample oil is distilled under prescribed conditions. Two cut fractions with same boiling ranges have different compositions.

③ 馏程指在规定的条件下蒸馏 100mL 试油所建立的 V 和 t 之间的馏分组成关系。沸程相同的两油馏分组成并不相同。

④ Differences between distillation range and boiling range: Distillation range is the relation of cut fraction composition between V and t, while boiling range is the temperature range between initial temperature and final temperature.

④ 馏程与沸程的区别：馏程指 V、t 之间的馏分组成关系；沸程是从 t_0 至 $t_{终}$ 的温度范围。

The distillate range data of the same oil varies with the measuring instruments and methods. In the quality standard for petroleum products, conditional test method is generally applied—Engler distillation (ASTM distillation, GB/T 6536—2010).

同一油品的馏程因测定仪器和测试方法不同，其馏程数据也有差别。在油品的质量标准中，大都采用条件性的馏程测定法——恩氏蒸馏（ASTM 蒸馏，GB/T 6536—2010）。

Pouring 100 mL specimen to a standard distillation flask, and heating at prescribed conditions. The temperature of liquid phase at which the first drop of condensate flow out is called the initial distillation point, and the temperature at which 10%, 20%...90% of distillate is produced is called 10%, 20%...90% distillation point respectively. The highest temperature at which the distillation reaches end point is called the final distillation point or dry point.

将100mL油品放入标准的蒸馏瓶中，按规定条件加热，流出第一滴冷凝液时的气相温度称为初馏点，馏出物为10%、20%……90%时的气相温度别称为10%馏出温度、20%馏出温度……90%馏出温度，蒸馏到最后所能达到的最高气相温度称为终馏点。

(2) Significance of determination of distillation range

（2）测定馏程的意义

① Evaluate the composition of petroleum fractions, determine process scheme and process techniques.

① 判断石油馏分组成，确定加工方案和加工工艺。

② Evaluate the evaporation, make primary appraisal of the fuel type and quality change.

② 判断蒸发性并初步鉴定燃料的种类及质量变化。

(3) Methods for determination *Standard Test Method for Distillation of Petroleum Products at Atmospheric Pressure* (GB/T 6536—2010). The related apparatus assemblies are shown in fig.7-8, fig.7-9. The position of thermometer in distillation is shown in fig.7-10. The test equi pment for distillation of petroleum products is shown in fig.7-11.

（3）测定方法 《石油产品常压蒸馏特性测定法》(GB/T 6536—2010)。相关试验装置见图7-8、图7-9。蒸馏瓶中温度计的位置见图7-10。石油产品馏程测定装置见图7-11。

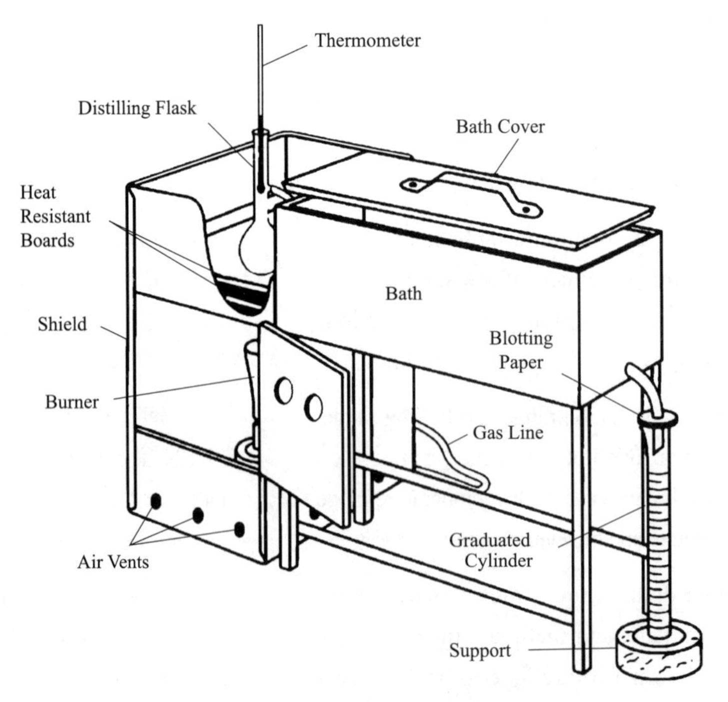

Fig.7-8 Apparatus assembly using gas burner flask

图7-8 利用气体和热的馏程测定仪

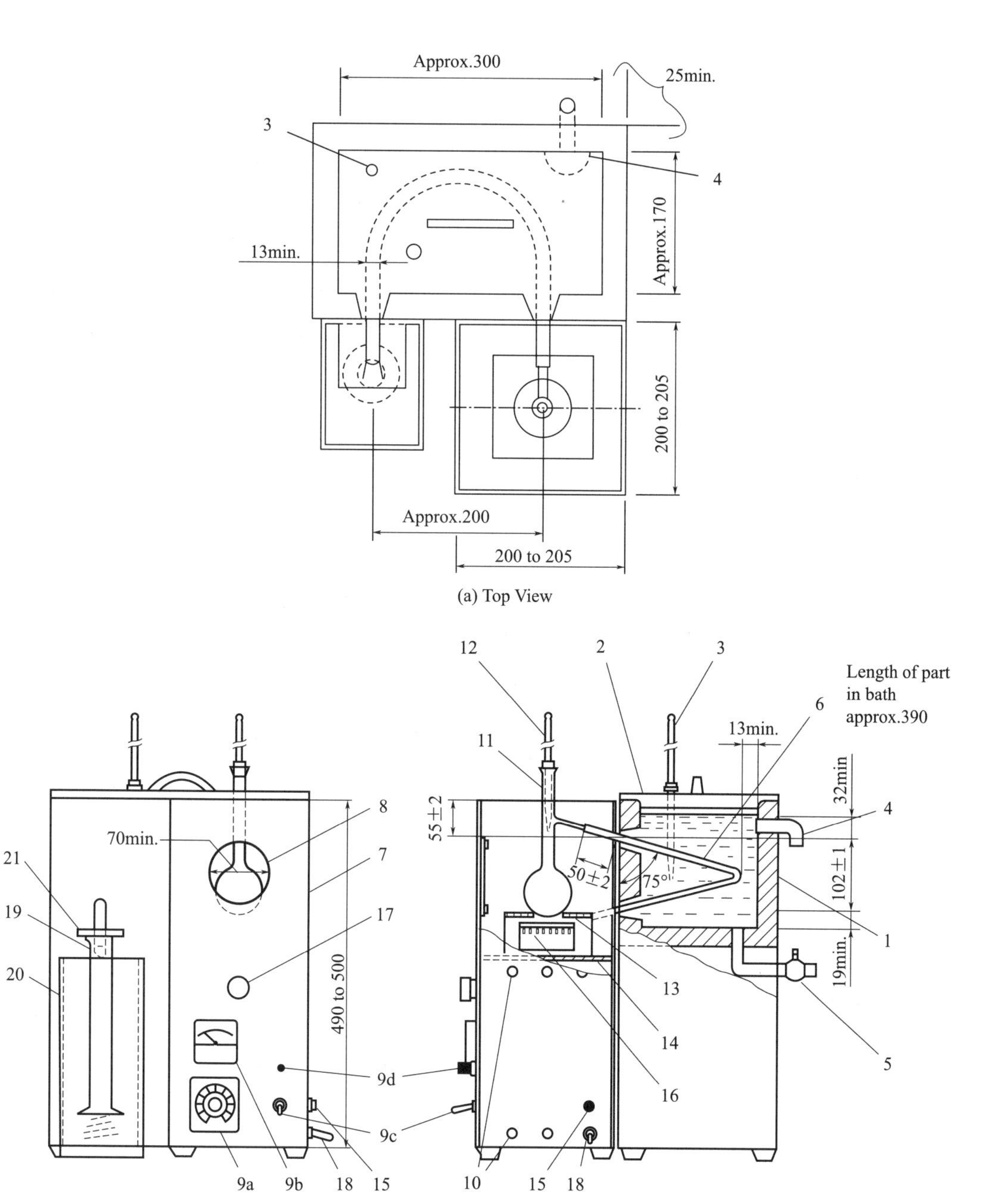

(a) Top View

(b) Front View

(c) Side View

Fig.7-9 Apparatus assembly using electric heater (mm)

1—Condenser bath; 2—Bath cover; 3—Bath temperature sensor; 4—Bath overflow; 5—Bath drain; 6—Condenser tube; 7—Shield; 8—Viewing window of support platform; 9a—Voltage regulator; 9b—Voltmeter or ammeter; 9c—Power switch; 9d—Power light indicator; 10—Vent; 11—Distillation flask; 12—Temperature sensor; 13—Flask support board; 14—Flask support platform; 15—Ground connection; 16—Electric heater; 17—Knob for adjusting level; 18—Power source cord; 19—Receiver cylinder; 20—Receiver cooling bath; 21—Receiver cover

图 7-9 电加热馏程测定仪（单位：mm）

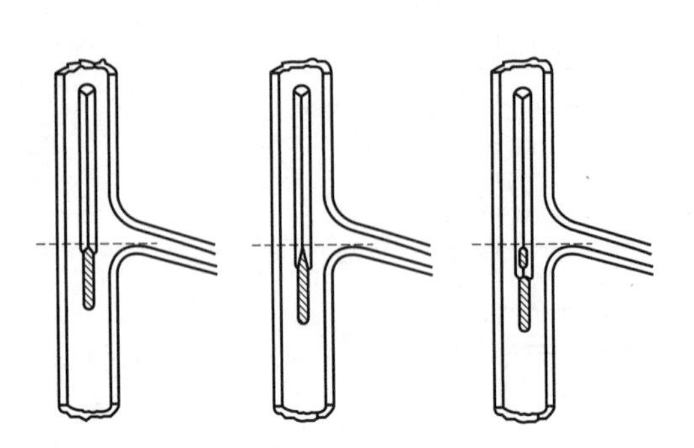

Fig7-10 Position of thermometer in distillation flask

图 7-10 蒸馏瓶中温度计的位置

Fig.7-11 Test equipment for distillation of petroleum products

图 7-11 石油产品馏程测定仪

Section 4 Determination of Cryogenic Flowability of Petroleum Products

第 4 节 油品低温流动性能的测定

1. Cloud Point, Crystallization Point, Freezing Point

Cloud point, crystallization point and freezing point are the indexes indicating the cryogenic properties of kerosene, aviation fuel and jet fuel. They have significant influence on the storage, transportation and use conditions. Each petroleum product has different cryogenic properties.

(1) Concept

① Cloud point is the highest temperature at which the sample oil precipitates giving the fluid a cloudy appearance under prescribed experimental conditions, unit: ℃ .

② Crystallization point is the highest temperature at which a liquid specimen when observable cluster of wax crystals first appears upon cooling under prescribed conditions, unit: ℃ .

1. 浊点、结晶点、冰点

浊点、结晶点、冰点是表征煤油、航空汽油和喷气燃料的低温性能指标。它显著影响油料输运、储存和使用条件，不同的石油产品低温流动性能有不同的评定指标。

（1）概念

① 在试验条件下把试油冷却到出现雾状或混浊时的最高温度叫浊点，单位：℃。

② 结晶点是在规定条件下冷却油品，出现用肉眼可以分辨的结晶时的最高温度，单位：℃。

③ Freezing point is the lowest temperature at which the petroleum products remain free of solid hydrocarbon crystals after being heated under prescribed conditions, unit: ℃ .

For a petroleum product, the freezing point is 1-3 ℃ higher than the crystallization point, cloud point > freezing point > crystallization point.

(2) Significance of determination

① The Crystallization point and the freezing point are quality indexes for evaluating the cryogenic properties of aviation fuels and jet fuels.

② The cloud point is a quality index for evaluating the cryogenic properties of kerosene.

(3) Methods for determination

① *Determination of Cloud Point and Crystallizing Point for Light Petroleum Products* (NB/SH/T 0179—2013).

② *Standard Test Method for Freezing Point of Aviation Fuels* (GB/T 2430—2008).

2. Solidification Point, Pour Point, Cold Filter Plugging Point

The solidification point, pour point and cold filter plugging point are important indexes reflecting the use of crude oil, diesel oil, lubrication oil and heavy fuel oil. Nowadays, there is tendency that the pour point replaces the solidification point and the cold filter plugging point replaces the solidification point of diesel oil in the domestic.

(1) Concept

① Pour point is the lowest temperature at which movement of the test specimen is observed under prescribed conditions of test, also called the movement limit. It can better reflect the cryogenic properties of petroleum products than the condensation point. It is stipulated as ISO standard.

③ 冰点是在规定条件下冷却油品到出现结晶后，再使其升温，使原来形成的结晶消失时的最低温度，单位：℃。

对于同一油品，冰点比结晶点高 1 ～ 3℃，浊点 > 冰点 > 结晶点。

（2）测定意义

① 结晶点和冰点是评定航空汽油和喷气燃料低温性能的质量指标。

② 浊点主要是煤油的低温性能指标。

（3）测定方法

①《轻质石油产品浊点和结晶点测定法》（NB/SH/T 0179—2013）。

②《航空燃料冰点测定法》（GB/T 2430—2008）。

2. 凝点、倾点和冷滤点

凝点、倾点和冷滤点是原油、柴油、润滑油和重质燃料油的重要使用性能指标。目前国内正逐步采用以倾点代替凝点、用冷滤点代替柴油凝点。

（1）概念

① 倾点是指油品在试验规定的条件下冷却时，能从规定仪器中流出的最低温度，也称为流动极限，它比凝点能更好地反映油品的低温性能，被规定作为 ISO 标准。

② Solidification point or solidifying point is the highest temperature at which a specimen cools down and the liquid surface does not move.

② 凝点又称凝固点，是指油品在试验规定的条件下冷却时液面不移动时的最高温度。

③ Cold filter plugging point is the highest temperature at which 20 mL of diesel oil cannot pass through a filtration device in 60 s under given pressure and at certain cooling speed.

③ 冷滤点是在规定的压力和冷却速度下，柴油试样在 60s 内不能通过过滤器 20mL 时的最高温度。

The cold filter plugging point could better reflect the pumping property and filtering property of diesel oil by good correlation with the practical use. Therefore, it tends to replace the condensation point.

冷滤点能较好地反映柴油的泵送和过滤性能，与实际使用情况有较好的对应关系，所以目前用冷滤点替换凝点指标。

(2) Significance of determination

（2）测定意义

① They are listed in the petroleum oil specification as the quality inspection standards for the production, storage and transportation of petroleum products.

① 列入油品规格，作为石油产品生产、贮存和运输的质量检测标准。

② Set the end-use temperatures of petroleum products.

② 确定油品的使用温度。

③ Evaluate the wax content and guide the production.

③ 估计石蜡含量，指导油品生产。

(3) Methods for determination

（3）测定方法

① *Petroleum Products—Determination of Pour Point* (GB/T 3535—2006).

①《石油产品倾点测定法》（GB/T 3535—2006）。

② *Determination of Solidification Point for Petroleum Products* (GB/T 510—2018).

②《石油产品凝点测定法》（GB/T 510—2018）。

③ *Determination of Cold Filter Plugging Point for Petroleum Products* (SH/T 0248—2006).

③《石油产品冷滤点测定法》（SH/T 0248—2006）。

Section 5 Determination of Combustion Properties of Petroleum Products

第 5 节 油品燃烧性能的测定

1. Antiknock Performance of gasoline

1. 汽油的抗爆性

(1) Phenomenon of knockings of gasoline engine The phenomenon of knockings of gasoline engine appears when there is a significant amount of low-combustion-

（1）汽油机爆震现象的产生 汽油机爆震现象的产生是因为汽油中含有较多的自燃点低易氧化的烃类，使烃类自

point and easy-oxidation hydrocarbons in the gasoline, the self-combustion point of hydrocarbons decrease and excess peroxides are produced in the unburned zone. If the cylinder temperature surpasses the self-combustion point, a few of burning centers form in the cylinder, resulting in self-combustion and knocking, then the metal slaps and black exhausting gas appears. That is, the detonation is related to the compositions of gasoline.

燃点下降，未燃区产生过多的过氧化物，气缸温度升高超过自燃点，便在气缸内产生多个燃烧中心而自燃，从而引起爆震。结果产生金属敲击声、排气冒黑烟。即爆震的产生与汽油的组成有关。

(2) Evalution Method of antiknock performance of gasoline—octane number Definition: the method utilizes a standard single-cylinder engine to compare the sample oil with the standard fuel. When the antiknock performance of the specimen is identical to that of a standard fuel of certain compositions, the volume percentage of isooctane in the standard fuel is the octane number of the specimen.

（2）汽油抗爆性的评价方法——辛烷值　定义：在实验用标准单缸发动机中将试油与标准燃料比较，当试油抗爆性与某一组成的标准燃料抗爆性相等时，标准燃料中所含异辛烷的体积分数即为试样的辛烷值。

(3) Significance of determination of antiknock performance

（3）测定汽油抗爆性的意义

① Classify the motor gasoline.

② Evaluate the quality of gasoline.

① 划分车用汽油牌号。

② 评价汽油质量。

(4) Methods for determination *Determination of The Gasoline Octane Number—Test Method for Research Octane Number* (GB/T 5487—2015). The CFR F-2u test equipment for research octane number is shown in fig.7-12.

（4）测定方法　《汽油辛烷值的测定　研究法》（GB/T 5487—2015）。CFR F-2u 型研究法辛烷值测定仪见图 7-12。

Fig.7-12　CFR F-2u test equipment for research octane number

图 7-12　CFR F-2u 型研究法辛烷值测定仪

2. Antiknock Performance of Diesel Oil

(1) Phenomenon of knockings of diesel engine The phenomenon of knockings of diesel engine appears when there is a little amount of low-combustion-point and easy-oxidation hydrocarbons in the diesel oil, the self-combustion delays and the ignition prolongs because the peroxides are not sufficient. If the injected diesel oil accumulates too much in the cylinder, they ignite at the same time upon self-combustion, the temperature and pressure rise sharply in the cylinder and knocking happens.

(2) Evalution Method of antiknock performance of diesel fuel oil—cetane number Definition: the cetane value is determined by comparing a specimen of sample oil with the standard fuel. The mixture of them at different ratio can give a series of standard fuel with different antiknock ratings. When the antiknock performance of the specimen is identical to that of a standard fuel with a certain volume ratio, the volume percentage of cetane in the standard fuel is the cetane number of the specimen.

(3) Significance of determination of antiknock performance

① Evaluate the combustion properties of diesel oil.
② Understand the correlation between the antiknock performance with the chemical composition of diesel oil.

(4) Methods for determination *Standard Test Method for Cetane Number of Diesel Fuel Oil* (GB/T 386—2010).

2. 柴油的抗爆性

（1）柴油机爆震现象的产生 柴油机爆震现象的产生是因为柴油中含有自燃点低易氧化的烃类少，过氧化物准备不足，迟迟不能自燃，使滞燃期增长，从而使喷入气缸内的柴油积累过多，一旦自燃，这些积累过多的柴油同时燃烧，而使温度急剧上升，压力急剧升高，故而产生爆震。

（2）柴油抗爆性的评价方法——十六烷值 定义：把试油与标准燃料对比的方法来测定十六烷值。二者以不同比例混合，可得一系列不同抗爆性等级标准燃料，当试油抗爆性与某一体积比的标准燃料抗爆性相等时，试油十六烷值就等于该标准燃料中正十六烷的体积分数。

（3）测定柴油抗爆性的意义

① 判断柴油燃烧性能。
② 了解柴油抗爆性与化学组成的关系，指导油品生产。

（4）测定方法 《柴油十六烷值的测定法》（GB/T 386—2010）。

Section 6 Determination of Corrosive Properties of Petroleum Products

第6节 油品腐蚀性能的测定

1. Determination of Acidity and Acid Number

(1) Concept

① Acidity Acidity and acid number are indexes

1. 酸度、酸值的测定

（1）概念

① 酸度 酸度和酸值都是定量表

indicating quantitatively the contents of acidic constituents in petroleum products, expressed as milligrams of potassium hydroxide required to neutralize the acidic constituents in 100 mL of petroleum product sample. The unit is mgKOH/100 mL.

示油品中酸性物质（主要是有机酸）含量的指标。用中和100mL石油产品中酸性物质所需要的KOH毫克数来表示酸度，单位为mgKOH/100mL；

② Acid number Acid number is expressed as milligrams of potassium hydroxide required to neutralize the acidic constituents in 1 g of petroleum product sample. The unit is mgKOH/g.

② 酸值　中和1g石油产品中的酸性物质，所需的氢氧化钾质量，称为酸值，单位为mgKOH/g。

Acidic substances affect the stability of oil and corrode equipment, which should be removed. Acidity is determined by light oils such as gasoline, kerosene and diesel oil. Acid number is determined by crude oil and lubrication oil.

酸性物质影响油品安定性，腐蚀设备，必须除去。汽、煤、柴油等轻质油品测定酸度，原油和润滑油等测定酸值。

(2) Significance in determination

（2）测定意义

① Determine the quantity of acidic constituents in petroleum products.

① 判断油品中所含酸性物质的数量。

② Evaluate the corrosion of petroleum products to metal materials.

② 判断油品对金属材料的腐蚀性。

③ Measure the degradation of petroleum products. The acidity and acid number will increase along with the degradation of oil in storage, so they are the important indexes to measure whether the oil is metamorphic.

③ 判断油品的变质程度。油品在储存中，因氧化变质其酸度和酸值会有所增大，因此它们也是衡量油品是否变质的重要指标之一。

2. Sulfur Content

2. 硫含量

The sulfur content is a key index for crude oil and oil products, which may lead to the corrosion, instability in storage and environment pollution in the use, so these indicators must be strictly controlled. The sulfur contents of different petroleum products are determined by different methods, such as X-ray spectrometer method and lamp method.

含硫量是原油和油品必需测定的重要指标，油品中的硫化物，使油品具有腐蚀性、储存的不安定性，使用时会造成环境污染，因此是必须加以严格控制的指标。不同油品含硫量的测定方法不同。如X射线光谱仪、燃灯法等。

Methods for determination:

测定方法：

① *Dark petroleum products-Determination of sulphur content(Tubular oven method)* (GB/T 387—90). The tester is shown in fig.7-13.

①《深色石油产品硫含量测定法（管式炉法）》（GB/T 387—90）。测定装置见图7-13。

② *Petroleum products-Determination of sulphur*

②《石油产品硫含量测定法（燃

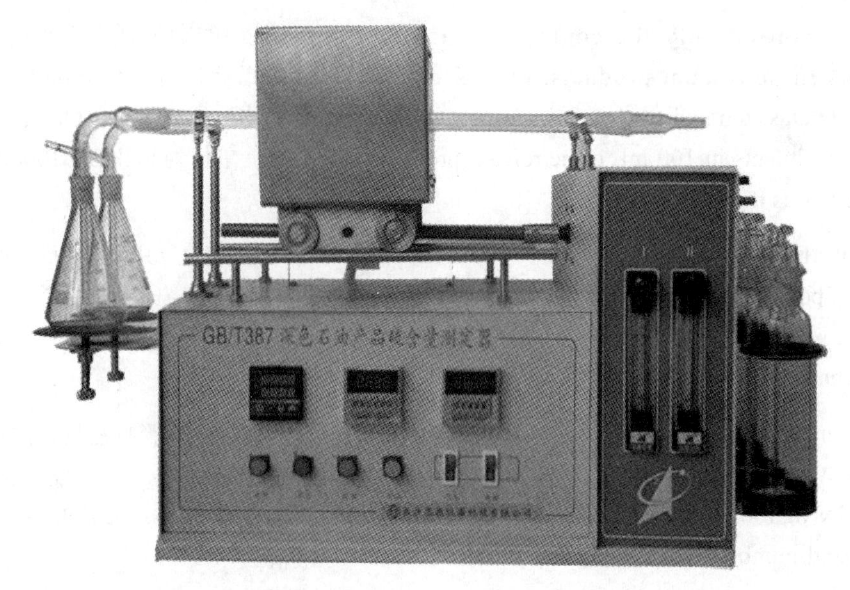

Fig.7-13 SC-387 Sulfur content tester for dark petroleum products (tubular furnace method)

图 7-13 SC-387 型重质石油产品硫含量测定仪（管式炉法）

content(*Lamp method*)(GB/T 387—77). The tester is shown in fig.7-14.

灯法）》（GB/T 380—77）。测定装置见图 7-14。

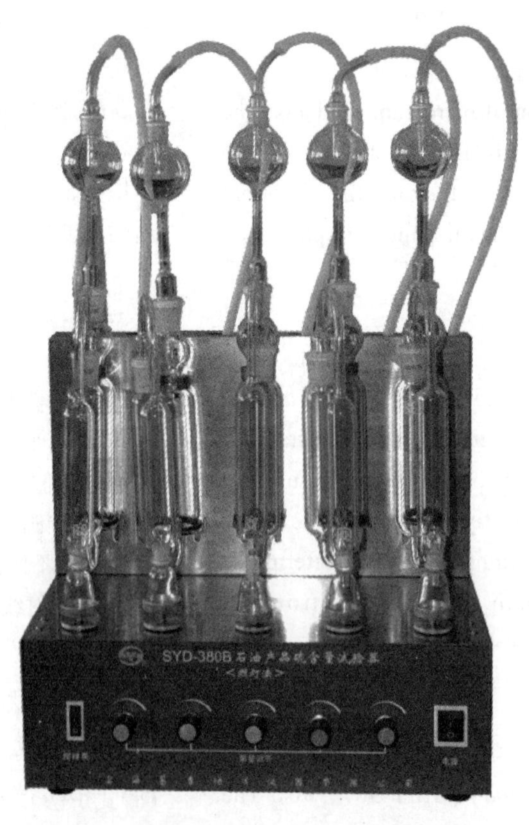

Fig.7-14 SYD-380B Sulfur content tester for dark petroleum products (Lamp method)

图 7-14 SYD-380B 型轻质石油产品硫含量测定仪（燃灯法）

Section 7 Determination of Stability of Petroleum Products

第 7 节 油品安定性的测定

1. Stability of Gasoline

The gasoline with good stability does not have apparent change in quality during storage and use. The gasoline with poor stability will oxidize and prone to acid formation in transportation, storage and use, producing acidic substances. The sticky gums and insoluble sediments darken the oil, decrease the octane number and corrode equipment. If the gum has a significant amount in oil, the oil way might be clogged resulting in insufficient oil supply, diluted fuel and air mixture and inability of air valve to close when engine works. Furthermore, the carbon deposit builds up, leading to poor heat radiation, knocking and preignition. The quality indexes for evaluating the stability of gasoline are existent gum and induction period.

Methods for determination: *Determination of Oxidation Stability of Gasoline—Induction Period Method* (GB/T 8018—2015). The tester is shown in fig.7-15.

1. 汽油安定性

安定性好的汽油，在贮存和使用过程中不会发生明显的质量变化；安定性差的汽油，在运输、贮存及使用过程中会发生氧化反应，易于生成酸性物质，黏稠的胶状物质及不溶沉渣，使油品颜色变深，导致辛烷值下降且腐蚀金属设备。汽油中生成的胶质较多，会使发动机工作时，油路阻塞、供油不畅、混合气变稀、气门被黏着而关闭不严，还会使积炭增加，导致散热不良而引起爆震和早燃等。评定汽油安定性的质量指标有实际胶质和诱导期。

测定方法：《汽油氧化安定性的测定 诱导期法》(GB/T 8018—2015)。测定仪器见图 7-15。

Fig.7-15 DSY-323A gasoline oxidation stability tester (induction period method)

图 7-15 DSY-323A 型汽油氧化安定性测定仪（诱导期法）

2. Stability of Diesel Oil

Similar to gasoline, the main factors affecting the stability of diesel oil are the unsaturated hydrocarbons, sulfur compounds or nitrogen compounds existing in the petroleum products. The influence of diesel oil stability to diesel engine is the same as that of gasoline stability on gasoline engine. The quality indexes for evaluating the stability of diesel oil are storage stability and thermal oxidation stability.

2. 柴油安定性

与汽油相似，影响柴油安定性的主要原因是油品中存在不饱和烃以及含硫、氮化合物等不安定组分。柴油的安定性对柴油机工作的影响与汽油的安定性对汽油机的影响基本相同。评定柴油安定性的指标主要有贮存安定性和热氧化安定性。

3. Iodine Number, Bromine Number and Bromine Index

(1) Concept

① Iodine number is the number of grams of iodine that will react with 100g of specimen under given conditions, expressed as $gI_2/100$ g.

② Bromine number is the number of grams of bromine that will react with 100g of specimen under given conditions, expressed as $gBr_2/100$ g.

③ Bromine index is milligrams of bromine that will react with 100 g of specimen under given conditions, expressed as $mgBr_2/100$ g.

(2) Significance in determination

The content of unsaturated hydrocarbons in petroleum products is not only one of the technical bases in the process control of oil refinery, but also an important index in evaluate the quality of oil. The content of unsaturated hydrocarbons and its influence on the stability of petroleum products can be indirectly expressed. The bigger the iodine number, bromine number and bromine index, the poorer the stability of oil. These indexes of aviation gasoline and aviation kerosene are strictly controlled to ensure the safety of flights.

3. 碘值、溴值及溴指数

（1）概念

① 碘值指在规定的条件下，100g试样所能消耗碘的质量，以 $gI_2/100g$ 表示。

② 溴值指在规定的条件下，100g试样所能消耗溴的质量，以 $gBr_2/100g$ 表示。

③ 溴指数指在规定的条件下，与100g 油品起反应时所消耗溴的质量，以 $mgBr_2/100g$ 表示。

（2）测定意义

不饱和烃类在石油产品中的含量是原油炼制过程中工艺控制技术依据之一，也是衡量油品质量好坏的重要指标。油品中不饱和烃含量及其对油品安定性的影响，可用碘值、溴值及溴指数来间接表示。碘值、溴值及溴指数越大，油品的安定性就越差，为了确保飞机的飞行安全，航空汽油、航空煤油对此类指标有严格的控制。

Section 8 Determination of Impurities in Petroleum Products

第8节 油品中杂质的测定

1. Water

(1) Forms of water existing in oil

① Suspending water suspending in oil as water drops, discernible to the naked eye.

② Emulsified water dispersing in oil as tiny water droplets, presenting in the form of colloidal solution in the oil. The emulsified water is stable, and its protective membranes are naphthenates in the form of water in oil.

③ Dissolved water existing in oil with the status of dissolution. In general, the amount of dissolved water is from several tens μL/L to two hundred μL/L. The heavier the oil, the more water it dissolves. (Heavy oil contains aromatic hydrocarbon, so the content of dissolved water is high because the content of polar substances is high).

Generally, it is unavoidable to have dissolved water existing in product oils or crude oils. The anhydrous oil in petroleum analysis is regarded as being free of suspending water and emulsified water.

(2) Risks of water existence

① Water in lubricant oil aggravating the equipment corrosion; when water is vaporized at the contact with the lubricated parts that temperature higher than 100 ℃, the lubricating oil film is damaged, and aggravates corrosion.

② Water in fuel oil decreasing the heat value of fuel oil; clogging the oil way by the ice particles deposited from oil at low temperature. Especially in aviation fuel and diesel, If the water contains, it will cause the blockage of the oil route, and the consequences are serious.

1. 水分

（1）油中水存在形式

① 悬浮水 以水滴形式悬浮于油中，肉眼可辨。

② 乳化水 以极小的水滴形式分散于油中。在油中呈胶体溶液，乳化水稳定，其保护膜是环烷酸盐类，形成油包水。

③ 溶解水 以溶解状态存在于油中，一般溶解水在几十到200μL/L，油越重溶解水越多（因重油含芳烃，极性物多，所以溶解水多）。

一般溶解水在成品油中乃至原油中是不可避免的，石油分析中把无水视为无悬浮水和乳化水。

（2）水分存在的危害

① 润滑油中含水 不但会增加机件腐蚀；而且会由于水同大于100℃的被润滑部件接触生成水蒸气，而破坏润滑油膜，进而增大磨损。

② 燃料油含水 一是降低燃料油热值，二是能因油在低温下使用析出冰粒而堵塞油路。尤其是航煤、柴油中含水，会造成堵塞油路，后果严重。

③ Water in crude oil increasing freight volume and making it difficult to the processing of crude oil. For example, the processing capacity of crude oil in China is about 150 million tons per year, in which 750 thousand tons of water has to be transported (water content <0.5% in crude oil).

④ Water in sample oil influencing the accuracy of tests. For example, the measurement of crystallization point, freezing point and cold filter point will be higher due to the presence of water.

⑤ Water in feed stocks poisoning catalysts. The reason is that water occupies the acid center of the catalyst and destroys the balance between the acid center and the metal center. Therefore, the reforming material should be distilled and dehydrated before entering the reactor. Pt-Rt reforming requires water content <5 μg/g, Pt reforming requires water content <30 μg/g. Therefore, the water content in petroleum and petroleum products must be strictly controlled.

(3) Determination method of water

① Distillation method *Test Method for Water in Petroleum Products—Distillation Method* (GB/T 260—2016). The distillation equipment is shown in fig. 7-16.

② Karl Fischer method The water measurement equipment for Karl Fischer method is shown in fig. 7-17。

③ 原油中含水 一是增大运输量，二是给原油加工带来困难（冲塔、腐蚀）。如：我国现原油生产能力约为1.5亿吨，按1.5亿吨计（原油含水要求<0.5 %），所以一年要多运75万吨水，即750列车（每列1000t）。

④ 试验用油含水 影响结果准确性。如测量结晶点、凝点、冷滤点会由于水的存在使结果偏高。

⑤ 重整原料含水 会使催化剂中毒。原因是水占据催化剂酸性中心，破坏酸性中心与金属中心的平衡使催化剂活性下降。所以重整原料在进入反应器前要蒸馏脱水。Pt-Rt重整要求含水 $<5\mu g \cdot g^{-1}$，Pt重整要求含水 $<30\mu g \cdot g^{-1}$。所以，石油及石油产品中含水量必须严格控制。

（3）水分测定方法

① 蒸馏法 《石油产品水含量的测定 蒸馏法》（GB/T 260—2016）。蒸馏法水分测定装置见图7-16。

② 卡尔费休法 卡尔费休法水分测定仪见图7-17。

2. Ash

(1) Definition the inorganic substance obtained by calcining the residue of carbonized petroleum products under prescribed conditions. That is the incombustible substance left after the ignition of oil under given conditions.

(2) Composition mainly metal oxides, e.g. CaO, MgO, Fe_2O_3, Al_2O_3, SiO_2 and a little amount of the oxides of V, Ni, Na, Mn.

(3) Source Organic acid salts, inorganic salts and organometallic oxides in petroleum are usually

2. 灰分

（1）定义 在规定条件下，油品被炭化后的残留物经煅烧所得的无机物。即油品在规定条件下灼烧后所剩的不燃物质。

（2）组成 灰分主要是金属氧化物（有CaO、MgO、Fe_2O_3、Al_2O_3、SiO_2及少量V、Ni、Na、Mn等金属氧化物）。

（3）来源 石油中的有机酸盐、无机盐和有机金属氧化物通常集中在

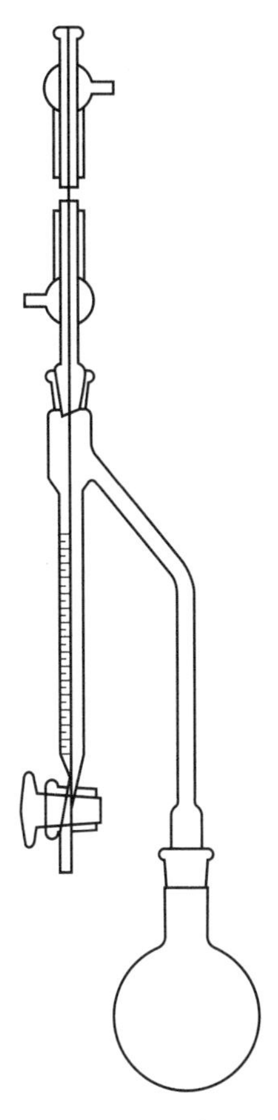

Fig.7-16 Distillation equipment for water measurement

图 7-16 蒸馏法水分测定装置

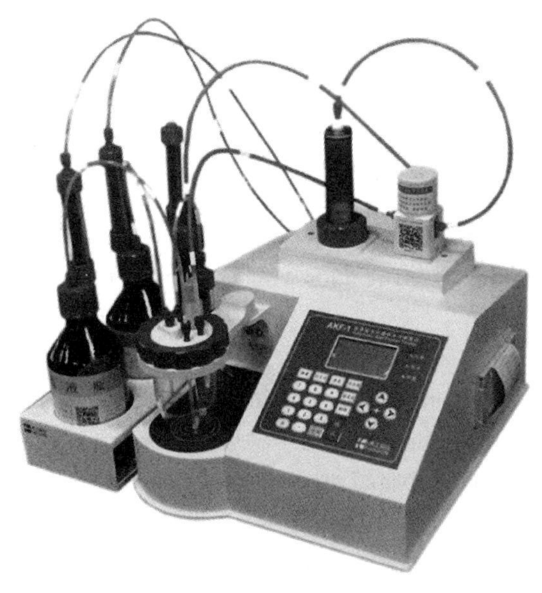

Fig.7-17 Water measurement equipment for Karl Fischer method

图 7-17 卡尔费休法水分测定仪

concentrated in the residue. The salt content in distillate oil is very small, it may be brought in by entered external mixing, corrosion or additives.

渣油中，馏分油中这些盐类含量极少，是外界混入、腐蚀时进入或加入的添加剂带入。

3. Mechanical Impurities

3. 机械杂质

(1) Definition All the precipitating or suspending substances insoluble in the solvent used (gasoline, benzene).

（1）定义 指存在于油品中所有不溶于所用溶剂（如汽油、苯等）的沉淀状或悬浮状物质。

(2) Compositions Compositions of mechanical impurities: sand, iron scurf, clay, mineral salts and carboids.

（2）组成 沙子、铁屑、黏土和矿物盐（如 Fe_2O_3）等及炭青质。

(3) Source blended into oil during processing, storage, transportation, or entrapped in the organic metal salt additives.

（3）来源 主要是在加工精制时、贮存运输过程中混入的，以及油品中加入某些有机金属盐类添加剂带入的。

(4) Determination method *Petroleum, Petroleum Products and Additives—Method for Determination of Machanical Admixtures* (GB/T 511—2010).

（4）测定方法 《石油和石油产品及添加剂机械杂质测定法》（GB/T 511—2010）。

Exercises

7-1 For the task of oil analysis, which of the following statements are false? (   )

A. Monitor the production process and provide data for controlling process conditions.

B. Inspect the quality of ex-factory oil products.

C. Evaluate oil performance.

D. Questionable samples cannot be used for oil quality arbitration.

7-2 Ignition point refers to the lowest temperature at which a petroleum product is heated under specified conditions and continuously burned by a contact flame (   ).

A. 10 s B. 5 s C. 15 s D. 20 s

7-3 Which of the following is self-ignition without ignition? (   )

A. flash point B. ignition point

C. spontaneous ignition point D. cloud point

7-4 The index for evaluating the tendency of oil to produce coke at high temperature is (   ).

A. carbon residue B. ash content C. mechanical impurities D. colloid

7-5 For a pure hydrocarbon, the vapor pressure increases with temperature (   ).

A. increasing B. decreasing C. no change D. rising and then falling

7-6 When measuring reed vapor pressure, the water bath temperature should be controlled at (   ).

A. 37.8 ℃ ±0.1 ℃ B. 37.8 ℃ ±3 ℃ C. 38℃ ±0.1 ℃ D. 38 ℃ ±3 ℃

7-7 In order to ensure all condensation of oil & gas and reduce distillation loss, the cold bath temperature should be controlled as ( ).

A. 0-10 ℃ B. 13-18 ℃

C. 0-1 ℃ D. not higher than room temperature

7-8 Which one indicates the average evaporation of automotive unleaded gasoline, it directly affects the acceleration and working stability of the engine? ( )

A. 50% evaporation temperature B. 10% evaporation temperature

C. end boiling point D. residual quantity

7-9 When oil is cooled under specified conditions, the highest temperature at which crystallization can be distinguished by naked eyes is called ( ).

A. freezing point B. crystallization point

C. cloud point D. condensation point

7-10 Under the specified pressure and cooling rate, the highest temperature of diesel sample can not pass through the filter 20 mL within 60 seconds is called ( ).

A. condensation point B. pour point

C. cold filter plugging point D. cloud point

7-11 The evaluation method of gasoline anti-knock property is ( ).

A. cetane number B. cetane index C. induction period D. Octane number

7-12 The ignition of diesel oil is closely related to its chemical composition and fraction composition. The longest ignition delay period of the following hydrocarbons is ( ).

A. n-alkanes B. alkene

C. naphthenic hydrocarbon D. aromatic hydrocarbon

7-13 Which is not an indicator to evaluate the corrosivity of light diesel and vehicle diesel? ( )

A. sulfur content B. acid value C. induction period D. copper corrosion

7-14 The quality indexes for evaluating the stability of gasoline are ( ).

A. actual colloid and induction period B. carbon residue and ash content

C. copper corrosion and sulfur content D. induction period and ash content

7-15 Unsaturated hydrocarbon content in oil and its influence on oil stability can be expressed indirectly by ( ), etc.

A. iodine value B. glue C. acidity D. sulfur content

7-16 If the oil contains water, at the freezing point determination, will it influence the determination result? ( ).

A. Little B. No impact C. Great D. Little and then great

7-17 All precipitated or suspended substances present in the oil that are insoluble in the solvent used (e.g. gasoline, benzene, etc.) are called ( ).

A. ash content B. colloid C. carbon residue D. mechanical impurities

7-18 For determination of moisture in oil, which of the following methods is true ? ( )

A. Light burning B. Distillation C. Filtration D. Indicator

7-19 If oil contains water, it is suspended in oil in the form of water droplets, which can be recognized by the naked eye, it is called ( ).

A. free water B. dissolved water C. emulsified water D. suspended water

7-20 As for the stability of gasoline, which of the following statements is false ? ( )

A. Oxidation reaction will occur in the process of transportation, storage and use, and it is easy to generate acidic substances.

B. More colloid is generated in gasoline.

C. It will increase carbon accumulation, resulting in poor heat dissipation and cause detonation and premature combustion.

D. It will make the oil darker and increase the octane number.

7-21 What is flash point? What are the measurement methods of flash point?

7-22 Briefly describe the cause and hazard of gasoline engine knocking.

7-23 Briefly describe the factors affecting the stability of oil products, and explain which methods can improve the stability of oil products.

7-24 Briefly describe the method of kinematic viscosity measurement of oil products.

7-25 What is the significance of the determination of acid substances in oil products?

Reference

参考文献

[1] 高职高专化学教材编写组 . 分析化学 . 4 版 . 北京：高等教育出版社，2014.

[2] 丁保君 . 分析化学（双语版）. 2 版 . 大连：大连理工大学出版社，2017.

[3] 高金波，吴红 . 分析化学 . 北京：中国医药科技出版社，2016.

[4] Douglas A. Skoog, F. James Holler, Stanley R. Crouch.Principles of Instrumental Analysis. 7th Edition. Stamford: Cengage Learning, 2017.

[5] 王英健，尹兆明 . 无机与分析化学 . 北京：化学工业出版社，2018.

[6] 王嗣岑，朱军 . 分析化学 . 北京：科学出版社，2019.

[7] 赵世芬，闫冬良 . 仪器分析技术 . 北京：化学工业出版社，2016.

[8] 曹国庆 . 仪器分析技术 . 2 版 . 北京：化学工业出版社，2018.

[9] 甘黎明，王海超 . 石油产品分析 . 2 版 . 北京：化学工业出版社，2019.

[10] 甘黎明，李锐 . 油品分析实训 . 北京：中国石化出版社，2015.

[11] 邢梅霞，夏德强 . 光谱分析 . 北京：中国石化出版社，2012.

[12] Daniel C. Harris. Quantitative Chemical Analysis. 9th Edition. New York: W. H. Freeman and Company, 2015.

[13] 丁瑞芳，段煜 . 分析化学双语实验 . 北京：中国医药科学出版社，2019.

[14] David S. Hage, James D. Carr. 分析化学和定量分析（英文版）. 北京：机械工业出版社，2012.

[15] F. W. Fifield, D. Kealey. Principles and Practice of Analytical Chemistry. 5th Edition. Oxford: Blackwell Science Ltd., 2000.